Springer Undergraduate Mathematics Series

Springer

London
Berlin
Heidelberg
New York
Barcelona
Hong Kong
Milan
Paris
Singapore
Tokyo

John Haigh

Probability Models

With 15 Figures

 Springer

John Haigh. MA, PhD
Centre for Statistics and Stochastic Modelling, School of Mathematical Sciences,
University of Sussex, Falmer, Brighton, BN1 9QH, UK

Cover illustration elements reproduced by kind permission of:
Aptech Systems, Inc., Publishers of the GAUSS Mathematical and Statistical System, 23804 S.E. Kent-Kangley Road, Maple Valley, WA 98038, USA. Tel: (206) 432 - 7855 Fax (206) 432 - 7832 email: info@aptech.com URL: www.aptech.com
American Statistical Association: Chance Vol 8 No 1, 1995 article by KS and KW Heiner 'Tree Rings of the Northern Shawangunks' page 32 fig 2
Springer-Verlag: Mathematica in Education and Research Vol 4 Issue 3 1995 article by Roman E Maeder, Beatrice Amrhein and Oliver Gloor 'Illustrated Mathematics: Visualization of Mathematical Objects' page 9 fig 11, originally published as a CD ROM 'Illustrated Mathematics' by TELOS: ISBN 0-387-14222-3, German edition by Birkhauser: ISBN 3-7643-5100-4.
Mathematica in Education and Research Vol 4 Issue 3 1995 article by Richard J Gaylord and Kazume Nishidate 'Traffic Engineering with Cellular Automata' page 35 fig 2. Mathematica in Education and Research Vol 5 Issue 2 1996 article by Michael Trott 'The Implicitization of a Trefoil Knot' page 14.
Mathematica in Education and Research Vol 5 Issue 2 1996 article by Lee de Cola 'Coins, Trees, Bars and Bells: Simulation of the Binomial Process' page 19 fig 3. Mathematica in Education and Research Vol 5 Issue 2 1996 article by Richard Gaylord and Kazume Nishidate 'Contagious Spreading' page 33 fig 1. Mathematica in Education and Research Vol 5 Issue 2 1996 article by Joe Buhler and Stan Wagon 'Secrets of the Madelung Constant' page 50 fig 1.

British Library Cataloguing in Publication Data
Haigh, John, Dr
  Probability models. - (Springer undergraduate mathematics
  series)
  1. Probabilities 2. Mathematical models 3. Probabilities –
  Problems, exercises, etc. 4. Mathematical models – Problems,
  exercises, etc.
  I. Title
  519.2
ISBN 1852334312

Library of Congress Cataloging-in-Publication Data
Haigh, John, Dr
  Probability models / John Haigh.
    p.   cm. -- (Springer undergraduate mathematics series)
  Includes bibliographical references and index.
  ISBN 1-85233-431-2 (alk. paper)
  1. Probabilities. I. Title. II. Series.
  QA273.H316 2002
  519.2—dc21   2001054261

Springer Undergraduate Mathematics Series ISSN 1615-2085
ISBN 1-85233-431-2 Springer-Verlag London Berlin Heidelberg
a member of BertelsmannSpringer Science+Business Media GmbH
http://www.springer.co.uk

Typesetting: Camera ready by the author
Printed and bound at the Athenæum Press Ltd., Gateshead, Tyne & Wear
12/3830-543210 Printed on acid-free paper SPIN 10791815

Preface

The purpose of this book is to provide a sound introduction to the study of real-world phenomena that possess random variation. You will have met some ideas of probability already, perhaps through class experiments with coins, dice or cards, or collecting data for projects. You may well have met some named probability distributions, and be aware of where they arise. Such a background will be helpful, but this book begins at the beginning; no specific knowledge of probability is assumed.

Some mathematical knowledge is assumed. You should have the ability to work with unions, intersections and complements of sets; a good facility with calculus, including integration, sequences and series; an appreciation of the logical development of an argument. And you should have, or quickly acquire, the confidence to use the phrase "Let X be ..." at the outset, when tackling a problem.

At times, the full story would require the deployment of more advanced mathematical ideas, or need a complex technical argument. I have chosen to omit such difficulties, but to refer you to specific sources where the blanks are filled in. It is not necessary fully to understand *why* a method works, the first time you use it; once you feel comfortable with a technique or theorem, the incentive to explore it further will come. Nevertheless, nearly all the results used *are* justified within this book by appeal to the background I have described.

The text contains many worked examples. All the definitions, theorems and corollaries in the world only acquire real meaning when applied to specific problems. The exercises are an integral part of the book. Try to solve them before you look at the solutions. There is generally no indication of whether an exercise is expected to be straightforward, or to require some ingenuity. But all of them are "fair", in the sense that they do not call on techniques that are more advanced than are used elsewhere within the book.

The text chapters have their natural order, but it is not necessary to assimilate *all* the material in Chapter n before embarking on Chapter $n + 1$. I suspect that many readers will find Chapter 6 markedly harder than the earlier ones; so that has a Summary, for ease of recall. You can come to grips with the subtleties of the different modes of convergence later. My excuse for including some of the results about fluctuations in random walks in Chapter 7 is their sheer surprise; some proofs here are more sophisticated.

No-one writes a textbook from scratch. We rely on our predecessors who have offered us their own insights, found a logical pathway through the material, given us their examples. Every serious probability text since 1950 has been influenced by William Feller's writings. In addition, it is a pleasure to acknowledge in particular the books by Geoffrey Grimmett and David Stirzaker, by Charles Grinstead and Laurie Snell, and by Sheldon Ross, that are cited in the bibliography. I also value *The Theory of Stochastic Processes*, by David Cox and Hilton Miller, and Sam Karlin's *A First Course in Stochastic Processes*, which were standard texts for many years.

I thank many people for their witting and unwitting help. Without the guidance of David Kendall and John Kingman, I might never have discovered how fascinating this subject is. Among my Sussex colleagues, conversations with John Bather and Charles Goldie have cleared my mind on many occasions. I also appreciate the discussions I have had with successive cohorts of students; listening to their difficulties, and attempting to overcome them, has helped my own understanding. Springer's referees have made useful comments on my drafts. I gave what I fondly hoped was the final version to Charles Goldie, who not only reduced the number of blunders, but made copious suggestions, almost all of which I have taken up. Mark Broom too found errors and obscurities, and the final text owes much to both of them. All the errors that remain are my responsibility. Without the help from two local TeX gurus, James Foster and James Hirschfeld, I would never have embarked on this project at all. Springer production staff (Stephanie Harding, Stephanie Parker and Karen Borthwick) have always responded helpfully to my queries. The patience and understanding of my wife Kay, during those long periods of distraction while I fumbled for the right phrase and the right approach, have been beyond what I could reasonably expect.

An updated file of corrections, both to the text and to the solutions, will be maintained at http://www.sussex.ac.uk/Staff/JH/ProbabilityModels.html. For each correction, the first person to notify me at J.Haigh@sussex.ac.uk will be offered the glory of being named as detector. I welcome feedback, from students and teachers; changes made in any subsequent editions because of such comments will be acknowledged.

John Haigh, Brighton, November 2001

Contents

1
Probability Spaces

1.1 Introduction

Imagine that some experiment involving random chance takes place, and use the symbol Ω to denote the set of all possible outcomes. For example, if you throw an ordinary die, then $\Omega = \{1, 2, 3, 4, 5, 6\}$. Or you might switch on the television set and ascertain what proportion of the current programme remains to be broadcast. Here Ω would be the continuum of real numbers from zero to unity.

If you toss a coin twice, and record the outcome, you might write $\Omega = \{HH, HT, TH, TT\}$, or, if you will simply count the number of Heads, then $\Omega = \{0, 1, 2\}$. In the game of Monopoly, the total score on two dice determines how far we move, and $\Omega = \{2, 3, \ldots, 12\}$. Ranging wider, in a dispute over the authorship of an article, Ω would be the list of all persons who could possibly have written it. "Experiment" here has a very wide meaning.

An *event* is a collection of outcomes whose probability we wish to describe. We shall use capital letters from the front of the alphabet to denote events, so that in throwing one die we might have $A = \{1, 2, 3, 4\}$ or, equivalently, we could say "A means the score is at most four". You can specify an event in any unambiguous way you like. It is enough to write $B = \{2, 3, 5\}$ without going to the tortuous lengths to find a phrase such as "B means we get a prime number". Any individual outcome is an event, but most events we shall be interested in will correspond to more than one outcome. We say that an event *occurs* when any of the outcomes that belong to it occur.

1

1.2 The Idea of Probability

We can get an intuitive understanding of probability notions by thinking about a simple experiment that is repeatable indefinitely often under essentially the same conditions, such as tossing an ordinary coin. Experience tells us that the proportion of Heads will fluctuate as we repeat the experiment but, after a large number of tosses, we expect this proportion to settle down around one half. We say "The probability of Heads is one half". This notion is termed "the *frequency interpretation* of probability".

I emphasise the phrase "a large number of tosses". No-one assumes the first two tosses will automatically balance out as one Head and one Tail, and the first six tosses might well all be Heads. So how many tosses are needed for the proportion of Heads to reach, and remain within, a whisker of 50%? Try it yourself. With just ten tosses, although five Heads and five Tails are more likely than any other specific combination, as few as two or as many as eight Heads should cause no alarm bells. On the other hand, with 100 tosses, the extremes of 20 or 80 Heads would be quite astonishing – we should be reasonably confident of between about 40 and 60 Heads. For 1000 tosses, the same degree of confidence attaches to the range from 470 to 530. In absolute terms, this range is wider than from 40 to 60, but, as a proportion of all the tosses, it is narrower. Were we prepared to make a million tosses, we could expect between 499 000 and 501 000 Heads with the same degree of confidence.

Probability is not restricted to repeatable experiments, but a different way of thinking is needed when the conditions cannot be recreated at will. Company executives make investment decisions based on their assessment of trading conditions, politicians are concerned with the probability they will win the next election. An art expert may claim to be 80% certain that Canaletto did not paint a particular picture, a weather forecaster suggests the chance of rain tomorrow is 50%. Here probability is being used to describe a *degree of belief*. To discover your own degree of belief in some event A, you could perform the following experiment, either for real, or in your imagination.

Take a flat disc, shaded entirely in black, fixed horizontally. At its centre is a pivot on which an arrow is mounted. When you spin the arrow, it will come to rest pointing in a completely random direction. A neutral observer colours a segment consisting of one quarter of the disc green, and poses the question:

which is more likely, the event A you are considering, or that

the arrow will come to rest in the green section?

If you think that A is more likely, the observer increases the size of the green segment, perhaps to one half of the disc, and poses the question again. If you think A is less likely, he similarly reduces the green region. This series of questions and adjustments continues until you cannot distinguish between the chances of the event A and the arrow settling in the green segment. Your degree of belief is then the proportion of the disc that is coloured green.

You can also use this idea to assess the probability you attach to repeatable events, such as selecting an Ace from a shuffled deck, or winning at Minesweeper or Solitaire on your home computer. For the Ace problem, most people quickly agree on the figure of one in thirteen without any auxiliary aid, but different people can legitimately have quite different ideas of good answers to the other two probabilities, even if they have identical experience.

1.3 Laws of Probability

There are three fundamental laws of probability. In the ordinary meaning of language, the words "impossible" and "certain" are at the extremes of a scale that measures opinions on the likelihoods of events happening. If event A is felt to be impossible, we say that the probability of A is zero, and write $P(A) = 0$, whereas if B is thought sure to occur, then $P(B) = 1$. If you select a card at random from an ordinary deck, it is certain the suit will be either red or black, it is impossible it will be yellow. Probabilities are real numbers and, whatever the event A, its probability $P(A)$ satisfies

Law 1 Always, $0 \leq P(A) \leq 1$.

Any values of a probability outside the range from zero to one make no sense, and if you calculate a probability to be negative, or to exceed unity (you will, it happens to all of us), your answer is wrong.

The second law is there to give assurance that our list of possible outcomes Ω really does include everything that might happen. If, for example, we write $\Omega = \{H, T\}$ when we toss a coin, we are excluding the possibility that it lands on its edge. This leads to

Law 2 $P(\Omega) = 1$.

To motivate the final law, note first that if we are interested in event A, then we are automatically interested in the complementary event, A^c, which consists precisely of those outcomes that do not belong to A. Exactly one of the events A and A^c will occur, and $(A^c)^c = A$. Further, if we are interested in events A and B separately, it is natural to be interested in both $A \cup B$ and $A \cap B$, respectively the ideas that *at least one of* A, B, and *both of* A, B occur. These notions extend to more than two events, so our collection $\mathcal{F}$ of events will have three properties:

1. Always, $\Omega \in \mathcal{F}$ (this means there is at least one event to talk about).

2. Whenever $A \in \mathcal{F}$, then also $A^c \in \mathcal{F}$.

3. If all of $A_1, A_2, A_3, \ldots \in \mathcal{F}$, then $\cup A_n \in \mathcal{F}$.

Since $\cap A_n = (\cup A_n^c)^c$, the last two properties ensure that whenever all of $A_1, A_2, A_3, \ldots \in \mathcal{F}$, then also $\cap A_n \in \mathcal{F}$, without the call to require this separately. The three properties of $\mathcal{F}$ describe what is known as a σ-field.

If we select just one card at random from an ordinary deck, the two events "Get a Heart" and "Get a Club" cannot both occur. Over a long series of experiments, we expect either event to occur about one quarter of the time, and so altogether half the experiments will lead to "Get either a Heart or a Club"; we add the frequencies together. On the other hand, although we expect a Red card half the time, we cannot similarly add the frequencies to deduce that we "Get a Heart or a Red card" three quarters of the time, since the Hearts are Red cards, and so would be counted twice. Again, since the Ace of Hearts is both an Ace and a Heart, we would not expect $P(\text{Ace or Heart}) = P(\text{Ace}) + P(\text{Heart})$. It is when events have *no overlap* that we should add their respective probabilities to find the chance that either occurs.

Formally, when $A \cap B = \varnothing$, the empty set, we say that A and B are *disjoint* or *mutually exclusive*. When one of them occurs, the other cannot. A sequence of events $A_1, A_2, A_3, \ldots$ is said to be *pairwise disjoint* if $A_i \cap A_j = \varnothing$ whenever $i \neq j$.

Law 3 Given pairwise disjoint events $A_1, A_2, A_3, \ldots$, then $P(\cup A_n) = \sum P(A_n)$.

Collecting these ideas together, we have

Definition 1.1

A *probability space* consists of three elements, $(\Omega, \mathcal{F}, P)$:

1. Ω is the set of outcomes.

2. $\mathcal{F}$ is a collection of subsets of Ω called events, that form a σ-field.

3. P is a function defined for each event, satisfying the three Laws.

It is usually not necessary to go to the lengths of spelling out every component of a probability space before you embark on any calculations, but it is useful to assure yourself you could do so, if necessary.

If A and B are such that $A \subset B$, this relation between sets is read as "A implies B" in terms of events. When A occurs, it is automatic that B occurs. (It is traditional that the statement "$A \subset B$" includes the possibility that $A = B$ in this corner of mathematics.) With a deck of cards, if A = Heart and B =

Red card then plainly $A \subset B$. In general, we might have either, both or neither of the statements $A \subset B$ and $B \subset A$ holding. A sequence of events such that $A_1 \subset A_2 \subset A_3 \subset \ldots$ (or with $\supset$ replacing $\subset$ throughout) is termed a *monotone sequence*.

1.4 Consequences

The three Laws have far-reaching consequences, among them

Theorem 1.2

For any events,

1. $P(A^c) = 1 - P(A)$.

2. If $A \subset B$, then $P(A) \leq P(B)$.

3. $P(A \cup B) = P(A) + P(B) - P(A \cap B)$.

Proof

1. Since A and A^c are disjoint, and also $A \cup A^c = \Omega$, then
$$1 = P(\Omega) = P(A \cup A^c) = P(A) + P(A^c)$$
from which the result follows.

2. We can write
$$B = (B \cap A) \cup (B \cap A^c) = A \cup (B \cap A^c)$$
as a union of disjoint events. Thus
$$P(B) = P(A) + P(B \cap A^c).$$
As the last term is a probability, it cannot be negative, so the result follows.

3. $A \cup B = A \cup (B \cap A^c)$ expresses $A \cup B$ as a union of disjoint events, so Law 3 shows that
$$P(A \cup B) = P(A) + P(B \cap A^c).$$
Moreover, $B = (B \cap A) \cup (B \cap A^c)$ is also a disjoint union, so Law 3 now leads to
$$P(B) = P(B \cap A) + P(B \cap A^c).$$
Subtracting one equation from the other gives our result.

$\square$

It is hard to overestimate the usefulness of the innocent expression that is the first part of this theorem. In order to find the probability that event A occurs, turn the question round and find the probability that A^c occurs, i.e. that A does not occur. This can be especially useful when A is of the form "At least one of such and such alternatives" – directly finding the chance that *none* of the alternatives occur is often much simpler.

Although the second statement of this theorem may look trivial, and hardly worth noting, consider the following information: "Linda is 31 years old, single, outspoken, and very bright. She majored in philosophy. As a student, she was deeply concerned with issues of discrimination and social justice, and also participated in anti-nuclear demonstrations." Psychologists Amos Tversky and Daniel Kahneman constructed eight possible pen-portraits of Linda, two of which were

- $A \equiv$ Linda is a bank teller active in the feminist movement.

- $B \equiv$ Linda is a bank teller.

They asked a group of subjects to rank these eight descriptions by the extent to which Linda would fit it: and 85% considered A to be more likely than B. But plainly $A \subset B$, so $P(A) \leq P(B)$. Any sensible ranking must have B more likely than A. Another possible application is when A is some complicated event, while B is much simpler, with $A \subset B$. This result then gives an upper bound on $P(A)$.

The third part is illustrated by the formula

$$P(\text{Ace or Heart}) = P(\text{Ace}) + P(\text{Heart}) - P(\text{Ace of Hearts}).$$

Whenever probabilities are constructed, they must be consistent with the results of this theorem. A decision to expand output may depend on judgements about events A = Interest rates will be low, and B = Demand will be high. Perhaps you believe that $P(A) = 65\%$ and that $P(B) = 80\%$. Then, unless you also believe that $45\% \leq P(A \cap B) \leq 65\%$, either the second or the third part of the theorem will be violated.

It is plain from the third part of the theorem that $P(A \cup B) \leq P(A) + P(B)$, whatever the events A and B. The exercises ask you to generalise this to prove *Boole's inequality,*

$$P\left(\bigcup_{n=1}^{N} A_n\right) \leq \sum_{n=1}^{N} P(A_n). \tag{1.1}$$

We now extend the third part of the theorem to give an expression for the chance that at least one of a collection of events occurs.

Corollary 1.3

$$P\left(\bigcup_{i=1}^{n} A_i\right) = S_1 - S_2 + S_3 - \cdots + (-1)^{n+1} S_n,$$

where

$$S_1 = \sum_{i=1}^{n} P(A_i), \quad S_2 = \sum_{i<j} P(A_i \cap A_j), \quad \ldots, \quad S_n = P\left(\bigcap_{i=1}^{n} A_i\right)$$

i.e. S_k is the sum, taken over all collections of precisely k events, of the probabilities that all these k events occur.

Proof

We use induction. The last part of the theorem shows this result holds when $n = 2$, so we suppose it holds for all collections of N events, and look at the collection $\{A_1, A_2, \ldots, A_{N+1}\}$. The main idea is to make use of the case $n = 2$ by writing

$$A_1 \cup A_2 \cup \cdots \cup A_{N+1} = (A_1 \cup A_2 \cup \cdots \cup A_N) \cup A_{N+1}.$$

Considering the right side as the union of two events, we have

$$P\left(\bigcup_{i=1}^{N+1} A_i\right) = P\left(\bigcup_{i=1}^{N} A_i\right) + P(A_{N+1}) - P\left(\left(\bigcup_{i=1}^{N} A_i\right) \cap A_{N+1}\right).$$

On the right side of this expression, apply the inductive hypothesis twice. First, use it as it stands on the term $P\left(\bigcup_{i=1}^{N} A_i\right)$, then rewrite the last term as $P\left(\bigcup_{i=1}^{N} (A_i \cap A_{N+1})\right)$ and apply the inductive hypothesis to *this* union of N events. Carefully collect together terms involving the joint probabilities of the same numbers of events and, lo and behold, the required expression emerges.   □

Example 1.4

In an investors' club, 38 members wish to buy automobile shares, 33 favour banks, and 39 favour construction shares. Moreover, 16 favour both automobiles and banks, 16 go for both banks and construction, 24 favour both construction and automobiles, while 6 want to buy all three types. How many club members want to buy at least one share in these categories?

To solve this, suppose there are N club members, and let A, B and C denote the events that a randomly chosen member favours automobiles, banks and construction respectively. The probability that an investor belongs to a given class is taken as the fraction of club members in that class. Using Corollary 1.3,

$$
\begin{aligned}
P(A \cup B \cup C) &= P(A) + P(B) + P(C) - P(A \cap B) - P(B \cap C) \\
&\quad - P(A \cap C) + P(A \cap B \cap C) \\
&= \frac{1}{N}(38 + 33 + 39 - 16 - 16 - 24 + 6) = \frac{60}{N}.
\end{aligned}
$$

Hence 60 members wish to buy at least one share in these categories.

Another universal result is

Theorem 1.5

Suppose $A_1 \subset A_2 \subset A_3 \subset \cdots$, and write $A = \bigcup_{n=1}^{\infty} A_n$. Then $P(A) = \lim P(A_n)$.

Proof

In order to use Law 3, we shall express A as a union of *disjoint* events, as follows: let $A_0 = \varnothing$, $B_1 = A_1$, $B_2 = A_2 \cap A_1^c$, ..., $B_n = A_n \cap A_{n-1}^c$, ... Plainly, the events $\{B_n\}$ are pairwise disjoint and, for all values of n, $A_n = \bigcup_{i=1}^{n} B_i$. Hence also $A = \bigcup_{n=1}^{\infty} B_n$, and so

$$
P(A) = P\left(\bigcup_{n=1}^{\infty} B_n\right) = \sum_{n=1}^{\infty} P(B_n) = \lim \sum_{n=1}^{N} P(B_n) \tag{1.2}
$$

by the definition of an infinite sum. But $A_n = B_n \cup A_{n-1}$ is a disjoint union, and so $P(B_n) = P(A_n) - P(A_{n-1})$; thus, for any N,

$$
\sum_{n=1}^{N} P(B_n) = \sum_{n=1}^{N}(P(A_n) - P(A_{n-1})) = P(A_N).
$$

Using Equation (1.2), the result follows. $\square$

Corollary 1.6

If $P(B_i) = 0$ for all values of i, then $P\left(\bigcup\limits_{i=1}^{\infty} B_i\right) = 0$.

Proof

Write $A_n = \bigcup\limits_{i=1}^{n} B_i$. Boole's inequality (1.1) shows that $P(A_n) \le \sum\limits_{i=1}^{n} P(B_i) = 0$. Plainly the events $\{A_n\}$ satisfy the conditions of the theorem, so $P(A) = \lim P(A_n) = 0$. But $A = \bigcup\limits_{n=1}^{\infty} A_n = \bigcup\limits_{n=1}^{\infty} \bigcup\limits_{i=1}^{n} B_i = \bigcup\limits_{i=1}^{\infty} B_i$, which establishes the result. □

EXERCISES

1.1. Find an expression in terms of $P(A)$, $P(B)$ and $P(A \cap B)$ for the probability that exactly one of the events A, B occurs.

1.2. To win the championship, City must beat both Town and United. They have a 60% chance of beating Town, a 70% chance of beating United, and an 80% chance of at least one victory. What is the chance they win the championship?

1.3. Use induction to prove Boole's inequality, as stated above in expression (1.1). Deduce that

$$P\left(\bigcap_{i=1}^{n} A_i\right) \ge 1 - \sum_{i=1}^{n} P(A_i^c).$$

1.4. Show that, if $P(B_n) = 1$ for $n = 1, 2, \ldots$, then $P\left(\bigcap\limits_{n=1}^{\infty} B_n\right) = 1$.

1.5. Use Theorem 1.5 to prove that, if $B_1 \supset B_2 \supset B_3 \supset \cdots$, then $P\left(\bigcap\limits_{n=1}^{\infty} B_n\right) = \lim P(B_n)$.

1.6. In the notation of Corollary 1.3, prove *Bonferroni's inequalities*, i.e.

$$S_1 - S_2 + \cdots - S_{2k} \le P\left(\bigcup_{i=1}^{n} A_i\right) \le S_1 - S_2 + \cdots + S_{2k-1}$$

for $k = 1, 2, 3, \ldots$.

1.5 Equally Likely Outcomes

Many experiments with finitely many outcomes have such a degree of symmetry that it is reasonable to assume that all these outcomes are equally likely. If there are K equally likely outcomes, since they form a collection of K disjoint events whose total probability is unity, it follows from Laws 2 and 3 that each outcome has probability $1/K$. Here the σ-field $\mathcal{F}$ of events can be taken as the collection of all possible subsets of Ω, the set of outcomes, and if A is an event that consists of r of these outcomes, then $P(A) = r/K$. In these experiments, questions of probability reduce to questions of counting.

If the outcomes are labelled as $\{1, 2, 3, \ldots, K\}$, we refer to the *discrete uniform* distribution over $[1, K]$, and denote it by $U[1, K]$. Many examples will be familiar.

Example 1.7

Throw an ordinary die once. To say it is fair is to say that all six outcomes are equally likely, so we have the discrete uniform distribution $U[1, 6]$, i.e. $P(i) = 1/6$ for $i = 1, 2, \ldots, 6$.

Now throw two fair dice, and record the total score. Plainly, this will be in the range from 2 to 12, and clearly these values are not all equally likely. As an intermediate step, we can list the outcomes as $\Omega = \{(i, j) : 1 \le i \le 6, 1 \le j \le 6\}$, where i is the score on the blue die, and j that on the red die. "Fairness" means we have the discrete uniform distribution over the 36 outcomes in Ω, so we can treat the distinct scores as events, and find their respective probabilities by counting how often each score can arise.

Score	2, 12	3, 11	4, 10	5, 9	6, 8	7
Probability	1/36	2/36	3/36	4/36	5/36	6/36

The answers are in the table.

Most counting problems can be resolved by the careful application of one principle: that if there are M ways to do one thing, and then N ways to do another, there are $M \times N$ ways to do both things, in that order. Applying that principle in various circumstances, we collect together some useful results, which I assume you already know, or can quickly justify.

1. If experiment j has n_j outcomes for $j = 1, 2, \ldots, m$, then the composite experiment of performing all these experiments in sequence has $n_1 \times n_2 \times \cdots \times n_m$ outcomes.

2. There are $(n)_k = n(n-1)(n-2)\cdots(n-k+1)$ ways of selecting k objects from a collection of n objects, without replacement but taking notice of the order of selection.

3. There are $\binom{n}{k} = \frac{n!}{k!(n-k)!}$ ways of selecting k objects from a collection of n objects, without replacement and ignoring the order of selection.

4. Given n_j objects of type j for $j = 1, 2, \ldots, m$, and so $n = n_1 + n_2 + \cdots + n_m$ objects altogether, there are $\frac{n!}{n_1! n_2! \cdots n_m!}$ ways of arranging them in order, if objects of the same type are not distinguished.

Example 1.8

The Art Gallery owns 6 Old Masters, 9 works by French Impressionists, and 5 modern pieces.

1. It can select one object from each category in $6 \times 9 \times 5 = 270$ ways.

2. In choosing pictures to hang in four spaces, there are $\binom{6}{4} = 15$ ways to select four Old Masters, and $(6)_4 = 360$ arrangements to display four.

3. There are $\binom{6}{2} \times \binom{9}{2} \times \binom{5}{2} = 5400$ ways to choose two representatives from each category.

4. There are $20! \approx 2.43 \times 10^{18}$ ways of *ranking* the pieces in the entire collection. If, in such a ranking, only the *type* of painting is specified, there are $\frac{20!}{6!9!5!} = 77\,597\,520$ different rankings.

A result that surprises many people the first time they meet it is the size of a group of randomly selected people that gives a good chance – say at least 50% – that two of them share a birthday. A common guess is 183. (Why?) Although more people have birth dates in the summer than in the winter, a reasonable initial model is to say that all possible birth dates are equally likely. If there are K possible dates, then the birth date of any one person is taken as uniformly distributed over $[1, K]$. If we make n selections from this distribution, ignoring previous results each time, there are $(K)_n = K(K-1)\cdots(K-n+1)$ ways in which all the birth dates are *different*. And as there are K^n ways of selecting birth dates for this sample of n people, our model leads to the expression $\frac{(K)_n}{K^n}$ as the probability that these n people all have different birth dates. Subtract this from unity to give the chance that our group has at least one pair with a common birthday (an easy use of the first part of Theorem 1.2).

Ignoring February 29, we take $K = 365$; it then transpires that so long as $n < 23$ it is more likely than not that all the group have different birthdays, but as soon as 23 or more are present, it becomes more likely that some pair share a birth date. Taking $K = 366$ leads to the same answer. Of course, our model is far from perfect as a representation of reality, but when we allow for the fact that some dates are slightly more likely than others to be birth dates, this only *increases* the chances of there being coincident birth dates. Groups of at least 23 randomly selected humans are more likely than not to contain a pair with a common birth date.

One place where it is intended that the discrete uniform distribution be an accurate model of reality is in organised lotteries. The first modern UK National Lottery asked players to select six different numbers from the list $\{1, 2, \ldots, 49\}$, so Ω consists of the $\binom{49}{6} = 13\,983\,816$ ways they can do so. An inanimate machine was designed to select six numbers from the same list "at random", so assuming it worked as intended, any player buying one ticket had chance $1/13\,983\,816$ of matching the machine's selection. To find the corresponding chances of matching exactly k of the winning numbers, we embark on a counting exercise. There are $\binom{6}{k}$ ways of choosing k of the winning numbers, and these are to be combined with the $\binom{43}{6-k}$ ways in which the remaining selection is among the losing numbers. That means the chance of matching k numbers in a 6/49 Lottery is $\binom{6}{k} \times \binom{43}{6-k} \div \binom{49}{6}$. The values (rounded) are shown in the table.

Number correct	Probability
6	7.2×10^{-8}
5	1.84×10^{-5}
4	0.000967
3	0.0177
2	0.1324
1	0.4130
0	0.4360

Nearly 85% of tickets bought match no more than one of the winning numbers.

Example 1.9

A secretary types n letters and their corresponding envelopes, but then ignores the addresses when putting the letters in the envelopes. What can be said about the number of letters that are in the correct envelopes?

This points to the model in which Ω is the list of the $n!$ ways of distributing the letters among the envelopes, and all these outcomes are to be equally likely. Let A_i be the event that letter i is in its correct envelope. We first use Corollary 1.3 to find $P\left(\bigcup_{i=1}^{n} A_i\right)$, the chance that at least one letter is in the correct envelope.

It is easy to see that $P(A_i) = 1/n$ for all values of i, and so $S_1 = n \times (1/n) = 1$. If the letters labelled $\{1, 2, \ldots, r\}$ are placed in their correct envelopes, there are $(n-r)!$ ways to distribute the other letters. Thus the probability that those r letters are in their correct envelopes is $(n-r)!/n!$. The same argument applies to every collection of precisely r letters, and since there are $\binom{n}{r}$ such collections, then $S_r = \binom{n}{r} \times (n-r)!/n! = 1/r!$. Hence

$$P\left(\bigcup_{i=1}^{n} A_i\right) = 1 - 1/2! + 1/3! - \cdots + (-1)^{n+1}/n!.$$

By the first part of Theorem 1.2, the probability that *no* letter is in its correct envelope is

$$P(0) = 1 - 1 + 1/2! - 1/3! + \cdots + (-1)^n/n! \qquad (1.3)$$

which you will recognise as the beginning of the usual expansion for e^{-1}.

Let $P(r)$ denote the probability that exactly r letters are in their correct envelopes. If it is the letters labelled $\{1, 2, \ldots, r\}$ that are inserted correctly, then all the remaining $(n-r)$ letters must be in the wrong envelopes. Since there are $(n-r)!$ ways of placing these letters somewhere, then Equation (1.3) shows that there are

$$(n-r)! \times (1 - 1 + 1/2! - 1/3! + \cdots + (-1)^{n-r}/(n-r)!) \qquad (1.4)$$

ways of having exactly the first r letters correctly inserted. But there are $\binom{n}{r}$ ways to choose which r letters are to be in their correct envelopes, so multiplying expression (1.4) by $\binom{n}{r}$ and then dividing by $n!$, the total number of arrangements, we see that

$$P(r) = (1 - 1 + 1/2! - 1/3! + \cdots + (-1)^{n-r}/(n-r)!)/r! \qquad (1.5)$$

for $r = 1, 2, \ldots$. If you compare Equations (1.5) and (1.3), you will see that this last expression also holds in the case $r = 0$ – maths is often kind to us in this fashion.

In Equation (1.5), fix r and let $n \to \infty$; then the expression for the probability that exactly r letters are correctly inserted, $P(r)$, converges to $e^{-1}/r!$. This collection of probabilities for $r = 0, 1, 2, \ldots$ is an example of the *Poisson distribution* which will appear again in Chapter 3.

After a probability calculation, it is often a good idea to pause and check that the answers make sense, to help avoid gross blunders. In this example, when $r = n - 1$ in Equation (1.5), the right side evaluates to zero – which, on reflection, is correct; it is not possible that *exactly* $n-1$ letters are in the correct envelopes. For another check, in the special cases when $n = 1, 2$ or 3, we can easily list all the outcomes and confirm the answers. Another check would be to sum all the values of the right sides in expression (1.5), for $r = 0, 1, 2, \ldots, n$, and confirm that the total is unity.

Sheldon Ross (1998) has given an ingenious example. Here is a simplified version. Imagine that it is one minute to midnight, and you have an empty urn and an infinite number of balls labelled $\{1, 2, 3, \ldots\}$. You are supremely dextrous, taking zero time to drop a ball into the urn, or to extract a ball from it. First, you drop balls labelled $\{1, 2\}$ into the urn, and withdraw ball number 2; thirty seconds later, you drop in balls $\{3, 4\}$, and take out ball 4; fifteen seconds later, in go $\{5, 6\}$, out comes 6; 7.5 seconds later, in go $\{7, 8\}$, out comes 8, and so on. How many balls are in the urn at midnight?

Plainly, there are infinitely many, as the balls labelled $\{1, 3, 5, 7, \ldots\}$ are dropped in, and left alone. Now suppose the process is altered slightly: we drop balls in exactly as before, but this time withdraw the lowest numbered of all the balls in the urn. How many are now in the urn at midnight? Well, whatever ball number you consider, be it $3, 42, 25\,506$, or whatever, that ball will *not* be in the urn, as it will have been withdrawn. So the urn is empty, even though you have inserted and withdrawn balls in the same numbers, and at the same times, as before!

It is not *how many* balls are withdrawn, it is *which* of them are withdrawn that leads to different answers. So far this has appeared as a warning to be careful when dealing with the infinite, and with limits: let us bring some probability into the process. Drop balls in as before, but now the ball withdrawn is to be selected "at random", i.e. all those then in the urn have the same chance of being withdrawn. What happens here?

Concentrate first on the ball labelled 1, and let A_n denote the event that it remains in the urn after n have been withdrawn. At the time of the rth withdrawal, there are $r + 1$ balls in the urn, so the total number of choices up to the time the nth ball has been withdrawn is $2 \times 3 \times \cdots \times (n + 1)$. The total number of ways in which event A_n can occur is $1 \times 2 \times \cdots \times n$, so

$$P(A_n) = \frac{1 \times 2 \times \cdots \times n}{2 \times 3 \times \cdots \times (n + 1)} = \frac{1}{n + 1}. \tag{1.6}$$

The sequence $\{A_n\}$ is monotone decreasing, so the result of Exercise 1.5 shows that $P\left(\bigcap_{n=1}^{\infty} A_n\right) = \lim(1/(n+1)) = 0$. But the event $B_1 = \bigcap_{n=1}^{\infty} A_n$ corresponds to "Ball 1 is in the urn at midnight", so we have just shown that $P(B_1) = 0$.

Write $B_K = $ "Ball K is in the urn at midnight", and apply the same reasoning. When K is either $2m - 1$ or $2m$, there are $m + 1$ balls in the urn immediately after it has been dropped in. Write C_n as the event that ball K remains in the urn after the withdrawal linked to the insertion of balls labelled $2n - 1$ and $2n$. Corresponding to Equation (1.6), we find that, when $n \geq m$,

$$P(C_n) = \frac{m \times (m + 1) \times \cdots \times n}{(m + 1) \times (m + 2) \times \cdots \times (n + 1)} = \frac{m}{n + 1}.$$

Since $B_K = \bigcap\limits_{n=m}^{\infty} C_n$, the same reasoning shows that $P(B_K) = 0$, and this is true for all values of K. Let $D_N = \bigcup\limits_{K=1}^{N} B_K$, i.e. D_N is the event that at least one of the balls labelled $\{1, 2, \ldots, N\}$ is in the urn at midnight. Boole's inequality (Exercise 1.3) shows that

$$P(D_N) = P\left(\bigcup_{K=1}^{N} B_K\right) \leq \sum_{K=1}^{N} P(B_K) = 0.$$

The events $\{D_N\}$ form a monotone sequence, and $D = \bigcup\limits_{N=1}^{\infty} D_N$ is the event that at least one ball is in the urn at midnight. Now Theorem 1.5 shows that $P(D) = 0$, i.e. the probability that there are any balls in the urn at midnight is zero. We can be certain the urn is empty, if balls are withdrawn at random.

Example 1.10

(TV game show addicts will recognise the Showcase Showdown as inspiring this account.) A device is capable of selecting one of the numbers $\{1, 2, \ldots, K\}$ at random, i.e. all of them are equally likely. Alex can profess himself satisfied with the first number generated ("stick"), or he may ask for a second number to be selected ("spin"). If he spins, his score is the sum of the two numbers, unless that sum is greater than K, in which case his score becomes zero. What are the consequences of the decision to stick or to spin?

If he sticks, his score has the uniform distribution $U[1, K]$. Were he to spin from an initial score of K, he is doomed to score zero. Suppose he spins from a score of n, where $1 \leq n < K$. If his second score exceeds $K - n$, which occurs with probability n/K, his total will be reduced to zero. Otherwise, his score will be among the list $\{n + 1, n + 2, \ldots, K\}$ and, by symmetry, all these scores are equally likely. To ensure the total probability of all scores is unity, each of these scores will have probability $1/K$.

Now suppose Alex's final score is r, where $0 \leq r \leq K$, and introduce his oppponent Bella, who sees him perform and operates to the same rules. She will win if she *beats* his score, he wins with a tie. What is the chance she wins?

Write $A =$ She wins at her first turn, and $B =$ She wins on her second turn. These are disjoint events, so we shall sum their probabilities. Event A occurs when her first score is among $\{r + 1, r + 2, \ldots, K\}$ and she sticks, so $P(A) = (K - r)/K$. Otherwise, her first score is among $\{1, 2, \ldots, r\}$, all equally likely, and she will spin. When she does so, whatever her initial score, there are exactly $K - r$ scores that allow her to overtake Alex, without being reduced to zero (think about it). Hence the total number of score combinations that lead to her winning on her second turn is $r(K - r)$, and as there are K^2 possible combinations from two spins, $P(B) = r(K - r)/K^2$. Thus Bella's winning chance is the sum

$$(K - r)/K + r(K - r)/K^2 = 1 - (r/K)^2.$$

There are some quick checks on the plausibility of this answer. It decreases from unity when $r = 0$ (plainly sensible) to zero when $r = K$ – also obviously correct. Good.

1.6 The Continuous Version

Now turn to the case when, instead of there being finitely many outcomes that are deemed equally likely from symmetry considerations, there are infinitely many. This gives an immediate problem – equally likely outcomes have the same probability, but passing to the limit as $K \to \infty$ means that each outcome has probability zero. And a whole lot of zeros still adds up to zero.

The continuous version of an experiment with equally likely outcomes is exemplified by the notion of selecting a single point, "at random", on a straight rod. Although any individual point will have zero probability, divide the rod into K equal intervals by marking $K - 1$ points appropriately. Selecting a *random* point means it should be equally likely to fall into any one of these intervals. Indeed, the chance a random point falls in any *interval* should be proportional to the length of that interval, whatever its position.

Definition 1.11

Suppose $\Omega = (a, b)$, the continuous interval from a to b (we do not care whether or not the endpoints are included, as all individual points have zero probability). Then the *continuous uniform distribution* over Ω is defined by the property that, whenever $a < c < d < b$, the probability of the interval (c, d) is $(d - c)/(b - a)$. We write $U(a, b)$, and allow the context to distinguish the discrete and continuous versions.

In particular, for the continuous uniform distribution over $(0, 1)$, the probability of any given interval is just the length of that interval.

What should the σ-field of events be? We have defined the probabilities for intervals, so it should certainly include all intervals, and all events obtained from sequences of intervals by the operations of complementation, union and intersection. The collection of events so generated is huge. Every individual point x is an event, as $x = \bigcap_{n=1}^{\infty}(x - 1/n, x + 1/n)$. Unions of disjoint intervals, with sequences of individual points also thrown in, are included. In fact, you have to work very hard to define a subset of Ω that does not arise in this fashion! And then Theorem 1.2 and its consequences enable us to calculate the probabilities of all these events, knowing the probabilities of intervals. This last point holds whenever we have a recipe for the probability of an interval, using this uniform model or not. So we make the following convention when the set of outcomes Ω is some interval (a, b): the σ-field of events will always be the smallest possible σ-field that contains all subintervals of Ω. With this convention, once we have described how to find the probability of an arbitrary interval within Ω, the probability space $(\Omega, \mathcal{F}, P)$ is specified. This really is a substantial simplification of the work otherwise required. We shall return to this remark in Chapter 3.

Imagine a darts player whose aim is so poor that, whenever an arrow hits the dartboard, it does so at a random point. The appropriate model will be the two-dimensional version of choosing a random point on a rod. For this player, Ω will be a disc and, by analogy with the rod, if we draw a rectangle within that disc, the chance a dart lands in it will be the ratio of the area of the rectangle to the area of the disc. $\mathcal{F}$ will consist of all shapes that can be constructed by the usual set operations on rectangles and, as with the one-dimensional rod, this is a vast collection. Figure 1.1 shows the beginnings of building up a circle as a union of a sequence of rectangles, and plainly there is an analogous construction for other shapes within the disc. As with intervals in one dimension, so knowing how to find the probability of a rectangle in two dimensions leads to the probability for any event in $\mathcal{F}$.

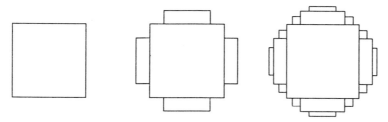

Figure 1.1 Building up a circle from a union of rectangles.

Exactly the same idea applies to choosing a "random point" in other two-dimensional bodies with a finite area, such as squares, ellipses or the surface of the Earth. The probability of any rectangle will be the ratio of the area of the rectangle to that of the whole body, events will be any shapes formed out of set operations on sequences of rectangles, and the probability space is sufficiently specified. We will apply these ideas to a particular problem.

Example 1.12

If a stick is broken at two points chosen at random on it, what is the probability that the three lengths of stick formed can make a triangle? Plainly, a triangle can be formed whenever the longest piece is shorter than the sum of the other two lengths or, equivalently, when none of the pieces is more than half the original length. If we were choosing just one point at random, the natural model would be to take the stick to have unit length, and select according to a $U(0,1)$ distribution. But choosing two points, the second selected without reference to the first, suggests a two-dimensional model: write

$$\Omega = \{(x,y) : 0 \le x \le 1, 0 \le y \le 1\},$$

where x and y are the distances of the break points from the left end of the stick. Thus our set of outcomes is the unit square, and choosing the two breaking points at random corresponds to selecting a random point in this unit square.

Formally, the probability space $(\Omega, \mathcal{F}, P)$ is given by

- Ω is the unit square, $\{(x,y) : 0 \le x \le 1, 0 \le y \le 1\}$.

- $\mathcal{F}$ consists of those subsets of Ω generated from sequences of rectangles by set operations.

- P is defined by $P(A) =$ The area of A.

To find the probability that a triangle can be formed, consider the cases $x < y$ and $y < x$ separately. When $x < y$, the three pieces have lengths x, $y - x$, and $1 - y$, and they will make a triangle so long as each of these lengths is less than one half, i.e.

$$x < 1/2, \quad y - x < 1/2, \quad 1 - y < 1/2$$

which is the same as

$$x < 1/2, \quad y - x < 1/2, \quad y > 1/2.$$

Similarly, when $x > y$, the condition becomes

$$x > 1/2, \quad x - y < 1/2, \quad y < 1/2$$

and so the point (x,y) must fall within the shaded area of Figure 1.2.

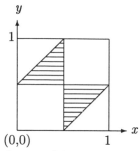

Figure 1.2 Where (x, y) must fall to enable a triangle to be formed.

Plainly, the shaded region occupies one quarter of the total area. There is one chance in four that the three pieces will form a triangle.

What can be said about the length of a chord, chosen at random in a circle? Again we need a model, but the exact meaning of "at random" is not so clear as before. One possible meaning is to choose two random points on the circumference of the circle (just like choosing two random points on a stick) and join them. Another is to select two points P and Q at random in the interior of the circle, then extend the line PQ to a full chord. Or select just one point, P, at random in the interior, and draw the chord through P perpendicular to the radius joining P to the centre of the circle. Not only are these descriptions different in wording, they also lead to different answers to such questions as the average length of a chord, or the chance the chord is longer than the radius of the circle. These are just three of the many ways of giving a reasonable interpretation to the phrase "a random chord". That phrase, alone, is too vague to work with. We might just as well ask how long is a piece of string.

EXERCISES

1.7. Set up a probability space for tossing a fair coin three times. What is the chance all three tosses give the same result?

Find the flaw in the argument: "Plainly, at least two of the tosses are the same, both H or both T. As the third coin is equally likely to be H or T, there is a 50% chance all three are alike."

1.8. Assume that the last two digits on a car number plate are equally likely to be any of the one hundred outcomes $\{00, 01, 02, \ldots, 98, 99\}$. Peter bets Paul, at even money, that at least two of the next n cars seen will have the same last two digits. Does $n = 16$ favour Peter or Paul? What value of n would make this a pretty fair bet?

1.9. Show that, for $0 < x < 1/2$,

$$x + \frac{x^2}{2} \le -\ln(1-x) \le x + x^2.$$

We have seen that if we make K random selections among n equally likely objects, with replacement each time, the chance they are all different is $(n)_K/n^K = p_K$, say. Deduce that, if $K < n/2$, then

$$\frac{K(K-1)(2K-1)}{12n^2} \le -\frac{K(K-1)}{2n} - \ln(p_K) \le \frac{K(K-1)(2K-1)}{6n^2}.$$

Hence show that, when n is large and we make about $\sqrt{-2n\ln(p)}$ selections, the chance they are all different is close to p.

1.10. An examiner sets twelve problems, and tells the class that the exam will consist of six of them, selected at random. Gavin memorises the solutions to eight problems in the list, but cannot solve any of the others. What is the chance he gets four or more correct?

1.11. A poker hand consists of five cards from an ordinary deck of 52 cards. How many poker hands are

(a) Flushes (i.e. all five cards are from the same suit).

(b) Four of a Kind (e.g. four Tens, and a Queen).

(c) Two Pairs (e.g. two Kings, two Fours, and a Seven).

1.12. (a) Write $x(r) = (43)_r/(49)_r$. Explain why $x(r)$ is the probability that none of the numbers $\{1, 2, \ldots, r\}$ are in the winning combination in a given draw of a 6/49 Lottery.

(b) Let A_i be the event that the number i does not appear in the winning combination in n given Lottery draws. Find $P(A_i)$ and $P(A_i \cap A_j)$ when $i \ne j$ in terms of $x(1)$ and $x(2)$. Hence use Corollary 1.3 to get an expression for the probability that *some* number does not appear as a winning number among the next n draws. Use Bonferroni's inequalities (Exercise 1.6) to evaluate this when $n = 50$.

1.13. Let $P = (x, y)$ be a point chosen at random in the unit disc, centre $(0, 0)$ and radius 1. Describe the model of this experiment, and evaluate the probabilities that P is within 0.5 of the centre; that $y > 1/\sqrt{2}$; and that both $|x - y| < 1$ and $|x + y| < 1$.

1.14. A breakdown vehicle cruises along the straight road of unit length that links Newtown to Seaport; help for stranded motorists is also available in both towns. Steve's car runs out of petrol. If x and y denote the distances of his car and the breakdown vehicle from Newtown, the distribution of (x, y) can reasonably be taken as uniform over the unit square, i.e. identical to that in Example 1.12. Show that, for $0 \leq t \leq 1/2$, the probability that Steve's nearest help is within distance t is $4t(1 - t)$.

1.15. In a fairground game, a flat surface is ruled with a grid of lines into squares of side s. A coin of diameter d ($d < s$) is rolled down a channel, and eventually topples over to come to rest on this grid. A prize is awarded if the whole of the coin is in the interior of a marked square. Set up a model, and find the chance of winning a prize.

1.7 Intellectual Honesty

Why have we been so cagey about what subsets of Ω are allowed to be described as events? We have gone out of our way to avoid saying that an event is *any* subset of Ω. In fact, there would be no problem in saying that any subset of Ω would qualify as an event, so long as we could label the outcomes as a sequence, e.g. $(0, 1, 2, \ldots)$. In these cases, as Chapter 3 will show, each outcome is assigned a probability, from which the probability of any subset of Ω can be calculated.

But suppose Ω were the circumference of a circle, and we wanted the probability of any interval J on that circumference to be proportional to the length of J. That corresponds to the idea of choosing a point at random on the circumference. For any subset A of Ω (not necessarily an interval) let $A(\theta)$ be the set obtained by rotating A through the angle θ.

By using great ingenuity (the details are in the Appendix of Capiński and Kopp, 1999), it is possible to construct a set E, thinly smeared around Ω, with the following properties:

1. Whenever θ and ϕ are different rational numbers, representing angles in the range $[0, 2\pi)$, then $E(\theta) \cap E(\phi) = \varnothing$.

2. $\cup E(\theta) = \Omega$, the union being taken over all rational angles θ with $0 \leq \theta < 2\pi$.

Since the sets $\{E(\theta)\}$ are just rotations of each other, we shall want them all to have the same probability as the original set E. Call this probability x.

Then we have

$$1 = P(\Omega) = P\left(\bigcup_\theta E(\theta)\right) = \sum_\theta P(E(\theta)) = \sum x.$$

And now we have a problem. Either $x = 0$, which gives the contradiction $1 = 0$, or $x > 0$, which leads to $1 = \infty$. Neither of these statements are tenable, but one of them must hold. The only thing that has gone wrong is to allow ourselves to suppose that $P(E)$ were a meaningful statement. The only escape is to accept that there can be subsets of Ω that do not qualify as events, to which probabilities can be attached.

There you have it. It is not laziness, unwillingness to do more work than is strictly necessary, that makes us place restrictions on what can be events, it is mathematical necessity.

2

Conditional Probability and Independence

2.1 Conditional Probability

If you throw a fair die, with no clues given about the outcome, the chance of getting a six is 1/6. But maybe you have been told that this *blue* die, and another *red* die were thrown, and their total score was four. Then you could be sure that the score on the blue die was not six. Similarly, to be told that the total score was ten makes it more likely that the blue die scored six, as all scores lower than four are eliminated. Information about the total score on the two dice can change your original opinion about the chance of a six.

"Conditional probability" addresses this sort of question. You are interested in some event A, whose probability is $P(A)$, and you learn that event B has occurred. What is your new opinion about the chance that A has occurred? The notation for what you seek is $P(A|B)$, read as "The probability of A, conditional upon B", or as "The probability of A, given B".

To see how this might relate to our previous ideas, consider a repeatable experiment with K equally likely outcomes. The probabilities of events A and B are then found by counting; how to assess the probability of A, given that B occurs? Plainly we can ignore all those experiments in which B does not occur. In the experiments where B does occur, we need to know how often A also occurs, i.e. how often $A \cap B$ occurs. If B occurs in $n(B)$ experiments, and A occurs in $n(A \cap B)$ of these, the ratio

$$n(A \cap B)/n(B)$$

23

should be close to the conditional probability of A given B, when $n(B)$ is large. But note that

$$\frac{n(A \cap B)}{n(B)} = \frac{n(A \cap B)}{N} \Big/ \frac{n(B)}{N}$$

where N is the total number of experiments. Since $n(A \cap B)/N \approx P(A \cap B)$ while $n(B)/N \approx P(B)$ when N is large, we make the following definition:

Definition 2.1

$P(A|B) = P(A \cap B)/P(B)$, provided that $P(B) > 0$.

It is an immediate consequence of this definition that

$$P(A \cap B) = P(A|B)P(B).$$

Examples often help to settle ideas. Suppose the experiment is selecting one card at random from a well-shuffled deck, so in our model Ω has 52 equally likely outcomes. Let events A, B, C be that the selected card is, respectively, an Ace, a Black card, or a Club. Then $P(A) = 1/13$, $P(B) = 1/2$ and $P(C) = 1/4$, as you should verify. We will check that use of the definition of conditional probability is in tune with our intuitive ideas. Simply using the definition, we see that

$$P(C|B) = P(C \cap B)/P(B) = P(C)/P(B) = 1/2,$$

which makes sense – given we have a Black card, there is a 50% chance it is a Club. The other way round leads to

$$P(B|C) = P(B \cap C)/P(C) = P(C)/P(C) = 1,$$

which is equally unsurprising – given we have a Club, we can be certain we have a Black card. And

$$P(A|C) = P(A \cap C)/P(C) = (1/52)/(1/4) = 1/13$$

i.e. given we have a Club, there is one chance in thirteen it is the Ace. The definition is working as we would anticipate.

The table in Example 1.7 shows the distribution of the total score when two fair dice are thrown. Write

- A = Total is at least seven, so that $P(A) = 21/36$

- B = Blue die scores at most four, so that $P(B) = 4/6$

- C = Blue die scores at most three, so that $P(C) = 3/6$

(it seldom helps to simplify fractions at this stage). By counting, $P(A \cap B) = 10/36$, and $P(A \cap C) = 6/36$, so the definition leads to

$$P(A|B) = (10/36)/(4/6) = 5/12, \quad P(A|C) = (6/36)/(3/6) = 4/12.$$

The information that the score on the Blue die is at most three, rather than at most four, makes it less likely that the total score is at least seven - as it should.

Theorem 2.2

Given a probability space $(\Omega, \mathcal{F}, P)$, let B be some event with $P(B) > 0$. With the same Ω and $\mathcal{F}$, define the function Q by

$$Q(A) = \frac{P(A \cap B)}{P(B)}.$$

Then $(\Omega, \mathcal{F}, Q)$ is also a probability space.

Proof

The only things to check are that Q satisfies the three Laws for a probability. First, since $A \cap B \subset B$, then $P(A \cap B) \leq P(B)$, so that $0 \leq Q(A) \leq 1$. Also, it is clear from the definition that $Q(\Omega) = 1$. Finally, let $\{C_i\}$ be a collection of pairwise disjoint events in $\mathcal{F}$; then

$$Q\left(\bigcup_i C_i\right) = \frac{P((\cup C_i) \cap B)}{P(B)} = \frac{P(\cup(C_i \cap B))}{P(B)} = \frac{\sum_i P(C_i \cap B)}{P(B)} = \sum_i Q(C_i)$$

which completes the proof. □

This means that any conditional probability $P(.|B)$ is also an ordinary probability. We do not have to use Definition 2.1 to compute a conditional probability, it is sometimes best to work directly.

As an illustration of this, we will compute the probability that a Bridge hand of 13 cards has at least two Aces, conditional on it containing the Ace of Spades. Using this direct approach, our model is that the hand consists of the Ace of Spades, along with 12 other cards that are selected at random from the remaining 51 cards, which comprise three Aces and 48 non-Aces. The complete hand will have at least two Aces when these 12 other cards contain at least one Ace.

This observation has removed the need to use the definition of conditional probability, we just work on the probability space for selecting 12 cards at random from 51. There are $\binom{51}{12}$ outcomes in this space, and $\binom{48}{12}$ of them have no Aces at all. Thus the ratio $\binom{48}{12}/\binom{51}{12}$ gives the chance of no Ace. We find the chance of at least one Ace by the usual trick of subtracting this value from unity, which gives the answer of 0.5612. The chance we have at least two Aces, given we have the Ace of Spades, is about 56%.

Suppose, on the other hand, that we want to assess the chance that our hand has at least two Aces, given it has at least one Ace. Many people will expect the answer to be the same as we have just computed. "After all", they say, "given that the hand has at least one Ace, the Spade Ace is just as likely as any other." An accurate observation, but not one from which the intended deduction can be drawn. Let's do it properly.

Write A as the event that the hand has at least two Aces, and B that it has at least one Ace. To find their probabilities, it is easiest to find $P(0)$ and $P(1)$, the chances that a hand has respectively no Aces and just one Ace. Counting again gives the answers. There are $\binom{52}{13}$ possible hands, and $\binom{48}{13}$ of them have no Ace, while $\binom{48}{12}\binom{4}{1}$ have one Ace. This leads to

$$P(0) = 0.3038 \quad \text{and} \quad P(1) = 0.4388.$$

Thus

$$P(B) = 1 - P(0) = 0.6962 \quad \text{and} \quad P(A) = 1 - P(0) - P(1) = 0.2573.$$

Since $A \subset B$, then
$$P(A|B) = P(A)/P(B) = 0.3696$$

when the dust has settled. Given we have at least one Ace, the chance we have at least two Aces is about 37%.

Intuitively, why should these two answers be so different? Think along the following lines: knowing that the hand has a particular Ace, each of the remaining 12 cards has chance 3/51 of being an Ace so there will be, on average, $12 \times 3/51 = 36/51 \approx 0.7$ other Aces. Since this comfortably exceeds one half, there is a pretty good chance of at least one Ace. Now look at hands in general. Symmetry dictates that they have one Ace, on average, so if we eliminate the Aceless hands, just one Ace seems rather more likely than two or more. But the statement "Given there is at least one Ace" does no more nor less than eliminate the Aceless hands. Perhaps the result is no real surprise.

Explicit use of the formula $P(A \cap B) = P(A|B)P(B)$ simplifies many calculations. Take an urn that initially has six white balls and four red ones. Balls are withdrawn at random, all balls in the urn being equally likely, and after each drawing, the ball is replaced, along with another of the same colour. What is the probability that the first two balls drawn are red?

Let A mean that the *second* ball drawn is red, and B that the first one is red. We shall find $P(A \cap B)$ using this formula. Plainly $P(B) = 4/10$ and, conditional on B, the urn has six white balls and five red ones when the second ball is selected. Hence $P(A|B) = 5/11$, and the answer to our problem is $(4/10) \times (5/11) = 2/11$.

This notion extends to more than two events:

Theorem 2.3 (Multiplication Rule)

Suppose $\{A_1, A_2, \ldots, A_n\}$ are events in the same probability space. Then

$$P\left(\bigcap_{i=1}^{n} A_i\right) = P(A_1)P(A_2|A_1)P(A_3|A_1 \cap A_2) \cdots P(A_n|A_1 \cap A_2 \cap \cdots \cap A_{n-1}),$$

provided, of course, that $P(A_1 \cap A_2 \cap \cdots \cap A_{n-1}) > 0$.

Proof

Just apply the definition of conditional probability to every term on the right side, and watch the expression collapse to that on the left side. □

To illustrate this result, we will calculate the chance that each of the four players in a game of Bridge is dealt an Ace. Our model is that the 52 cards are randomly distributed, each player getting 13 of them. Let A_i denote the event that player i gets an Ace, and use the theorem to find $P(A_1 \cap A_2 \cap A_3 \cap A_4)$. The first player receives 13 cards, chosen at random from 48 non-Aces and 4 Aces, so $P(A_1) = \binom{48}{12}\binom{4}{1}/\binom{52}{13}$.

To find $P(A_2|A_1)$, we work with the reduced pack of 36 non-Aces and 3 Aces, which is what remains, conditional on A_1 having occurred. Thus $P(A_2|A_1) = \binom{36}{12}\binom{3}{1}/\binom{39}{13}$. Continuing in this fashion, we find $P(A_3|A_1 \cap A_2)$ by working from a pack with 24 non-Aces and 2 Aces. You should write down the expression. Finally, there is no work at all to find $P(A_4|A_1 \cap A_2 \cap A_3)$, as the answer is plainly unity. (If you do not see this, you have been reading too quickly.)

Theorem 2.3 tells us to multiply together all the probabilities we have just found; there is massive cancellation among the factorial terms, and the final expression shows that the probability that the Aces are equidistributed is $\frac{6 \times 13^3}{51 \times 50 \times 49} \approx 0.1055$.

The next result shows another way in which using conditional probability can simplify a calculation. The essence of the bargain is to replace one fairly complex calculation by a number of easier ones. The key is finding a *partition* of the set of outcomes that is relevant to the calculation in question.

Definition 2.4

$\{B_1, B_2, \ldots\}$ are said to form a *partition* of Ω if

(i) $B_i \cap B_j = \varnothing$ if $i \neq j$

(ii) $\bigcup B_i = \Omega$.

Theorem 2.5 (Law of Total Probability)

If A is any event, and $\{B_1, B_2, \ldots\}$ form a partition of Ω, then

$$P(A) = \sum_i P(A \cap B_i) = \sum_i P(A|B_i)P(B_i).$$

Proof

Using the definition of a partition, we have

$$P(A) = P(A \cap \Omega) = P(A \cap (\cup B_i)) = P(\cup(A \cap B_i)) = \sum P(A \cap B_i)$$

because the events $\{B_i\}$ are pairwise disjoint. The final expression arises from the definition of conditional probability.   □

There is the potential problem that some of the terms $P(A|B_i)$ are not yet defined when $P(B_i)$ is zero. But this is easily overcome. Whatever the expression $P(A|B_i)$ might mean, it is a probability, and so lies between zero and one; and since it gets multiplied by $P(B_i)$, the product will be zero. So we can use this result without worrying about what that conditional probability might mean.

To see this theorem in action, recall Example 1.10, and seek to decide the best strategy for Alex when his initial total is n. We know that if he sticks, his winning chance is $(n/K)^2$.

Suppose he spins; let $A =$ Alex wins, and let $B_i =$ Alex scores i with his second spin. Plainly these events $\{B_i\}$ form a partition, and each has probability $1/K$. For $1 \leq i \leq K - n$,

$$P(A|B_i) = P(\text{Alex wins from } n + i) = (n+i)^2/K^2.$$

For $i > K - n$, then $P(A \mid B_i) = 0$, as Alex's score gets reduced to zero. The theorem shows that

$$P(A) = \sum_{i=1}^{K-n} \frac{(n+i)^2}{K^2} \times \frac{1}{K} = \frac{1}{K^3} \sum_{r=n+1}^{K} r^2.$$

So the best tactics, assuming he wants the fun of spinning when the chances are the same, are to spin whenever $(n + 1)^2 + (n + 2)^2 + \cdots + K^2 \geq Kn^2$.

Given any K, there will be some unique value n_0 such that this inequality holds if, and only if, $n \leq n_0$. Theorem 2.5 also helps us evaluate each player's winning chance, if they play optimally. Again $A = $ Alex wins, and let $C_i = $First spin scores i. By the above work,

1. if $i > n_0$, then Alex will stick and $P(A|C_i) = i^2/K^2$;

2. if $i \leq n_0$, Alex will spin and $P(A|C_i) = \frac{1}{K^3} \sum\limits_{r=i+1}^{K} r^2$.

Since $P(C_i) = 1/K$, Theorem 2.5 shows that the chance Alex wins is

$$P(A) = \frac{1}{K^3} \sum_{i=n_0+1}^{K} i^2 + \frac{1}{K^4} \sum_{i=1}^{n_0} \sum_{r=i+1}^{K} r^2.$$

Table 2.1 gives some illustrative values. Recall that Alex wins if the scores are tied, but that Bella has the advantage of going second. The table makes sense   Alex's advantage counts for less as K, the number of different scores, increases, and Bella is soon favourite.

Table 2.1 For various values of K, the optimum tactics for Alex and his winning chances.

Value of K	10	20	35	50	75	100	200
Spin if at most	5	10	18	26	40	53	106
Prob. Alex wins	0.512	0.482	0.470	0.465	0.461	0.460	0.457

This law of total probability was the basis, in 1654, for Pascal's method of solving the "problem of points". This asks how a prize should be divided between two players if the series of games in which they compete has to be abandoned before the outcome is settled. When the game began, the winner would be the first to win N games, but the series stops when Clive still needs m wins, and Dave needs n.

There is no unique answer, but the approach developed by Pascal and Fermat is generally accepted as satisfactory. They suggested treating each player as having a 50% chance of winning any of the unplayed games. Let A denote the event that Clive would win the contest, and let B mean that Clive wins the next game. Then

$$P(A) = P(A|B)P(B) + P(A|B^c)P(B^c).$$

If $p(m, n)$ is the probability that Clive would win the contest when the respective targets are m and n, this means that

$$p(m, n) = \frac{1}{2}p(m - 1, n) + \frac{1}{2}p(m, n - 1).$$

Since $p(m,0) = 0$ when $m \geq 1$, and $p(0,n) = 1$ when $n \geq 1$, we can apply the above equation as often as necessary to build up to, and calculate, $p(m,n)$. The prize is then apportioned proportional to each player's chance of victory. Table 2.2 shows some sample values.

Table 2.2 Clive's winning chances $p(m,n)$ when he requires m further victories and Dave requires n.

m/n	1	2	3	4
1	1/2	3/4	7/8	15/16
2	1/4	1/2	11/16	13/16
3	1/8	5/16	1/2	21/32
4	1/16	3/16	11/32	1/2

EXERCISES

2.1. An urn initially has six red balls and eight blue ones. Balls are chosen one at a time, at random, and not replaced. Find the probabilities that

(a) the first is blue, the second is red;

(b) the first is blue, the sixth is red;

(c) the first three are all the same colour.

2.2. Show that

$$P(A|B) = P(A|B \cap C)P(C|B) + P(A|B \cap C^c)P(C^c|B).$$

For each of the following statements, either prove it valid, or give a counterexample, using events from the experiment of one throw of a fair die.

(a) $P(A|B) + P(A^c|B^c) = 1$

(b) $P(A|B) + P(A|B^c) = 1$

(c) $P(A|B) + P(A^c|B) = 1$.

2.3. Three cards of the same size and shape are placed in a hat. One of them is red on both sides, one is black on both sides, the third is red on one side and black on the other. One card is selected at random, and one side is exposed: that side is red. What is the chance the other side is red?

Explain what is wrong with the following argument: "Seeing the red side eliminates the black–black card; since all the cards were initially equally likely, the remaining two cards are equally likely, so there is a 50% chance the other side is red."

2.4. Show that, if $P(A|C) > P(B|C)$ and $P(A|C^c) > P(B|C^c)$, then $P(A) > P(B)$.

2.5. (Snell's problem) You have w white balls and b black balls to distribute among two urns in whatever way you like. Your friend will then select one urn at random, and withdraw one ball at random from that urn. What should you do to maximise the chance of getting a black ball?

2.6. In a TV quiz show, the star prize is in one of three boxes, A, B or C. When the contestant selects one of these boxes, the host (who knows which box contains the star prize) opens one of the other two boxes, showing it to be empty. He invites the contestant to stick with her original choice, or to swap to the third, unopened, box. Should she change? Or does it make no difference to her winning chances whether she sticks or swaps?

2.2 Bayes' Theorem

With the same set-up as Theorem 2.5, we can look at matters the other way round: *given* that event A has occurred, what can be said about the chances of the different events $\{B_i\}$? This is exactly the situation facing a doctor when she has examined a patient: event A is the set of symptoms and the general background information, while the events $\{B_i\}$ are the various diagnoses possible – measles, beri-beri, hypochondria or whatever. The value of i such that $P(B_i|A)$ is largest will point to the diagnosis that is most likely.

It is also the situation faced by a jury in a criminal trial. Here A represents all the evidence, while $B_1 = $ Guilty and $B_2 = $ Not Guilty are the possible verdicts, forming a *partition*, since the jury must select one of them. Formally, the jury is asked to evaluate $P(B_1|A)$, and to return a verdict of Not Guilty unless this conditional probability is very close to 1 – "beyond reasonable doubt". Each juror will have a personal threshold value (80%? 99%? Higher? Lower?) that they use to interpret this time-honoured phrase.

Theorem 2.6 (Bayes' Theorem)

If A is any event with $P(A) > 0$ and $\{B_i\}$ form a partition, then

$$P(B_j|A) = \frac{P(A|B_j)P(B_j)}{P(A)} = \frac{P(A|B_j)P(B_j)}{\sum_i P(A|B_i)P(B_i)} \qquad (2.1)$$

Proof

Because $P(A \cap B_j)$ can be written both as $P(B_j|A)P(A)$ and as $P(A|B_j)P(B_j)$, the first equality is immediate. The second comes from using Theorem 2.5 to give an alternative form of $P(A)$. $\square$

Take the special case when there are just two alternatives, B_1 and B_2 in this partition, so that $B_2 = B_1^c$. The ratio $P(B_1)/P(B_2)$ is then termed the *odds ratio* of B_1, as it compares the chance B_1 happens to the chance it does not happen. Using Bayes' theorem for $j = 1$ and $j = 2$, and dividing the two expressions, we have

$$\frac{P(B_1|A)}{P(B_2|A)} = \frac{P(A|B_1)}{P(A|B_2)} \times \frac{P(B_1)}{P(B_2)} \qquad (2.2)$$

In a Court, if B_1 represents a Guilty verdict, this formula tells us how each piece of evidence A should influence our opinions. The left side is the odds ratio, taking account of this evidence, and it is found by multiplying the odds ratio $P(B_1)/P(B_2)$ before the evidence (the *prior* odds) by the so-called *likelihood ratio*, i.e. the ratio of the chances of the evidence arising when the accused is Guilty, or Not Guilty. As each new piece of evidence is introduced, the odds ratio is updated by multiplying by this likelihood ratio. In particular, it follows that the *order* in which the evidence is introduced makes no difference to the final answer. A major problem – and it is of crucial sensitivity – is to give a convincing initial value to $P(B_1)$ before any evidence is given.

Example 2.7 (Attributed to J.M. Keynes)

There are $2\,598\,960$ different poker hands of five cards for an ordinary deck, of which just four are Royal Flushes, the highest hand possible. Thus the chance of a Royal Flush is about 1 in $650\,000$. You play one hand of poker against the Archbishop of Canterbury, who promptly deals himself a Royal Flush! What is the probability he is cheating?

Let B_1 denote the event that he is cheating, and let B_2 mean he is acting honestly. Let A be the evidence, this Royal Flush on the first deal. We know from the above counting that $P(A|B_2) = 1/650\,000$, and we may as well take

$P(A|B_1) = 1$, on the grounds that if he is cheating, he can certainly deal himself a Royal Flush. In Equation (2.2), the first term on the right side is 650 000, and we seek the value of the left side – the odds ratio of cheating, given the evidence.

As the equation makes clear, the answer depends entirely on the initial odds ratio. Taking the view that the Archbishop is a man of enormous integrity, our initial feeling may be that $P(B_1)$ is tiny, say 0.000 000 01. In that case $P(B_2) = 0.999\,999\,99$, and the initial odds ratio is effectively the tiny value of $P(B_1)$. When we apply the equation, the odds ratio becomes about 0.0065, so we retain a very strong belief in the integrity of our opponent.

But suppose he deals himself another Royal Flush next hand? After we have multiplied this last odds ratio by the factor 650 000 again, we now believe the odds are about 4000 to one that cheating is taking place.

If our opponent had been someone else, a suspected card sharp, we might have begun with the opinion $P(B_1) \approx 0.1$, and then even after the first deal, the odds ratio has shifted to over 70 000 to one in favour of cheating.

Robert Matthews has pointed out that, in certain circumstances, confession evidence may *weaken* the case against a suspect.

- G = The accused is guilty of a terrorist offence
- I = The accused is innocent of that offence
- C = Confession, extracted after long interrogation.

Equation (2.2) shows that the change in the odds of Guilt comes from the ratio $P(C|G)/P(C|I)$. Provided that $P(C|G) > P(C|I)$, then the confession does shift the odds more towards Guilt. But actual terrorists may have been trained to resist prolonged interrogation, while an innocent person may make a confession simply to end his ordeal. In these circumstances, the above inequality may be reversed, and confession would *reduce* the odds of Guilt. (A real terrorist may reach the same conclusion, and try the double bluff of confessing.)

The general point is that drawing attention to the actions of an accused only strengthens the case against him if those actions are *more likely* when he is guilty that when he is innocent. If a bus journey took the accused past both the crime scene and the shopping centre he said he was visiting, then evidence that tended to prove he took that journey would not add to the case against him, as the two values of $P(\text{Journey}|\text{Innocent})$ and $P(\text{Journey}|\text{Guilty})$ would have the same value.

Example 2.8 (Daniel Kahneman and Amos Tversky)

In a given city, 85% of the taxicabs are Green, 15% are Blue. A witness to a hit-and-run accident identified the perpetrator as a Blue taxicab. Tests under similar lighting conditions showed the witness would correctly identify the colour 80% of the time, and be wrong 20%. What is the chance that the cab was indeed Blue?

Let B =Cab was Blue, G =Cab was Green, and W =Witness says the cab was Blue. Then

$$P(B) = 0.15, \quad P(G) = 0.85, \quad P(W|B) = 0.80 \text{ and } P(W|G) = 0.20.$$

We require $P(B|W)$, so Equation (2.2) can be transformed into

$$\frac{P(B|W)}{P(G|W)} = \frac{P(W|B)}{P(W|G)} \times \frac{P(B)}{P(G)}.$$

Using the figures, the odds ratio for a Blue cab is $(0.80/0.20) \times (0.15/0.85) = 12/17$, i.e. despite the evidence of a witness who is 80% reliable claiming to have seen a Blue cab, it is more likely to have been Green!

For a different medical application of Bayes' theorem, suppose one person in 1000 suffers an adverse reaction to a drug, and a simple test for this reaction is on offer. The test is said to be 95% reliable, meaning that if the person would suffer a reaction, a positive result comes up 95% of the time, and if they would not have a reaction, a negative result occurs 95% of the time. What can we conclude from the knowledge that Susie tests positive?

It is far too tempting to conclude that, as the test is 95% reliable, there is a 95% chance she would suffer a reaction. This answer is quite wrong. Let $S =$ Susie tests positive, and let $R =$ She would suffer an adverse reaction. We seek $P(R|S)$, using R and R^c as the partition for Bayes' theorem. The background information can be expressed as

$$P(S|R) = 0.95 \quad \text{and} \quad P(S|R^c) = 0.05,$$

while we also know $P(R) = 1/1000$. Hence

$$P(S) = P(S|R)P(R) + P(S|R^c)P(R^c) = 0.95 \times \frac{1}{1000} + 0.05 \times \frac{999}{1000} = 0.0509.$$

By Bayes' theorem, $P(R|S) = P(S|R)P(R)/P(S) = 0.00095/0.0509 \approx 0.0187$! When Susie tests positive, the chance she would suffer the reaction is under 2% – the test is virtually useless, even though it can claim to be 95% reliable.

Here, and in Example 2.8, many people's intuitive answer will be wrong, because they have not properly taken into account the background information – the frequencies of Blue cabs, or the prevalence of the disease.

EXERCISES

2.7. Soccer player Smith has a good game two times in three, otherwise a poor game. His chance of scoring is 3/4 in a good game, 1/4 in a poor one. What is the chance he scores in a game? Given that he has scored, what is the chance he had a good game?

2.8. Of all the hats in a shop, one half come from Luton, one third from Milton Keynes, and the rest from Northampton. Two-thirds of hats from Luton are for formal wear, as are one half of those from Milton Keynes and one third of the Northampton hats. A hat is selected at random from the shop; what is the chance it is for formal wear? Given that it is for formal wear, what is the chance it originated in Northampton?

2.9. In the last medical example above, make one change in the parameters: it is n in 1000, not 1 in 1000, who would suffer an adverse reaction. Compute the chance that Susie would suffer a reaction, given that she tests positive, as a function of n ($1 \leq n < 1000$).

2.10. One coin amongst n in a bag is double-headed, the rest are fair. Janet selects one of these coins at random and tosses it k times, recording Heads every time. What is the chance she selected the double-headed coin?

2.11. Assume that if a woman carries the gene for haemophilia, any child has a 50% chance of inheriting that gene, and that it is always clear whether or not a son has inherited the gene, but the status of a daughter is initially uncertain. Karen's maternal grandmother was a carrier, the status of her mother in unknown; but Karen's sister Penny has one son, who is healthy.

   (a) Find the chance that Karen's first child inherits the haemophilia gene.

   (b) Penny now has a second healthy son; repeat the calculation in (a).

   (c) But Penny's third son is a haemophiliac; again, repeat (a).

2.12. The till in the pub contains 30 £20 notes and 20 £10 notes. There is a dispute about what denomination Derek used to pay his bill, and the initial assumption is that all 50 notes were equally likely. The barmaid, Gina claims he used a £10 note, Derek disagrees. Both are honest, but may make mistakes. Show that, using the information that Gina correctly identifies notes 95% of the time, the chance it was a £10 note is 38/41.

Derek, who correctly identifies £20 notes 80% of the time, and correctly identifies £10 notes 90% of the time, says he used a £20 note. Update your calculation.

2.3 Independence

Suppose $P(A|B) = P(A)$, i.e. knowing that B occurs makes no difference to the chance that A occurs. Since

$$P(A) = P(A|B)P(B) + P(A|B^c)P(B^c),$$

then

$$P(A|B^c) = \frac{P(A) - P(A)P(B)}{1 - P(B)} = P(A);$$

the information that B did *not* occur has also made no difference to the chance that A occurs. Further, we also have

$$P(B|A) = P(B \cap A)/P(A) = P(A|B)P(B)/P(A) = P(B),$$

and similarly $P(B|A^c) = P(B)$. Whenever $P(A|B) = P(A)$, knowing whether or not either A or B occurs makes no difference to the chances of the other one. This is the intuitive notion that A and B are *independent* events.

Definition 2.9

Events A and B are said to be *independent* when $P(A \cap B) = P(A)P(B)$.

Making the definition in this format emphasises the symmetry, and removes any concern about whether either event has zero probability. You should check that, by this definition, any event whose probability is zero is automatically independent of any other event. Hence the same applies to any event with probability unity.

Consider the experiment of dealing two cards, at random, from a full deck. Write

$A =$ First card is an Ace, $B =$ First card is a Spade,

$C =$ Second card is an Ace, $D =$ Second card is a Spade.

We can check the working of the definition by verifying that $\{A, B\}$ are independent, as are $\{C, D\}$, $\{A, D\}$ and $\{B, C\}$, but neither $\{A, C\}$ nor $\{B, D\}$ are independent.

Suppose we have more than two events. The *idea* of independence is that knowing whether any of them occurs should make no difference to the chances of other events. For example, suppose $\Omega = \{1, 2, \ldots, 8\}$, with all outcomes having probability 1/8. Write $A = \{1, 2, 3, 4\}$, $B = \{1, 2, 5, 6\}$, $C = \{3, 4, 5, 6\}$. Each pair of these events are independent, so the information that A occurs, or the information that B occurs, makes no difference to the chance that C occurs. However, $P(A \cap B \cap C) = 0$, so if we know that *both* A and B occur, then it is impossible for C to occur.

On the same space, if also $D = \{1, 3, 4, 5\}$ and $E = \{1, 6, 7, 8\}$, we have $P(A \cap D \cap E) = 1/8 = P(A)P(D)P(E)$. A careless beginner might think that it will follow that A, D and E are independent. But simple calculations show that *none* of the pairs $\{A, D\}$, $\{A, E\}$ or $\{D, E\}$ satisfy the definition of pairwise independence. When looking at as few as three events, it is not enough to look at them just in pairs, or just as a triple, we have to look at all the pairs and the triple together. The general definition, for three or more events, is

Definition 2.10

Events $\{A_1, A_2, \ldots, A_n\}$ are *independent* if

$$P(A_{1'} \cap A_{2'} \cap \ldots \cap A_{r'}) = P(A_{1'})P(A_{2'}) \cdots P(A_{r'})$$

whenever $\{1', 2', \ldots, r'\} \subset \{1, 2, \ldots, n\}$.

An infinite collection of events is independent if every finite subcollection of them is independent.

Independence among a collection of events is a very strong property. To verify all the conditions of this definition with ten events would call for over 1000 calculations! Fortunately, this terrifying prospect seldom arises: usually either the events have been set up to be independent, or one of the conditions obviously fails.

How should we deal with "independent" repetitions of the same experiment, such as tossing a coin or rolling a die? More generally, how can $\{(\Omega_1, \mathcal{F}_1, P_1), (\Omega_2, \mathcal{F}_2, P_2), \ldots\}$ be probability spaces that correspond to independent experiments? Without constructing the whole of $(\Omega, \mathcal{F}, P)$ that corresponds to all these experiments in sequence, the key step is to have

$$P(A_1 \cap A_2 \cap \ldots \cap A_n) = P_1(A_1)P_2(A_2) \cdots P_n(A_n),$$

whenever each A_i belongs to $\mathcal{F}_i$, $i = 1, 2, 3, \ldots$, and for all values of n. In future work, we will continue to suppress reference to the underlying probability space, unless it is absolutely necessary, and just use this fundamental property without comment. Our first application yields a result that is no surprise, but you should reflect on the steps we have taken to prepare the ground for it.

Corollary 2.11

If $P(A) > 0$ and the experiment is repeated independently, indefinitely often, then A must occur sometime.

Proof

Let B_n be the event that A does not occur in the first n experiments. Then $B_1 \supset B_2 \supset B_3 \supset \cdots$, and $B = \bigcap_n B_n$ is the event that A never occurs. But Exercise 1.5 shows that $P(B) = \lim P(B_n)$; and

$$P(B_n) = (1 - P(A))^n \to 0 \text{ as } n \to \infty$$

since $P(A) > 0$. Hence $P(B) = 0$, so A is certain to occur sometime. □

Example 2.12

Let A and B be mutually exclusive, each with non-zero probabilities. What is the chance that A occurs before B, if the experiment is repeated indefinitely often?

Write C_n to mean that neither A nor B occur in the first n experiments, but one of them does occur in the $(n + 1)$th. These C_n form a partition, since they are plainly disjoint and exhaustive, and let D denote that A occurs before B. Then

$$P(D) = \sum_n P(D \cap C_n) = \sum_{n=0}^{\infty} (1 - P(A) - P(B))^n P(A) = \frac{P(A)}{P(A) + P(B)}$$

on summing the geometric series. To assess which of A or B is likely to appear first, look at the ratio of their probabilities.

The answer makes intuitive sense: perhaps there is a more direct way to obtain it? Try conditioning on the outcome of the first experiment. That leads to

$$P(D) = P(D|A)P(A) + P(D|B)P(B) + P(D|(A \cup B)^c)P((A \cup B)^c)$$

since A, B and $(A \cup B)^c$ partition the space. Plainly $P(D|A) = 1$ and $P(D|B) = 0$. Also, if neither A nor B occur in the first experiment, we can act as though that had never taken place, so that $P(D|(A \cup B)^c) = P(D)$. Hence

$$P(D) = P(A) + 0 + P(D)(1 - P(A \cup B)),$$

from which the previous result follows, as $P(A \cup B) = P(A) + P(B)$.

An application of this result is to the game of craps. Here the shooter throws two dice and wins immediately if the total is 7 or 11. She loses immediately with a total of 2, 3 or 12, and otherwise the game is not yet over: she throws the dice repeatedly until either she repeats her first total – she wins – or she throws a 7 – a loss. We will work out her winning chances.

The table in Example 1.7 shows the chances of the different totals on each throw. Leaving the fractions with their common denominator of 36 simplifies the steps: the chance she wins on her first throw is $6/36+2/36 = 8/36$. When the first throw is among $\{4, 5, 6, 8, 9, 10\}$ she embarks on a series of repetitions: for example, the chance she throws a 4 and subsequently wins is $3/36 \times 3/(3+6) = 1/36$. Making similar calculations, and using the symmetry about the midpoint of 7, her overall winning chance is

$$\frac{8}{36} + \left(\frac{1}{36} + \frac{4}{36} \cdot \frac{4}{10} + \frac{5}{36} \cdot \frac{5}{11}\right) \times 2 = \frac{244}{495}$$

about 0.493. As ever, the game favours the house.

Example 2.13

Suppose a fair coin is tossed repeatedly, and the outcomes noted. Which is more likely to appear first, HH or HT? What about HH and TH?

For any given pair of tosses, all four outcomes $\{HH, HT, TH, TT\}$ are equally likely, which suggests that we should use Example 2.12 to conclude that, for both problems, either sequence has a 50% chance of arising first. But that example may not necessarily apply here, as there the events A and B are within a single experiment, but here we are looking at pairs of experiments that have some *overlap*.

We will show that, for the first problem, it is true that either sequence is equally likely to appear first, but that TH arises before HH 75% of the time. For HH versus HT, let p be the probability that HH comes up first, and let A be that the first toss is a Head. Then

$$p = P(HH \text{ before } HT|A)P(A) + P(HH \text{ before } HT|A^c)P(A^c).$$

Now

$$P(HH \text{ before } HT|A) = P(H \text{ at second toss}) = 1/2,$$

and also

$$P(HH \text{ before } HT|A^c) = P(HH \text{ before } HT)$$

because, if the first toss is a Tail, it can be ignored. Thus

$$p = (1/2) \times (1/2) + p \times (1/2),$$

so $p = 1/2$ as we claimed.

But for HH versus TH, unless the first two tosses are both Heads, it is clear (yes?) that TH must arise first. So the chance HH is before TH is the chance of beginning HH, which is 1/4, and TH does have a 75% chance of arising before HH.

Of course for sequences of three successive tosses, there are eight outcomes to consider. It turns out that, whichever one of them is chosen, there is one among the remaining seven whose chance of arising before it is at least 2/3 ! You might imagine playing a game against an opponent, allowing her to select whichever of the eight sequences she wants; you will (carefully) pick from the seven that are left – the winner is the one whose sequence turns up first. This contest goes under the name of Penney-ante.

Suppose a complex piece of machinery is made up from linked components, each of which may fail independently of the rest. To give extra protection, some components may be duplicated and placed in parallel, so that only one of them need work; but for components in series, all must work or the system will fail. Consider the set-up shown in the figure on the left, where the numbers in the boxes are the probabilities that component will fail.

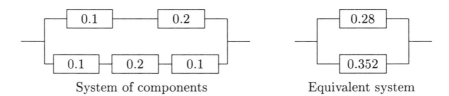

System of components   Equivalent system

The top row will *work* only if both components work, so the chance is $0.9 \times 0.8 = 0.72$, hence the probability of failure is 0.28. Similarly, the chance of failure in the bottom row is 0.352, so the original figure is equivalent to the new system on the right. That system has two components in parallel, so fails only when both fail, which has probability $0.28 \times 0.352 = 0.09856$, or about 10%. The whole system works with probability 0.90144, call it 90%.

Example 2.14 (Simpson's Paradox)

Suppose that 200 of 1000 males, and 150 of 1000 females, in University A read Economics. In University B, the figures are 30 of 100 males, and 1000 of 4000 females. In each institution, a higher proportion of males read Economics. But if the universities combine to form a super-university, it has $230/1100 \sim 21\%$ of males in Economics, and $1150/5000 = 23\%$ of females. Proportionally more females than males read Economics!

The possibility of this phenomenon was noted by E.H. Simpson (1951). To see how it might arise, write $E =$ Student reads Economics, $M =$ Student is male. Then, in each university separately, we have $P(E|M) > P(E|M^c)$. From the definition of conditional probability, this is equivalent to

$$P(E \cap M)/P(M) > P(E \cap M^c)/P(M^c),$$

which can be rearranged, using $P(E) = P(E \cap M) + P(E \cap M^c)$, as $P(E \cap M) > P(E)P(M)$. E and M are not independent, they are "positively associated" in each university.

But as the example shows, positive association in each institution on its own need not carry over to a combination. We have two probability spaces, Ω_1 with N_1 points and Ω_2 with N_2. In each space, all points are equally likely, and $P(E_i \cap M_i) > P(E_i)P(M_i)$ for $i = 1, 2$. Simpson's paradox notes that if $\Omega_3 = \Omega_1 \cup \Omega_2$, and all $N_1 + N_2$ points are equally likely, it does not necessarily follow that $P((E_1 \cup E_2) \cap (M_1 \cup M_2)) > P(F_1 \cup E_2)P(M_1 \cup M_2)$ in Ω_3.

In the first innings of a cricket match, bowler Eric takes 2 wickets for 36 runs, while Alec takes 4 for 80. In the second innings, Eric takes 1 for 9, Alec takes 6 for 60. In each innings, Eric's runs per wicket were lower, but overall, Eric took 3 for 45 at an average of 15, Alec took 10 for 140 at average of 14. Eric's average was better in each innings, Alec's match average was better. Amalgamating the two innings is appropriate.

60 of 900 people given drug A suffered a reaction (6.7%), while 32 of 400 given drug B (8%) suffered a reaction. Drug A appears safer. But there were 1000 healthy people, and 300 who were ill. 800 healthy people were given A, of whom 40 (5%) reacted, only 2 of 200 healthy people (1%) reacted to B. Among 100 ill people given A, 20 reacted (20%), while 30 of 200 ill people given B reacted (only 15%). Drug B is safer for both ill people and healthy ones! Amalgamating the data would give an unsound conclusion.

Within an organisation, it is perfectly possible that division A shows better than division B for each of ten subgroups of clients, but B shows better than A when the groups are combined. Management must be very careful when deciding which division deserves a bonus.

EXERCISES

2.13. Confirm the remark after Definition 2.9 that if $P(A) = 0$ or 1, then A is independent of any other event. (Such events A are called *trivial* events.)

2.14. Suppose events A and B are mutually exclusive. Show they can be independent only if one of them has probability zero. (Thus "mutually exclusive" and "independent" are normally far apart in meaning. But many students, unaccountably, confuse the two.)

2.15. Suppose Ω has n equally likely outcomes. If n is prime, show that non-trivial events A and B cannot be independent.

2.16. A spinner is a device for selecting one of the integers $\{1, 2, \ldots, n\}$ with equal probability. For what values of n are the events

$$A = \text{ Even number, } \quad \text{and } B = \text{ Multiple of three}$$

independent?

2.17. Show that, if

(a) a fair die is thrown four times independently, it is more likely than not that at least one six appears;

(b) a pair of fair dice are thrown 24 times independently, it is more likely than not that a double six does *not* appear.

(This pair of calculations has an honoured place in the history of the development of the formal study of probability. Some seventeenth century gamblers are said to have believed that, since (a) holds, then having six times as many throws ($4 \to 24$) "ought" to give the same chance of getting an event that was one sixth as likely (six $\to$ double six). It is very satisfying to see a loose argument give the wrong answer.)

2.18. In the diagram, the numbers are the (independent) chances the components will fail within ten years. Find the chance the system fails within ten years.

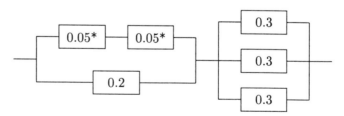

Given that the system has not failed in ten years, find the chance neither component marked * has failed.

2.19. Poker dice have six equally likely faces labelled $\{9, 10, J, Q, K, A\}$. When five such dice are thrown independently, what are the probabilities of the different types of "hand" which, in rank order are: Five of a Kind (aaaaa); Four of a Kind (aaaab); Full House (aaabb);

Threes (aaabc); Two Pairs (aabbc); One Pair (aabcd); and No Pair (abcde)? ("Runs" are not normally considered in this game.)

Given any of these hands, you are allowed to roll again any dice that are unmatched. Find your respective chances of improving your hand for the different initial holdings (ignoring "improvements" such as turning $(KKKQJ)$ to $(KKKA9)$).

2.4 The Borel–Cantelli Lemmas

Suppose $\{A_n\}$ is an arbitrary sequence of events. Write $B_N = \bigcup_{n=N}^{\infty} A_n$, and $C = \bigcap_{N=1}^{\infty} B_N$, so that $C = \bigcap_{N=1}^{\infty} \bigcup_{n=N}^{\infty} A_n$. We claim that event C is the event that infinitely many of the events $\{A_n\}$ occur.

For, suppose infinitely many of the $\{A_n\}$ occur. Then, no matter how large we take N, some A_n with $n \geq N$ occurs, i.e. B_N occurs for every N. Thus C occurs. The argument holds in reverse: if C occurs, so does B_N for all N, and hence no matter how large N is, some A_n with $n \geq N$ occurs. Thus infinitely many A_n occur.

What can we say about $P(C)$, knowing the probabilities of the events $\{A_n\}$?

Theorem 2.15 (Borel–Cantelli Lemmas)

1. If $\sum P(A_n)$ converges, then $P(C) = 0$.

2. If $\sum P(A_n)$ diverges, and the events $\{A_n\}$ are independent, then $P(C) = 1$.

Proof

Suppose the series is convergent. Since $P(C) = P\left(\bigcap_{N=1}^{\infty} B_N\right)$, so $P(C) \leq P(B_N)$ for every choice of N. But

$$P(B_N) = P\left(\bigcup_{n=N}^{\infty} A_n\right) \leq \sum_{n=N}^{\infty} P(A_n),$$

using Boole's inequality (Exercise 1.3). Whatever $\epsilon > 0$ is given, the convergence of the series means that we can find N such that $\sum_{n=N}^{\infty} P(A_n) < \epsilon$. For such values of N, we see that $P(B_N) < \epsilon$. But we established that $P(C) \leq P(B_N)$, which means that $P(C) < \epsilon$. This holds whatever positive value we take for ϵ, no matter how small, so the inevitable conclusion is that $P(C) = 0$.

For the other result, note that, for any N,

$$P(B_N^c) = P\left(\left(\bigcup_{n=N}^{\infty} A_n\right)^c\right) = P\left(\bigcap_{n=N}^{\infty} A_n^c\right) \leq P\left(\bigcap_{n=N}^{N+M} A_n^c\right)$$

for any $M > 0$. But when the events $\{A_n\}$ are independent, this becomes

$$P(B_N^c) \leq \prod_{n=N}^{N+M} (1 - P(A_n)) \leq \exp\left(-\sum_{n=N}^{N+M} P(A_n)\right),$$

using the simple inequality $1 - x \leq \exp(-x)$, valid when $0 \leq x \leq 1$.

If the series is divergent, whatever the value of N, we can then choose M so that the sum $\sum_{n=N}^{N+M} P(A_n)$ is as large as we like, and hence the value of $\exp(-\sum_{n=N}^{N+M} P(A_n))$ is as small as we like. Hence $P(B_N^c) = 0$, and so $P(B_N) = 1$. This holds for all values of N and so, using Exercise 1.4, $P\left(\bigcap_{N=1}^{\infty} B_N\right) = 1$, i.e. $P(C) = 1$. $\qquad\square$

The Borel–Cantelli lemmas are an example of a so-called Zero–One Law: we have an event C that is either impossible, or it is certain. The second part of the theorem has the extra condition that the $\{A_n\}$ are independent, and it is plain that some such condition will be required. For otherwise we could take $A_n = D$ for some fixed event D, with $0 < P(D) < 1$; and then $B_N = D$ for all N, so $C = D$ and $P(C)$ can be whatever we choose.

For a potential application of the theorem, let A_n be the event that, in year n, there is a catastrophic accident at some nuclear power station. Plainly, we shall require that $P(A_n)$ is small, but the Borel–Cantelli lemmas are much more demanding. It seems plausible to take the events $\{A_n\}$ as independent, and then we have:

1. Provided $\sum_{n=1}^{\infty} P(A_n)$ converges, it is impossible that there will be infinitely many catastrophic accidents (small mercies indeed). More alarmingly

2. If $\sum_{n=1}^{\infty} P(A_n)$ diverges (even if $P(A_n) \to 0$), it is certain there will be infinitely many catastrophes.

So be warned. It is not enough that $P(A_n)$ be tiny, the successive values really ought to be getting smaller year by year, *and* get smaller fast enough for the series to converge. Safety standards in dangerous industries must be perennially made stricter, or we are doomed.

3

Common Probability Distributions

Here we look at a number of experiments in which chance plays a role, set up plausible probability spaces, and describe their main properties. The general phrase "a probability distribution" means a description of the outcomes, and how to find the probabilities of the events.

3.1 Common Discrete Probability Spaces

A probability space is said to be discrete when the set of outcomes Ω can be put in one–one correspondence with the list of non-negative integers, $(0, 1, 2, \ldots)$, or a finite subset of them. As we found in the last chapter, it is sufficient here to list the outcomes, and to attach to outcome n its probability p_n, for then the probability of any event $A \subset \Omega$ is found as $P(A) = \sum_{n \in A} p_n$. So long as each $p_n \geq 0$, and also $\sum_{n \in \Omega} p_n = 1$, the probability space is specified, and all subsets of Ω are events. The phrase *probability mass function* means the list of outcomes, along with their associated probabilities.

We met the discrete uniform distribution in the last chapter. There is some finite number, K, of outcomes, each with probability $1/K$. This would seem to be an appropriate model for experiments such as drawing lots to decide which lane your team occupies in a relay race at the Olympic Games, or using a pin to make your choice of horse in the Derby.

Without naming it, we also met the *hypergeometric distribution*, when we sought the winning chances in a 6/49 Lottery. The general model arises when selecting n objects at random, without replacement, from a collection of M Blue objects and N Red objects, and we can assume that all remaining objects have the same chance of selection each time. The quantity of interest is k, the number of Blue objects in the selection. By counting, we see that

$$p_k = \frac{\binom{M}{k} \times \binom{N}{n-k}}{\binom{M+N}{n}} \tag{3.1}$$

for $k = 0, 1, 2, \ldots$. Note that k is at least $n - N$, and cannot exceed either M or n, but there is no need to insert these restrictions, so long as we maintain the convention that the value of the expression $\binom{a}{b}$ is taken as zero when $b > a$, or $b < 0$.

The hypergeometric distribution is frequently used in card games such as Bridge. The declarer finds that she and dummy have eight Spades between them, and she seeks the distribution of the other five Spades between her two opponents. Her left-hand opponent has 13 of the 5 Spades and 21 non-Spades that remain: so the chance this opponent has exactly k Spades is $\binom{5}{k}\binom{21}{13-k} / \binom{26}{13}$. The values are shown in the table.

Number of Spades	0	1	2	3	4	5
Probability	0.020	0.141	0.339	0.339	0.141	0.020

The Spades will split 3–2, 4–1 or 5–0 about 68%, 28% and 4% of the time respectively.

Another application is in capture–recapture sampling. An unknown number of fish swim in a pond; M of them are captured, tagged, and released back into the pond. A few days later, n are captured, and k of these are found to be tagged. What can we say about the number of fish in the pond?

All we know for certain is that there are at least $M+n-k$ fish. However, if we are prepared to believe that all the fish, tagged or untagged, have equal chances of being captured, then proceed as follows. Let N be the unknown number of untagged fish. The value of N that makes expression (3.1) as large as possible will be that which maximises the probability of the observed outcome of the experiment. In Statistics, this is known as *maximum likelihood estimation*, and is clearly a reasonable procedure.

The best approach is to refer to the right side of (3.1) as $L(N)$, the *likelihood* of N, and compute the ratio $L(N+1)/L(N)$. This collapses down to

$$\frac{(M+N+1)(N+1-n+k)}{(M+N+1-n)(N+1)}$$

which is less than or equal to unity so long as (after simplifying)

$$N + 1 \leq M(n-k)/k.$$

The right side here is fixed as N changes. Thus, for low values of N, $L(N)$ will be less than $L(N+1)$, and then that inequality reverses. Hence $L(N)$ increases initially, reaches its maximum value, and then decreases. If $X = M(n-k)/k$ is not an integer, the unique estimate of N, the unknown number of untagged fish, is the largest integer that does not exceed X. If it is an integer, then both X and $X-1$ would be equally good as estimates.

The simplest non-trivial experiment is one with just two outcomes, which might be Heads or Tails, Success or Failure, 0 or 1 according to context. The term *Bernoulli trials* is used to mean independent repetitions of such an experiment, the chances of the two outcomes remaining fixed. This name is in honour of James (also known as Jacob) Bernoulli, whose *Ars Conjectandi* was published in 1713.

About 51% of human births are male. We might regard the successive births on a maternity ward as a sequence of Bernoulli trials, with the probability of Success taken as either 0.51 or 0.49, according to the arbitrary convention of whether it is a male or a female birth that is labelled a Success.

Definition 3.1

The *Bernoulli distribution* with values $\{0, 1\}$ has $p_1 = p$ and $p_0 = q = 1 - p$ for some p, $0 \leq p \leq 1$. Take $0 \equiv$ Failure, and $1 \equiv$ Success.

Suppose we conduct a fixed number, n, of Bernoulli trials, and record k, the number of Successes. Then $0 \leq k \leq n$, but some work is needed to find the respective probabilities. Any outcome is a sequence of length n such as $FFS \ldots FS$, and there are 2^n possible sequences. By the independence of successive trials, the probability of such an outcome that contains k Successes and $n-k$ Failures will be $p^k q^{n-k}$, irrespective of the order. As there are $\binom{n}{k}$ ways of choosing where those k Successes occur, the probability of exactly k Successes is $\binom{n}{k} p^k q^{n-k}$.

Definition 3.2

The *Binomial distribution*, with symbol $\mathrm{Bin}(n, p)$, is defined by

$$p_k = \binom{n}{k} p^k q^{n-k}$$

for $0 \leq k \leq n$.

The $\mathrm{Bin}(10, 1/6)$ distribution would be a reasonable model for the number of sixes in ten throws of a fair die. Over a series of six Test Matches, the $\mathrm{Bin}(6, 1/2)$

distribution would indicate how often the home captain could expect to win the toss. The number of males in a family with five children might have a Bin(5, 0.51) distribution. Figure 3.1 illustrates the binomial distribution.

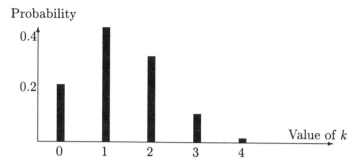

Figure 3.1 The values of the probabilities in a Bin(4, 1/3) distribution.

The ratio of successive probabilities in the binomial distribution simplifies to

$$\frac{p_{k+1}}{p_k} = \frac{(n-k)p}{(k+1)q}.$$

This ratio exceeds unity when $(n-k)p > (k+1)(1-p)$, i.e. when $k+1 < np+p$. Just as we argued when we looked at capture–recapture sampling above, this means that the successive probabilities $\{p_k\}$ initially increase, reach a maximum, then fall away. Whether that maximum is unique, or occurs at two successive values, depends entirely on whether $np+p$ happens to be an integer. The maximum probability in this binomial distribution is when k is at, or just below, $np + p$.

The expression for the ratio of successive terms shows a convenient algorithm for computing the probabilities with a pocket calculator. Plainly $p_0 = q^n$, and if we have found $p_0, \ldots, p_k$ then p_{k+1} comes from $p_{k+1} = p_k \times (p/q) \times (n-k) \div (k+1)$. You can get a check on your calculations by independently working out p_n as p^n; if that is correct, it is most unlikely that any intermediate values are wrong.

The binomial distribution might help an airline to investigate the economics of overbooking passengers onto planes. Suppose the plane's capacity is 100, and that any booked passenger has, independently of the rest, a 10% probability of not turning up for the flight. If tickets are freely transferable, an airline that never overbooks will lose considerable revenue. What are the consequences of booking n passengers, when $n > 100$?

On this model, the number who do show up has a Bin(n, 0.9) distribution. For $n = 105$, calculations show there is a chance in excess of 98% that all who turn up can be accommodated; for $n = 110$, this drops to 67%, and for $n = 115$ it is below 20%. Whatever value of n is chosen, the distributions of the number of empty seats, and the number of passengers who have to be bribed to transfer to another flight can be calculated. The financial consequences of that choice of n then follow.

We ought to look for weaknesses in our model, before basing an entire policy on some possibly dubious calculations. It is fundamental to the $\text{Bin}(n, p)$ model that all the n "trials" are independent, and have the same value of p. If groups of passengers – members of the same family, say – are booked together, then it is likely that either all or none of them will show up. This is a lack of independence. And passengers using the flight for business purposes may have a quite different probability of showing up from those travelling as tourists.

The binomial distribution is the basis of Fermat's method of solving the "problem of points" discussed earlier. We assume all games are independent, and each player has a 50% chance to win any one of them. Clive wins the contest if he wins m games before Dave wins n. Fermat neatly observed that the contest would certainly be decided during the next $m + n - 1$ games, and it did no harm to allow the contest to run to that length, as it was impossible for both players to reach their targets in that time.

Over this series of $m + n - 1$ games, the number Clive wins has a $\text{Bin}(m + n - 1, 0.5)$ distribution, and we require the probability that the outcome is at least m. You should verify that the entries in Table 2.2 do correspond to these calculations from a binomial distribution.

The binomial distribution applies when we are interested in the number of Successes in a fixed number of Bernoulli trials. But what is the distribution of the number of trials to achieve the first Success? This is the question faced by a childless couple considering the process of *in vitro* fertilisation (IVF), where Success on any attempt is far from guaranteed. For exactly k trials to be needed, the first $k - 1$ must result in Failure, the kth in Success, so this outcome has probability $q^{k-1}p$ for $k = 1, 2, 3, \ldots$. If, instead of asking for the number of *trials*, we were interested in the number of *Failures* before the first Success, the probability of exactly k Failures would be $q^k p$.

Definition 3.3

The *Geometric distribution* with symbol $G_1(p)$ is defined by $p_k = pq^{k-1}$ for $k \geq 1$. The distribution with $p_k = pq^k$ for $k \geq 0$ with symbol $G_0(p)$ is also known as the Geometric distribution.

The reason for the name is clear: successive probabilities form a geometric progression, each one being just q times the last. Thus the maximum probability is at the beginning. An optimistic couple embarking on IVF will note that the first attempt is more likely than any other named attempt to achieve the first conception; a pessimistic cricket batsman, who believes that the balls he faces until his dismissal form a sequence of Bernoulli trials, will know he is more likely to be out first ball than any other specified time.

Example 3.4

One white ball is placed in a bag. Sheena throws a fair die, dropping a black ball into the bag at every throw. She stops as soon as she has thrown a Six, and then selects one ball at random from those in the bag. With what probability is it the white ball?

To solve this, we note that the final number of black balls in the bag has a $G_1(p)$ distribution, with $p = 1/6$. If there are k black balls, the chance she selects the white one is $1/(k+1)$. As the distinct values of k form a partition, we have

$$P(\text{White ball}) = \sum_{k=1}^{\infty} P(\text{White ball}|k)P(k) = \sum_{k=1}^{\infty} \frac{1}{k+1}\frac{1}{6}(\frac{5}{6})^{k-1}.$$

For $0 < x < 1$, we can expand $-\ln(1-x)$ as the series $x + x^2/2 + x^3/3 + \cdots$, so $\sum_{k=1}^{\infty} \frac{1}{k+1}(\frac{5}{6})^{k+1} = -\ln(1-\frac{5}{6}) - \frac{5}{6}$. The probability she gets the white ball is $\frac{6}{25}(\ln(6) - 5/6) \approx 0.23$.

In a sequence of Bernoulli trials, to find the probability of no more than r Failures before the first Success, you could compute the sum $\sum_{k=0}^{r} pq^k$. Rather easier, note that this event occurs if, and only if, the first $r+1$ trials include at least one Success, which has probability $1 - q^{r+1}$.

The distributions of the number of trials to achieve r Successes, and the number of Failures before the rth Success, can be found in a similar fashion. For the latter, argue that there will be exactly k Failures if, and only if, the first $k+r-1$ trials include $r-1$ Successes, and the $(k+r)$th trial is a Success. Apply the Binomial distribution to the first $k+r-1$ trials, and use the independence of the next trial from its predecessors. This leads to

Definition 3.5

The *Negative Binomial distribution*, symbol $\text{NB}_0(r, p)$, is given, for $k = 0, 1, 2, \ldots$, by

$$p_k = \binom{k+r-1}{r-1}p^r q^k.$$

Similarly, the $\mathrm{NB}_r(r,p)$ distribution has, for $k = r, r+1, \ldots,$

$$p_k = \binom{k-1}{r-1} p^k q^{k-r}.$$

The $\mathrm{NB}_0(1,p)$ distribution coincides with the $G_0(p)$, and the $\mathrm{NB}_1(1,p)$ with $G_1(p)$. A nice application of this distribution is to Banach's matchbox problem. Tom has two boxes of matches, one in his left pocket and one in his right. Initially both have N matches, and when he needs a match, he is equally likely to select either pocket. What is the distribution of the number of matches in the other box when he first finds one box empty?

Let Success mean he selects the right pocket, so that $p = 1/2$, and find the chance that the $(N+1)$th Success occurs at trial number $(N+1) + (N-k) = 2N + 1 - k$, which leaves k matches in the left pocket. This chance is thus p_{2N+1-k}, evaluated from the $\mathrm{NB}(N+1, 1/2)$ distribution. The probability we seek, that Tom has k matches in the other pocket is just double this, i.e. $\binom{2N-k}{N}/2^{2N-k}$, for $k = 0, 1, 2, \ldots, N$ as there is just the same chance his left pocket will be found empty first.

In the $\mathrm{Bin}(n,p)$ distribution, think of n as being large, λ fixed, so that $p = \lambda/n$ is small. The probability

$$p_k = \binom{n}{k} \left(\frac{\lambda}{n}\right)^k \left(1 - \frac{\lambda}{n}\right)^{n-k}$$

can be rearranged as

$$p_k = \frac{\lambda^k}{k!} \left(1 - \frac{\lambda}{n}\right)^n \times \frac{(n)_k}{n^k} \times \left(1 - \frac{\lambda}{n}\right)^{-k}.$$

As $n \to \infty$, the second term converges to $e^{-\lambda}$ and the last two terms each converge to unity (recall that k is fixed). Thus the whole expression converges to $\lambda^k e^{-\lambda}/k!$.

Definition 3.6

The *Poisson distribution*, symbol $\mathrm{Poiss}(\lambda)$, is given by $p_k = e^{-\lambda}\lambda^k/k!$ for $k = 0, 1, 2, \ldots$.

It is almost impossible to exaggerate the importance of the Poisson distribution in models of random experiments. We met it in Example 1.9, when a secretary scattered letters at random into envelopes; the most notorious example is due to von Bortkewitsch (1898), who observed that the numbers of deaths by horsekick in the Prussian Cavalry Corps over a twenty-year period followed this distribution beautifully. (In any Corps, in any year, there would be a very large number of very tiny chances of such a death, exactly the template for our derivation of the Poisson.) William Feller gave data on flying bomb hits on London during the Second World War: with a grid half a kilometre square superimposed on the city, the numbers of hits in these different blocks also followed this distribution. Other applications include the number of telephone calls at an exchange over a specified quarter-hour; the number of misprints on each page of a book; the emission of radioactive particles from a source. Whenever things tend to occur at random in time or space, at some overall average rate, the number in a specified region will tend to follow this Poisson distribution. We will justify this statement in the next chapter.

Just as we saw with the binomial, the expressions for the ratios of successive probabilities in the Poisson simplify, and we can use this to calculate the individual probabilities. Initially $p_0 = \exp(-\lambda)$, and then $p_{k+1} = p_k \times \lambda \div (k+1)$.

EXERCISES

3.1. In the hypergeometric distribution given by expression (3.1), let n, k be fixed, while $M, N \longrightarrow \infty$ keeping the ratio $M/(M+N) = p$ constant. Show that this expression converges to the value p_k given in Definition 3.2 for the binomial distribution.

Why is this not a surprise?

3.2. A majority verdict of 10–2 or better may be permitted in a jury trial. Assuming each juror has probability 0.9 of reaching a Guilty verdict, and decides independently, what is the probability the jury decides to convict?

3.3. Assume that within a given service game at tennis, successive points form Bernoulli trials with $p = P(\text{Server wins}) > 1/2$. Tennis rules say that the service game ends as soon as either player has won at least four points, and is at least two points ahead of the other. Find the chances the server wins the game $4 - 0$, $4 - 1$ and $4 - 2$; find also the chance the game reaches $3 - 3$ ("Deuce").

Let D be the chance the server eventually wins the game, if the score is now Deuce. Show that $D = p^2 + 2pqD$, and hence deduce that the overall probability the server wins the game is $(p^4 - 16p^4q^4)/(p^4 - q^4)$.

3.4. Show that, as in the Binomial distribution, the successive probabilities in a Poisson distribution increase to a maximum, then decrease towards zero. Under what circumstances is there a unique maximum probability?

3.5. Model the number of attempts a jailer makes to locate the key to a cell, when he has a bunch of K keys, and has no idea which is the correct key. Give the answers for the cases when (a) he goes logically through the keys, one at a time, and (b) he is drunk, and has no idea whether he has tried any particular key before.

3.6. Let P_1 and P_2 be two discrete distributions over the same set of outcomes. Define $d(P_1, P_2) = \sup_A |P_1(A) - P_2(A)|$ as the *distance* between them, i.e. the largest difference between the probabilities that P_1 and P_2 attach to any event. Show that $d(P_1, P_2) = \sum_{n \in \Omega} |p_1(n) - p_2(n)|/2$.

Find the distance between the Bin(3, 1/3) distribution, and the Poiss(1).

3.2 Probability Generating Functions

Whenever all the outcomes are non-negative integers, a useful tool resides in the probabilist's kit-bag:

Definition 3.7

Suppose $(p_k : k = 0, 1, 2, \ldots)$ form a probability distribution. Then $g(z) \equiv \sum_{k=0}^{\infty} p_k z^k$ is the corresponding *probability generating function*, or *pgf*.

For example, the pgf of the Bin(n, p) distribution is $\sum_{k=0}^{n} \binom{n}{k} p^k q^{n-k} z^k$ which simplifies to $(pz + q)^n$, using the binomial theorem (Appendix 9.3A). The pgf of a Poiss(λ) is $\sum_{k=0}^{\infty} e^{-\lambda} \lambda^k z^k / k! = e^{-\lambda} \sum_{k=0}^{\infty} (\lambda z)^k / k!$, which reduces (standard series for exp) to $e^{-\lambda(1-z)}$. Because $p_k \geq 0$ and $\sum p_k = 1$, the pgf always converges when $|z| \leq 1$. The definition shows that a distribution determines a pgf. Indeed, the converse is also true, and justifies the terminology "probability generating function": if you expand a pgf as a power series, it *generates* the individual probabilities as the coefficients of the successive powers z^k.

A common reaction amongst students, on first meeting a pgf, is to ask "What is z?". That is a good question. One answer is to demonstrate the usefulness of the idea of a generating function in a different context. Suppose you are challenged to prove that, for all non-negative integers k, m, n,

$$\binom{m+n}{k} = \sum_{r=0}^{k} \binom{m}{r}\binom{n}{k-r}.$$

Plainly $(1+z)^{m+n} \equiv (1+z)^m(1+z)^n$, and the binomial theorem tells us that

$$(1+z)^{m+n} \equiv \sum_{k=0}^{m+n} \binom{m+n}{k} z^k.$$

Using similar expressions for both $(1+z)^m$ and $(1+z)^n$ leads to the identity

$$\sum_{k=0}^{m+n} \binom{m+n}{k} z^k \equiv \sum_{r=0}^{m} \binom{m}{r} z^r \sum_{s=0}^{n} \binom{n}{s} z^s.$$

On the right side, terms in z^k arise when a term in z^r from the first sum multiplies one in z^{k-r} in the second. Picking out the coefficients of z^k on both sides of the identity in this fashion proves our result. Introducing z allowed us to handle all the binomial coefficients together, in a single expression $(1+z)^{m+n}$.

So my answer to "What is z?" is that at first it merely gives an alternative way of specifying the distribution, but we shall be able to use $g(z)$ in unexpected ways to simplify arguments later. The development of Branching Processes would be grotesquely complex if pgfs were not available.

3.3 Common Continuous Probability Spaces

The times recorded in swimming sprints at the Olympic Games are announced to 100th of a second, but the times to 1000th of a second have been used to split close finishes. Even so, these times are only a close approximation to the actual times, which will be from the continuum of real numbers. The chance the winner took *exactly* the 52.004 seconds shown on the electronic clock is zero. At best we can say the time was between 52.0035 and 52.0045 seconds.

This is the essential difference between discrete probability spaces, and the continuous spaces we now study. The outcomes Ω will be the whole of the real line, or perhaps the non-negative reals, or a finite interval. Every individual outcome has zero probability, but, just as we saw in the first chapter with the continuous uniform distribution, tiny intervals within Ω can have non-zero probability.

In Chapter 1, we noted that for such an Ω, it was enough to describe how to find the probability associated with an interval within Ω.

Definition 3.8

Suppose $f(x)$ is defined for $x \in \Omega$, and satisfies

(i) $f(x) \geq 0$

(ii) $\int_\Omega f(x)dx = 1$.

Then f is a *probability density function*, and the probability of the interval (a, b) is $\int_a^b f(x)dx$.

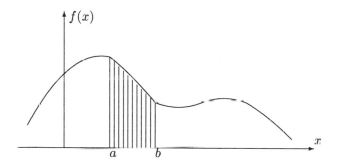

There is a close affinity between a density function, and the mass function for discrete probability spaces. Instead of individual probabilities being non-negative, the density is non-negative; and sums get replaced by integrals to obtain the probability of an event.

Definition 3.9

Given any density function $f(x)$, its corresponding *distribution function* is $F(x) = \int_{-\infty}^{x} f(u)du$. The probability of the interval (a, b) is then $F(b) - F(a)$.

Clearly, since a distribution function evaluates the probability of being not more than x, it is an increasing function, taking the values 0 at $-\infty$ and 1 at $+\infty$. We could have defined the distribution function for a discrete space, by a sum rather than an integral, but distribution functions tend to be less useful in discrete spaces. A distribution function and a density function determine each other, the density function being the derivative of the distribution function. Since a distribution function computes the probability of a definite event, while a density function gives the rate of change of probability, a more nebulous notion, a distribution function is often more convenient for theoretical work. On the other hand, a density function is better for a quick indication of the regions of greater or lesser probability.

We have already met the $U(a, b)$ distribution. Its density function is $f(x) = 1/(b-a)$ for $a < x < b$, and $f(x)$ is zero elsewhere. Its distribution function is

$$F(x) = \begin{cases} 0 & x \leq a \\ (x-a)/(b-a) & a \leq x \leq b \\ 1 & x \geq b. \end{cases}$$

How long should we wait for an α-particle to be emitted from a radioactive source? Ω here will be the interval $[0, \infty)$, and suppose the density function is $f(x)$. If $P(t)$ is the probability that we have to wait more than time t, then

$$P(t) = \int_t^\infty f(x)dx. \tag{3.2}$$

$P(t)$ can also be interpreted as the probability there are no emissions before t, and the idea that emissions occur at random means that the numbers in non-overlapping time intervals are independent. Hence, for any $u > 0$ and $v > 0$,

$$\begin{aligned} P(u+v) &= P(\text{No emissions in } (0, u+v)) \\ &= P(\text{None in } (0, u) \cap \text{None in } (u, u+v)). \end{aligned}$$

By independence, and assuming that the distribution of the number of emissions depends only on the length of the interval, not its position, this gives

$$P(u+v) = P(u)P(v).$$

Let $Q(t) = \ln(P(t))$. The last equation shows that $Q(u+v) = Q(u) + Q(v)$ whenever $u > 0$ and $v > 0$.

Interesting consequences tumble from this expression. First, take $u = v$ to see that $Q(2u) = 2Q(u)$, and then use this idea repeatedly to find $Q(nu) = nQ(u)$ for all $u > 0$ and positive integers n. In particular, take $u = 1/n$ to see that $Q(1/n) = Q(1)/n$. Thus, for any positive integer m, $Q(m/n) = (m/n)Q(1)$. Whenever t is a positive *rational* number, $Q(t) = tQ(1)$.

The definition of P shows that it is monotone decreasing, hence so is Q, and continuity considerations imply $Q(t) = tQ(1)$ for all $t > 0$. Now $P(1)$, being a probability, is less than unity, so $Q(1)$ will be negative: write $Q(1) = -\lambda$. Then $P(t) = e^{-\lambda t}$. Equation (3.2) gives $f(x) = \lambda e^{-\lambda x}$ when $x > 0$.

Definition 3.10

The *Exponential distribution* $E(\lambda)$ has density $\lambda e^{-\lambda x}$ on $x > 0$.

A similar argument indicates that this density function should be a useful model whenever we can persuade ourselves that events of interest are happening randomly, but at some overall average rate. It has been found to give sensible answers when modelling the times of goals in soccer matches, the arrivals of customers in queues, births in a population, cell division, light bulbs failing, and many more. The parameter λ is interpreted as the average rate at which these events of interest occur, so high values of λ are associated with short waiting times.

Real life seldom matches a simple mathematical model completely, we have to be content with good approximations – the usual mathematical trade-off between reality and tractability. A modeller would like a bank of probability densities, whose parameters can be chosen to give a fair fit to data. Here are two examples, each having two parameters.

Definition 3.11

Take $f(x) = Kx^{\alpha-1}e^{-\lambda x}$ on $x > 0$, where $K = \lambda^{\alpha}/\Gamma(\alpha)$ is chosen to ensure $\int f(x)dx = 1$. This is called the *Gamma distribution*, with symbol $\Gamma(\alpha, \lambda)$; the parameters α and λ can take any positive values.

This Gamma distribution has been found useful in modelling the service times of customers making transactions such as buying rail tickets, or in supermarkets. It can model quantities whose values are non-negative, and unbounded above; the density function rises to a single peak before falling away. Taking $\alpha = 1$ shows that the Exponential distribution belongs to this family.

Definition 3.12

Take $f(x) = Kx^{\alpha-1}(1-x)^{\beta-1}$ over $0 < x < 1$, where the constant K is given by $K = \Gamma(\alpha+\beta)/\{\Gamma(\alpha)\Gamma(\beta)\}$. This is the *Beta distribution*, with symbol $B(\alpha, \beta)$; the parameters α and β can take any positive values.

The Beta distribution can model quantities that range between zero and unity, so the Beta family is a candidate for describing our knowledge or belief about the value of an unknown probability. When $\alpha = \beta = 1$, it reduces to the $U(0,1)$ density, which we would use to express complete ignorance about a probability – any value from 0 to 1 seems just as good as any other. If we favour values near $2/3$, some $B(2m, m)$ model could be appropriate; the larger the value of m, the more concentrated around $2/3$ would our beliefs be. Choosing α in the range $0 < \alpha < 1$ allows us to give high weight to values of x that are close to zero. This distribution is very flexible.

Suppose that the quantity we wish to model lies in the finite range (a, b), rather than $(0, 1)$. We could adapt the Beta family for use here; first, a scale change to map the interval $(0, 1)$ to $(0, b - a)$, then a location change by adding a to all values.

There are many more standard continuous distributions, including the daddy of them all, the Normal or Gaussian distribution, but we will postpone their introduction until they arise naturally.

EXERCISES

3.7. Find the pgfs of the $G_1(p)$, $U[1, K]$ and $NB_0(r, p)$ distributions.

3.8. The *Logarithmic distribution* has been used in ecology to model how often any particular species appears in a habitat. Its pgf is of the form $c \ln(1 - qz)$, where c is a constant and $0 < q < 1$. Find c in terms of the parameter q, and write down the expression for p_k, the chance a species is located k times.

3.9. Suppose buses arrive at random, at average rate λ, so that the time to wait follows an Exponential distribution. Show that, conditional on you having already waited time T, without a bus, the remaining time you have to wait is independent of T.

3.10. For each function f, decide whether or not there is some constant k that makes f a density function. When such a k can be found, find it, evaluate the probability of getting a value between 1 and 2, and the distribution function $F(x)$.

(a) $f(x) = kx$ over $0 < x < 2$.

(b) $f(x) = -k \sin(x)$ over $0 < x < \pi/2$.

(c) $f(x) = k \cos(x)$ over $0 < x < 2$.

(d) $f(x) = k|x|$ over $-1 < x < 1$.

(In all cases, $f(x) = 0$ outside the indicated range.)

3.11. Assume that the proportion of commercial vehicles among users of the Humber Bridge varies randomly from day to day, with density $f(x) = cx(1 - x)^2$ over $0 < x < 1$, where c is a constant. Show that $c = 12$, find the distribution function, and sketch the density and distribution functions over $-1 < x < 2$.

On what fraction of days is the proportion of commercial vehicles between 20% and 50%?

3.12. Eleven horses enter a race. Mr C takes exactly 60 seconds to complete the course. Each of the other ten horses, independently, takes more than 60 seconds with probability p, or less than 60 seconds with probability $1 - p$. Give a model for Mr C's position in the race. Find the chances that he wins, and that he is placed (i.e. in the first three).

Find the value of p that makes all eleven horses equally likely to win. For this value of p, find the chance Mr C is placed. Is it a surprise that your last answer is not 3/11?

3.13. Verify that $f(x) = -\ln(x)$ for $0 < x < 1$ is a density function, and sketch it. Write $A = (1/4, 3/4)$ and $B = (0, 1/2)$. Use your density sketch to *assess* which is bigger, $P(A)$ or $P(A|B)$. Calculate both values; check that your assessment was correct.

3.14. For both the Gamma and Beta distributions, find the value of x that maximises the density function. When is there just one local maximum?

3.4 Mixed Probability Spaces

Not all probability spaces fit conveniently into the category of either discrete or continuous. Here are a few experiments where the outcomes naturally have both a discrete and a continuous component.

- Select an adult male at random, and measure his alcohol intake over a week. A proportion of the population will be teetotal, consuming no alcohol at all, while the consumption of the rest will vary over a continuous range.

- In a clinical trial over a three-month period, the patient either dies at some (continuous) time during those three months, or is recorded as having survived at least three months – the discrete component.

- A squirrel has secreted food at several locations in a tree, and either gambols around the tree, or visits one of these locations for a feast. At any random time, the distribution of its position in the tree will have a positive probability attached to each store, otherwise a continuous distribution over the rest of the tree.

- When you take your shopping trolley to the supermarket checkout, the time you wait until you are served may be zero with probability say, 20%, otherwise will follow some continuous distribution.

Such examples cause no real difficulty. If the total probability of the discrete components is p, then the density function for the outcomes in the continuous part of Ω will be non-negative, and its integral over that region will be $1 - p$.

However, be warned that there are probability spaces that are more complex than discrete, continuous, or such a mixture. As an illustration, we introduce a distribution function F based on *Cantor's singular function*.

First, let $F(x) = 0$ for $x \leq 0$, and $F(x) = 1$ for $x \geq 1$. For $0 \leq x \leq 1$, define F as follows:

(i) for $1/3 < x < 2/3$, then $F(x) = 1/2$;

(ii) for $1/9 < x < 2/9$ then $F(x) = 1/4$; for $7/9 < x < 8/9$ then $F(x) = 3/4$;

(iii) continue in this fashion indefinitely, using the middle third of each remaining interval. The next step is to take the four intervals $(1/27, 2/27)$, $(7/27, 8/27)$, $(19/27, 20/27)$, $(25/27, 26/27)$, and let $F(x)$ be respectively $1/8$, $3/8$, $5/8$, $7/8$ on them.

The figure shows F, so far as has been explicitly described.

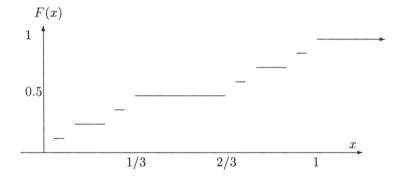

If you imagine completing the sketch of F, you will appreciate that it is continuous at all points, and that it satisfies all the conditions to be a distribution function. It is even differentiable, and its derivative is zero, in the interiors of these intervals of lengths $1/3$, $1/9$, $1/27$ etc., where we have defined its values.

However, in the corresponding probability space over the interval $(0, 1)$, there is no discrete component, as no individual point has non-zero probability. And where F is differentiable, its derivative is zero, so any density function would have to be zero everywhere. There is no continuous component either!

Such a probability space might arise only rarely as a model for a real-life phenomenon. But its existence is a warning against any belief that all probability spaces can be reduced to a mixture of discrete and continuous components, even though they are all we shall meet in this book.

4
Random Variables

Sometimes, in a random experiment, the sole items of interest are the individual outcomes and events. More often, the outcome is just the trigger for one or more consequences, which are the real focus of attention.

One example is in playing roulette. The natural probability space for one spin of the wheel takes the 37 outcomes $\Omega = \{0, 1, 2, \ldots, 36\}$ as equally likely. Suppose Christine has staked two units on the 12 numbers $\{3, 6, 9, \ldots, 36\}$ in a single column, while Sheila has risked one unit on the three numbers $\{7, 8, 9\}$ across one row. If Christine wins, she makes a profit of four units (her bet pays out at odds of 2 : 1), Sheila's bet has odds of 11 : 1 so she would win 11 units. Neither gambler cares particularly what the individual outcome is, all that matters is whether or not it favours her bet. Using X and Y to denote the changes in their fortunes, X takes the value $+4$ with probability 12/37 and the value -2 with probability 25/37, while Y is $+11$ with probability 3/37 and -1 with probability 34/37.

X and Y are known once the result of the spin is known, but we do not need a list of 37 outcomes to describe their properties. X has a space with outcomes $\{-2, 4\}$, Y has a space with outcomes $\{-1, 11\}$, and these simpler spaces are all that are necessary to study either X or Y on its own. But if we want to look at X and Y together, these separate spaces are not enough, since both quantities rest on the same spin. The original Ω remains useful.

4.1 The Definition

Definition 4.1

Given any probability space $(\Omega, \mathcal{F}, P)$, a *random variable* is a real-valued function X on Ω such that, for all real numbers t, the set $\{\omega \in \Omega : X(\omega) \leq t\}$ belongs to $\mathcal{F}$.

Just as we generally use capital letters from the front of the alphabet to denote events, so we will normally use capital letters near the end to denote random variables. We shall reserve the letter Z for a random variable having the Standard Normal distribution (see later).

A trivial example of a random variable is when X is a constant function, taking the same value whatever the outcome. The phrase "degenerate" variable is used in this case. It is plain that the definition is satisfied, as $\{\omega \in \Omega : X(\omega) \leq t\}$ is either the empty set, or the whole space, both of which always belong to $\mathcal{F}$.

The condition that $\{\omega \in \Omega : X(\omega) \leq t\}$ belongs to $\mathcal{F}$ ensures that we can always evaluate its probability, i.e. we can find the probability that $X \leq t$ for any real number t. When the probability space is discrete, that condition is automatically satisfied, as all subsets of Ω are events in $\mathcal{F}$. The condition will never be a serious obstacle, unless you try very hard to construct exotic functions on continuous probability spaces. You will not go far wrong if you look on a random variable as *any* real-valued function determined by the outcome of some random experiment.

In some experiments, it is the outcome itself that is of prime interest. When Ω is a list of real numbers, or an interval, then plainly the identity function, $X(\omega) = \omega$, is a random variable. In the last chapter, we introduced several probability spaces that might be models for random experiments. We use the same names – Binomial, Poisson, Exponential etc. to denote the random variables that are the identity functions on these spaces. (Of course, random variables having these distributions can also arise as suitable functions on other probability spaces.)

Definition 3.9 defined a distribution function in terms of a density function. Whatever the random variable X, its distribution function is given by $F(t) = P(X \leq t) = P(\omega : X(\omega) \leq t)$.

If, when we conduct an experiment, we shall be interested in only one random variable defined on it, little purpose is served by going through the two stages of setting up the probability space, and then defining X on it. It will be enough to describe the distribution of X directly, either as the mass function of a discrete variable, or the density function of a continuous one. That is the approach we shall use. It is when we look at more than one random variable, as in the roulette example above, that it can be useful to have explicit mention of the probability space.

The distribution function of a random variable gives a complete description of its properties. But the saying "You can't see the wood for the trees" forces its way irresistibly forward: on many occasions, we are better served by one or more quantities that summarise important facts about the variable. One candidate is what might be termed the middle value of X, so that X will be below or above that value half the time.

Definition 4.2

Suppose X is a random variable. Then any value m such that $P(X \leq m) \geq 1/2$ and $P(X \geq m) \geq 1/2$ is a *median* of X.

For a continuous random variable with distribution function F, we solve the equation $F(m) = 1/2$. For a discrete random variable, that equation may have no solution. For example, if X is Poiss(2.4), then $P(X \leq 1) = 0.3084$, $P(X \leq 2) = 0.5697$; for no value of m is $P(X \leq m) = 1/2$, and here the unique value $m = 2$ satisfies the definition of being a median. But for the discrete $U[1,6]$ distribution corresponding to the throw of a fair die, any real number m with $3 \leq m \leq 4$ satisfies the condition to be a median.

In a similar fashion, we could define the *quartiles* of a random variable, being the three places that split the range of X into four quarters with equal probability. The median is the second quartile, and the distance between the first and third quartiles will give an indication of the variability of X. Other useful summary measures will be considered below.

4.2 Discrete Random Variables

Definition 4.3

Given a discrete random variable X taking values $(x_1, x_2, \ldots)$ with respective probabilities $(p_1, p_2, \ldots)$, its *mean* or *expected value* is defined as

$$E(X) = \sum_i x_i p_i$$

provided this series is absolutely convergent.

This is the mathematical definition of what is often informally described as an *average*. Each value of X is weighted according to the theoretical frequency with which it will arise. If we did not insist that the series be absolutely convergent, we would have to face the possibility that it is conditionally convergent, which would lead to a major embarrassment. In any conditionally convergent series, the order of the terms can be rearranged to make the resulting series converge to any answer whatsoever! That would be absurd, if we wanted the sum to describe the "average" value of X, as that average value would depend on the arbitrary decision about the order of summing the terms.

Banning conditionally convergent series lets us escape that trap. And actually, if the series is not conditionally convergent, but diverges to either $+\infty$ or to $-\infty$, this also causes no trouble, as we can write $E(X) = +\infty$, or $E(X) = -\infty$, without fear of misinterpretation. It is therefore convenient to extend the definition to include these possibilities. The symbol μ is often used to denote $E(X)$.

A different problem arises when the series is not conditionally convergent, but does not diverge to $+\infty$ or $-\infty$ either. For example, suppose Emily and Marion takes turns to toss a fair coin, Emily first. If the first Head appears on the nth toss, the one who obtained that Head gives the other 2^n peanuts. From Emily's perspective, writing losses as negative amounts, she "wins" -2, $+4$, -8, $+16$, ... peanuts with respective probabilities $1/2$, $1/4$, $1/8$, $1/16$, ... Formally using Definition 4.3, the "mean" number of peanuts she wins is

$$-2 \times (1/2) + 4 \times (1/4) - 8 \times (1/8) + \cdots = -1 + 1 - 1 + 1 - \cdots,$$

a meaningless expression. We cannot simply apply $E(X) = \sum x_i p_i$ without checking that the series behaves in a friendly fashion.

Example 4.4

Suppose X has the binomial distribution $\text{Bin}(n, p)$. Then its mean is given by

$$\mu = E(X) = \sum_{k=0}^{n} k \times \binom{n}{k} p^k q^{n-k} = np \sum_{k=1}^{n} \binom{n-1}{k-1} p^{k-1} q^{n-k},$$

and since the sum reduces to $(p + q)^{n-1} = 1$, so $\mu = E(X) = np$.

This result should be no surprise. For a binomial distribution, we conduct n Bernoulli trials, each with a probability p of Success. To find that the mean number of Successes is np accords with intuition.

Theorem 4.5

Suppose X has pgf $g(z)$. Then $\mu = E(X) = g'(1)$, where g' is the derivative of g. (If g is not differentiable at $z = 1$, $g'(1)$ is taken as the limit as $z \uparrow 1$.)

Proof

Write $P(X = i) = p_i$ so that $E(X) = \sum i p_i$ and the pgf is $g(z) = \sum_i p_i z^i$, convergent certainly for $|z| \leq 1$. A power series can be differentiated term by term, so $g'(z) = \sum_i i p_i z^{i-1}$, convergent for $|z| < 1$. Now let $z \uparrow 1$. That useful analytical result known as Abel's Lemma allows us to interchange sum and limit to obtain $g'(1) = \sum i p_i$, proving the result. $\square$

As an illustration, we know that the pgf of the standard binomial distribution is $(pz + q)^n$. Differentiate, put $z = 1$, and quickly confirm the result $\mu = np$. Similarly, since a Poiss(λ) distribution has pgf $\exp(\lambda(z - 1))$, the same procedure shows that $E(X) = \lambda$. The parameter of a Poisson distribution is interpreted as its mean.

Where the pgf does not simplify, using Theorem 4.5 seldom helps, and we use Definition 4.3 directly. The mean outcome of Christine's roulette bet is $E(X) = -2 \times (25/37) + 4 \times (12/37) = -2/37$, and Sheila's mean winnings are $E(Y) = -1 \times (34/37) + 11 \times (3/37) = -1/37$.

As well as evaluating the mean value of X, we might be interested in some function of X, and in the mean of that function.

Definition 4.6

Using the notation of Definition 4.3, let $h(X)$ be some function of X. Then the expected value of $h(X)$ is given by

$$E(h(X)) = \sum_i h(x_i)p_i,$$

with the same convention about the absolute convergence of the series.

The very assiduous reader will have spotted a potential problem. We could use this last definition to find $E(h(X))$, but we might equally have started with $Y = h(X)$ as the random variable of interest, and used Definition 4.3. The next result assures us that the two routes lead to the same answer.

Theorem 4.7 (Law of the Unconscious Statistician)

Let X be a discrete random variable on the probability space $(\Omega, \mathcal{F}, P)$, and suppose $Y = h(X)$ is also a random variable. Then using Definition 4.3 for $E(Y)$ leads to the same answer as using Definition 4.6 for $E(h(X))$, provided both $E(h(X))$ and $E(Y)$ are well-defined.

Proof

We collect together like terms. For any value x_i taken by X, write $K_i = \{\omega : X(\omega) = x_i\}$. For any value y_j taken by $Y = h(X)$, write $L_j = \{i : h(x_i) = y_j\}$. Then $\{\omega : Y(\omega) = y_j\} = \bigcup_{i \in L_j} K_i$. Using Definition 4.3,

$$E(Y) = \sum_j y_j P(\omega : Y(\omega) = y_j) = \sum_j y_j \sum_{i \in L_j} P(K_i).$$

Since $P(K_i) = P(X = x_i)$, rearranging the order of summation leads to $E(Y) = \sum P(X = x_i)h(x_i)$, which proves the theorem. □

Suppose X has a well-defined mean, that $P(X = x_i) = p_i$, and that a, b are constants. Then

$$E(aX + b) = \sum(ax_i + b)p_i = a\sum x_i p_i + b\sum p_i = aE(X) + b.$$

To illustrate this, suppose Trixie stumbles into the wrong examination room, and finds herself facing a multi-choice exam with 25 questions, each with four possible answers. In each case, just one answer is correct. Trixie ought to walk out, but to cover her embarrassment, she stays and makes a random guess at each question. A correct answer scores five marks, an incorrect answer leads to a one-mark penalty. What is Trixie's mean score?

Let X denote the number of correct answers. Then there are $25 - X$ wrong answers, so her total score will be $5X + (-1)(25 - X) = 6X - 25$. We take X plausibly to have a $\text{Bin}(25, 1/4)$ distribution, and our earlier result shows that $E(X) = 25/4$. Thus her mean score is $E(6X - 25) = 6 \times 25/4 - 25 = 12.5$.

The mean μ of X is often the single most important quantity. Like the median, it gives some idea of a middle, perhaps a typical, value. But we are often very interested in the variability of X. One possible approach would be to calculate $E(|X - \mu|)$, but this is awkward to deal with in mathematical manipulations. More convenient is to use the following notion.

Definition 4.8

If X has finite mean μ, its *variance*, commonly denoted by σ^2, is defined as $\text{Var}(X) = E((X - \mu)^2)$.

Since a variance is the mean value of the non-negative quantity $(X - \mu)^2$, it cannot be negative. And if X is measured in feet, its variance is in square feet. It is thus natural to take the square root of a variance, thereby obtaining the so-called *standard deviation* (s.d.) of X, which is measured in the same units as X.

Note that

$$E((X - \mu)^2) = E(X^2 - 2\mu X + \mu^2) = E(X^2) - 2\mu^2 + \mu^2 = E(X^2) - \mu^2.$$

This gives an alternative, often easier, way of calculating a variance.

The knowledge that $E(aX + b) = a\mu + b$ shows that

$$\text{Var}(aX + b) = E((aX + b - a\mu - b)^2) = E(a^2(X - \mu)^2) = a^2\text{Var}(X).$$

Adding the constant b has made no difference to the variance – which makes sense if variance is to measure variability.

Theorem 4.9

Suppose X has pgf $g(z)$ and a finite mean. Then $\text{Var}(X) = g''(1) + g'(1) - g'(1)^2$.

Proof

Consider the meaning of the expression $E(z^X)$. Since z^X takes the value z^i when $X = i$, we see that

$$E(z^X) = \sum p_i z^i = g(z),$$

the pgf. Differentiate to obtain

$$g'(z) = E(Xz^{X-1}) \quad \text{and} \quad g''(z) = E(X(X-1)z^{X-2}).$$

Hence $g''(1) = E(X^2 - X)$; we know from previous work that $g'(1) = E(X)$, so $g''(1) + g'(1) = E(X^2)$. The result follows.   □

For example, in the binomial distribution we have $g'(z) = np(pz + q)^{n-1}$, so $g''(1) = n(n-1)p^2$ and the variance is $\sigma^2 = n(n-1)p^2 + np - n^2p^2 = npq$. For the Poisson distribution, you should verify that the variance, like the mean, evaluates to λ.

Example 4.10 (A Fair Game?)

How much would you pay to play the following game? There is a sequence of Bernoulli trials, whose probability of Success is p, with $0 < p < 1$. You are given a prize of $£X$, where

1. if the first trial is S, then X is the length of the first run of Fs;

2. if the first trial is F, then X is the length of the first run of Ss.

So, for instance, $X = 4$ if the sequence begins $SSFFFFSFS\ldots$.

To judge how much it is worth paying, we work out $E(X)$. To find the distribution of X, let $A = $ First trial is S. When A occurs, ignore the initial run of Ss, and start counting from the first F. Then $X = k$ means that there are exactly $k-1$ Fs after this first one, and the chance of this is $q^{k-1}p$, the Geometric distribution. Hence

$$P((X = k) \cap A) = P(X = k|A)P(A) = q^{k-1}p.p = p^2q^{k-1}.$$

Similarly, we find $P((X = k) \cap A^c)$, and then

$$P(X = k) = P((X = k) \cap A) + P((X = k) \cap A^c) = p^2q^{k-1} + q^2p^{k-1}.$$

The pgf of X is

$$g(z) = \sum_{k=1}^{\infty}(p^2q^{k-1} + q^2p^{k-1})z^k = p^2z\sum_{k=1}^{\infty}(qz)^{k-1} + q^2z\sum_{k=1}^{\infty}(pz)^{k-1}$$

which shows that $g(z) = p^2z/(1-qz) + q^2z/(1-pz)$. Thus $g'(z) = p^2/(1-qz) + p^2qz/(1-qz)^2 + q^2/(1-pz) + q^2pz/(1-pz)^2$. By Theorem 4.5, $E(X) = g'(1)$ evaluates as 2, *whatever the value of p*! You do not need to know the chance of Success to compute the fair entrance fee. A fee of £2 leads to a fair game, in the sense that your mean reward exactly balances the cost of one play.

The variance of X is found in a similar way using Theorem 4.9. You should check that the expression simplifies to $2/(pq) - 6$. Plainly, this is minimised, with value 2, when the chances of Success and Failure are equal. The variability of your winnings *does* depend on p.

Example 4.11 (Random Equivalence Relations)

The algebraic notion of an *equivalence relation* on a set S splits its members into disjoint blocks, or *equivalence classes*. For example, if $S = \{a, b, c\}$, there are just five distinct equivalence relations on it: $\{a, b, c\}$; $\{a, b\}, \{c\}$; $\{a, c\}, \{b\}$; $\{b, c\}, \{a\}$; and $\{a\}, \{b\}, \{c\}$. Given a set S with n members, if one of the possible equivalence relations is selected at random, all equally likely, what can be said about the number of equivalence classes?

Write $S(n, k)$ as the number of equivalence relations with exactly k non-empty classes. The listing above shows that $S(3, 1) = 1$, $S(3, 2) = 3$ and $S(3, 3) = 1$. It is plainly easy to count up $S(n, k)$ when n is small, and for larger values we can use the recurrence relation

$$S(n + 1, k) = kS(n, k) + S(n, k - 1). \tag{4.1}$$

To prove this formula, suppose we wish to split $n + 1$ objects into exactly k subsets. Either the first n objects are already split among k non-empty sets, in which case there are k choices for the last object, or the first n objects are divided among $k - 1$ non-empty sets, and the last object then forms the final subset. The $S(n, k)$ are called the Stirling numbers of the Second Kind. A table of their early values is

n/k	1	2	3	4	5
1	1				
2	1	1			
3	1	3	1		
4	1	7	6	1	
5	1	15	25	10	1

Summing along the rows gives the *Bell numbers* $B_1 = 1, B_2 = 2, B_3 = 5, B_4 = 15, B_5 = 52, \ldots$ B_n is the total number of equivalence relations on a set of n objects, and it is conventional to write $B_0 = 1$.

Given n, let X denote the number of equivalence classes when one of the equivalence relations is selected at random. Then $P(X = k) = S(n, k)/B_n$, and so $E(X) = \sum_k kS(n, k)/B_n$. By Equation (4.1), which holds even when $k > n$,

$$\sum_k kS(n, k) = \sum_k S(n + 1, k) - \sum_k S(n, k - 1) = B_{n+1} - B_n,$$

and so $E(X) = B_{n+1}/B_n - 1$.

To find the variance, multiply Equation (4.1) by k and sum, so that

$$\sum_k k^2 S(n, k) = \sum_k kS(n + 1, k) - \sum_k kS(n, k - 1) = B_{n+2} - 2B_{n+1}.$$

Hence $E(X^2) = (B_{n+2} - 2B_{n+1})/B_n$, from which $\text{Var}(X) = B_{n+2}/B_n - (B_{n+1}/B_n)^2 - 1$.

To compute this mean and variance, we need an efficient way to find the Bell numbers. We can show that

$$B_{n+1} = \sum_{k=0}^{n} \binom{n}{k} B_k.$$

For, take the first n items in a collection of size $n + 1$: there are $\binom{n}{k}$ ways to select k of these, and then there are B_k ways to partition these k items. Add the $(n + 1)$th item to the other $n - k$ to form the final block in the partition. It is easy to see that all partitions obtained in this fashion are different, and that every partition of $n + 1$ items arises somewhere in this process. The right side of our formula counts all B_{n+1} partitions according to this description. Exercise 4.9 shows a generating function for these Bell numbers.

One way to explore the properties of a random variable is through *simulation*. It is quite standard for a computer to have been programmed to generate a series of values that can be treated as coming from the continuous $U(0,1)$ distribution, so we assume we can call up those values at will. Let X be a discrete random variable, taking the values $0, 1, 2, \ldots$ with respective probabilities $p_0, p_1, p_2, \ldots$.

Define $\{q_n\}$ to be the partial sums of these probabilities, i.e. $q_0 = p_0$, $q_1 = p_0 + p_1$, $q_2 = p_0 + p_1 + p_2$ and so on. (These $\{q_n\}$ are all values taken by the distribution function of X.) Given any value u from a $U(0,1)$ distribution, there will be a unique value of n such that $q_{n-1} < u \leq q_n$ (taking q_{-1} as 0 by convention). The desired value of X is then this value of n. Exercise 4.10 asks you to verify that this procedure does what we desire.

For some of the discrete distributions we have looked at, there are ways of taking advantage of how they arise to simulate values in a more efficient manner. But to see this algorithm in practice, suppose X takes the values $0, 1, 4$ with respective probabilities $1/2, 1/3, 1/6$. Then $q_0 = 1/2$, $q_1 = 5/6 = q_2 = q_3$, and $q_n = 1$ for $n \geq 4$. If u, the value of the $U(0,1)$ variable, is in the range $(0, 1/2)$, we have $q_{-1} = 0 < u \leq q_0 = 1/2$, so the value generated is $X = 0$. For $1/2 < u \leq 5/6$, correspondingly $q_0 < u \leq q_1$, so $X = 1$; and if $u > 5/6$, then $q_3 < u \leq q_4$, so $X = 4$. It is clear that X has the distribution described.

EXERCISES

4.1. Use pgfs to find the mean and variance of the Geometric $G_0(p)$ and $G_1(p)$ distributions, and also for the Negative Binomial $\mathrm{NB}_r(r, p)$ distribution.

4.2. Find the mean and variance of the $U[1, K]$ distribution, without using pgfs.

4.3. Let X have the hypergeometric distribution defined in expression (3.1). Show that

$$E(X) = \frac{Mn}{M + N}, \quad \mathrm{Var}(X) = \frac{MNn(M + N - n)}{(M + N)^2(M + N - 1)}.$$

(a) Check that these values converge to the corresponding values for the binomial distribution, as $M, N \longrightarrow \infty$ with $M/(M + N) = p$ kept constant. (Recall Exercise 3.1.)

(b) Deduce the mean and variance of the number of winning numbers on a single ticket in a 6/49 Lottery.

4.4. Suppose X takes values among the non-negative integers only. Show that $E(X) = \sum_{n \geq 1} P(X \geq n)$.

4.5. Let X have a Poiss(λ) distribution. Find the mean value of $1/(1+X)$. Repeat the calculation when X has the Bin(n,p) distribution.

4.6. You buy 100 used computer monitors for a lump sum of £1500. Your basis is that, on past history, 60% will be serviced at essentially zero cost, and sold for £40 each, while the rest can be sold at £5 each for their components.

Model X, the number of monitors that can be serviced and sold. Write down your net profit, in terms of X. Hence deduce the mean and standard deviation of your net profit.

4.7. In order to test a vaccine, we have to find patients with a certain blood type, that is found in 20% of the population. Model W, the number of people sampled until we have found one with this blood type, X, the number sampled to find four with the blood type, and Y, the number with this blood type among 20 people. Find the means and standard deviations of W, X and Y.

4.8. It is desired to estimate θ, the mean number of a strain of bacteria per ml of a sample of seawater. 100 tubes, each holding 20 ml of seawater have a test agent added: the water will remain clear if there are no bacteria, but turn cloudy if there are any bacteria at all. Assuming the number of bacteria in a sample follows a Poisson distribution, estimate θ from the information that 30 tubes turned cloudy.

4.9. Given the recurrence relation

$$B_{n+1} = \sum_{k=0}^{n} \binom{n}{k} B_k$$

with $B_0 = 1$, show that

$$\sum_{n=0}^{\infty} \frac{B_n x^n}{n!} = \exp(e^x - 1).$$

4.10. Prove that the algorithm described immediately prior to these exercises does indeed generate values having the distribution of X.

4.3 Continuous Random Variables

Sums in discrete mathematics generally correspond to integrals in continuous mathematics, and dealing with random variables conforms to this rule. A continuous random variable X is one with a density function $f(x)$, and its properties can also be found from its distribution function $F(x) = \int_{-\infty}^{x} f(t)dt$.

Definition 4.12

Given a continuous random variable X, its *expected value*, or *mean* is $\mu = E(X) = \int_{-\infty}^{\infty} xf(x)dx$, provided that $\int_{-\infty}^{\infty} |x|f(x)dx < \infty$. When μ is finite, the *variance* of X is given by $\sigma^2 = E((X - \mu)^2) = \int_{-\infty}^{\infty} (x - \mu)^2 f(x)dx$.

The condition on the integral of $|x|$ is there for the same reason as in Definition 4.3 for a discrete distribution, to ensure the integral of $xf(x)$ can be evaluated without ambiguity. Just as in the discrete case, we could (and we will) extend this definition to other cases where there is no mathematical difficulty. For instance, if $\left| \int_{-\infty}^{0} xf(x)dx \right| < \infty$ and $\int_{0}^{\infty} xf(x)dx = \infty$, then $E(X) = \infty$, as $\int_{-\infty}^{\infty} xf(x)dx = \infty$ under any reasonable interpretation.

The argument showing the alternative calculation

$$\sigma^2 = E(X^2) - \mu^2$$

remains valid, and we can adapt the argument given for discrete variables to show we can continue to use

$$E(aX + b) = aE(X) + b.$$

Similarly, we define

$$E(h(X)) = \int_{-\infty}^{\infty} h(x)f(x)dx,$$

provided that $\int_{-\infty}^{\infty} |h(x)|f(x)dx < \infty$.

Whenever X is a random variable, and $h(x)$ is a reasonably well-behaved function, then $Y = h(X)$ will also be a random variable. The most systematic way to find the density function $g(y)$ of Y is via its distribution function, $G(y)$. We have

$$G(y) = P(Y \leq y) = P(h(X) \leq y).$$

The next step is to turn the statement "$h(X) \leq y$" into a statement of the form "$X \in A(y)$", for the appropriate set $A(y)$. When h is either strictly increasing or strictly decreasing – which is often the case – life is simple. The event "$h(X) \leq y$" is written $X \leq h^{-1}(y)$, or as $X \geq h^{-1}(y)$, according as h is increasing or decreasing. Hence we have

$$G(y) = \begin{cases} F(h^{-1}(y)) & \text{if } h \text{ is increasing;} \\ 1 - F(h^{-1}(y)) & \text{if } h \text{ is decreasing.} \end{cases}$$

We now differentiate, and find the formula $g(y) = f(h^{-1}(y))|\frac{d}{dy}(h^{-1}(y)|$, for all possible values y of Y.

I have deliberately chosen not to display this last formula as a theorem, useful though it can be. Experience tells me that too many students will seek to apply the formula, without checking the vital condition that the function h is monotone. I strongly recommend that you do not commit the formula to memory, and if you need to perform the calculation, then work from first principles, using the distribution function of Y as described.

For example, suppose X has a $U(-1, 2)$ distribution, so that its density is $1/3$ over the interval from -1 to $+2$, and we seek the density of $Y = X^2$. Plainly, $0 \leq Y \leq 4$, so suppose y is in that range. Following the notation above, $G(y) = P(Y \leq y) = P(X^2 \leq y)$. For $0 \leq y \leq 1$, $X^2 \leq y$ corresponds to $-\sqrt{y} \leq X \leq \sqrt{y}$, and this event has probability $2\sqrt{y}/3$. For $1 \leq y \leq 4$, $X^2 \leq y$ whenever $X \leq \sqrt{y}$, which has probability $(\sqrt{y} + 1)/3$. Having found the distribution function of Y, we now differentiate to obtain the density

$$g(y) = \begin{cases} 1/(3\sqrt{y}) & 0 < y < 1 \\ 1/(6\sqrt{y}) & 1 < y < 4. \end{cases}$$

Theorem 4.13 (Probability Integral Transform)

Suppose X has a continuous distribution with distribution function F. Then $Y = F(X)$ has the $U(0, 1)$ distribution.

Proof

Since a distribution function only takes values in the range from zero to unity, suppose $0 \leq y \leq 1$. We have

$$G(y) = P(Y \leq y) = P(F(X) \leq y).$$

Now F is an increasing function, so the event $F(X) \leq y$ corresponds to the event $X \leq F^{-1}(y)$, with one caveat: it is possible that F is not strictly monotone, so there may be intervals (a, b) with $a < b$ and $F(a) = F(b)$; in these circumstances, we shall take $F^{-1}(y)$ as the largest possible value, i.e. $\sup\{x : F(x) \leq y\}$.

Now F^{-1} is well-defined, so for all $0 \le y \le 1$,

$$G(y) = P(X \le F^{-1}(y)) = F(F^{-1}(y)) = y$$

which is indeed the distribution function of a $U(0,1)$ variable.   □

If X has a discrete component, this result cannot hold, as there will be values in the interval $(0,1)$ that F cannot take. But for continuous variables, this result suggests a way of using a computer to generate values from that distribution. We assume, as earlier, that the computer has been satisfactorily programmed to generate values from the $U(0,1)$ distribution.

Let F be the distribution function for the variable of interest. Given some value y from the $U(0,1)$ distribution, solve the equation $F(x) = y$ to obtain the value of x, using the convention in the proof of Theorem 4.13 in case of ambiguity. The theorem ensures that x has the desired properties.

For example, suppose we wish to simulate from the $E(\lambda)$ distribution. Here $F(x) = 1 - \exp(-\lambda x)$ on $x > 0$, so the solution of $F(x) = y$ comes from $1 - \exp(-\lambda x) = y$, i.e. $x = -\frac{1}{\lambda}\ln(1 - y)$. We can make a small simplification in this recipe: it is clear that the value of $1 - y$ is also $U(0,1)$ so the formula $x = -\frac{1}{\lambda}\ln(y)$ will also give a value with the correct statistical properties.

In Exercise 3.9, we saw that if X, the time to wait for a bus, had an exponential distribution, then even if we have already waited for time T, the residual waiting time still has the original distribution of X. This is referred to as the *memoryless* property of the exponential distribution. It would be very annoying if real buses routinely behaved in this fashion, but this particular property makes calculations in some models much easier. We will now establish that the exponential distribution is the only continuous distribution with this property.

Write $G(x) = P(X > x)$. Then, for $s > 0$ and $t > 0$,

$$P(X > s + t | X > t) = P(X > s + t \cap X > t)/P(X > t) = G(s + t)/G(t).$$

If X has the memoryless property, $P(X > s + t \mid X > t) = P(X > s)$, and so $G(s + t)/G(t) = G(s)$, i.e.

$$G(s + t) = G(s)G(t).$$

We have met this before. In Chapter 3, when modelling the time to wait for the emission of an α-particle, we established the equation $P(u + v) = P(u)P(v)$, and showed that the only monotone solution was of the form $P(t) = \exp(-\lambda t)$. Thus $G(t) = \exp(-\lambda t)$ for some $\lambda > 0$, and X has an Exponential distribution.

EXERCISES

4.11. Show that, if X has a $U(0,1)$ distribution, its mean is $1/2$ and its variance is $1/12$.

   Let $Y = a + (b - a)X$ where a, b are constants, $a < b$. What is the distribution of Y? What are its mean and variance?

4.12. Suppose X has the Exponential distribution $E(\lambda)$, i.e. its density is $\lambda \exp(-\lambda x)$ on $x > 0$. Show that its mean and s.d. are both $1/\lambda$.

4.13. For what value of c is $f(x) = c/x^3$ for $x > 1$ a density function? Show that, if X has this density, then $E(X) = 2$, but $\mathrm{Var}(X) = \infty$.

4.14. Suppose X has the Exponential distribution $E(\lambda)$. Find the density functions of $Y = aX$ $(a > 0)$, $W = X^2$ and $V = |X - 1|$.

4.15. Suppose X has density $f(x) = \exp(-|x|)/2$ for $-\infty < x < \infty$. Find its mean and variance. (This is known as the *Laplace* or the *double exponential* distribution.) What is the density of $|X|$?

4.16. X has density function $x/2$ for $0 < x < 2$. Sketch its density and distribution functions. Give a formula for simulating values of X, given a supply of $U(0,1)$ values.

   Find the distribution function, and hence the density function, of each of $2X$, $\ln(X)$, $1/X$ and $(X - 1)^2$.

4.17. X has the $U(-\pi/2, \pi/2)$ distribution, and $Y = \tan(X)$. Show that Y has density $1/(\pi(1 + y^2))$ for $-\infty < y < \infty$. (This is the *Cauchy* density function.) What can be said about the mean and variance of Y? How could you simulate values from this distribution, given a supply of $U(0,1)$ values?

4.18. Show that $E(X) = \int_0^\infty P(X > x)dx$ when $X \geq 0$ is continuous. (Compare Exercise 4.4.)

4.19. Find the mean and variance of random variables having the Gamma and Beta distributions, described in Definitions 3.11 and 3.12.

4.4 Jointly Distributed Random Variables

We began this chapter by looking at two random variables, X and Y, the changes in the fortunes of two players determined by the same spin of a roulette wheel. Denote the change in fortune of a third player, who bets one unit on the single number 6, by W. Then $W = -1$ with probability $36/37$, and $W = +35$ with probability $1/37$. If we knew that $W = 35$, we could be certain the outcome had been 6, and hence that $X = +4$ and $Y = -1$. In the intuitive meaning of the words, these three random variables are not independent.

To look at the joint properties of several random variables, they must all be defined on the same probability space. The notion of independence of random variables is that *any* information about some of them should give no clues whatsoever to any of the others. That strong requirement is expressed formally as follows.

Definition 4.14

Let (X_α) be an indexed collection of random variables, all defined on the same probability space. If, for every finite collection real numbers $\{t_1, t_2, \ldots, t_m\}$, and choice of $\{X_{1'}, X_{2'}, \ldots, X_{m'}\}$, the events $\{X_{1'} \leq t_1\}, \ldots, \{X_{m'} \leq t_m\}$ are independent, then $\{X_\alpha\}$ are independent.

It is easy to see that even the two original random variables X and Y in the roulette example are not independent. There are 23 outcomes which lead to both players losing, so $P(X \leq -2 \cap Y \leq -1) = 23/37$, whereas $P(X \leq -2)P(Y \leq -1) = (25/37) \times (34/37)$, which is plainly different. Just as with independence of events in general, we shall only rarely need to check this definition when considering independence of random variables. Usually, the random variables will have been set up to be independent, or to be dependent in a particular way.

Example 4.15

Suppose the number of telephone calls, X, made to an insurance company follows a Poiss(λ) distribution, and any call is about houses with probability p. What is the distribution of Y, the number of calls made about house insurance?

We assume the subjects of calls are independent. Then, when $X = m$, the distribution of Y is Bin(m, p). The joint distribution of X and Y is given by

$$P(X = m, Y = n) = P(Y = n | X = m)P(X = m) = \binom{m}{n} p^n q^{m-n} \frac{e^{-\lambda} \lambda^m}{m!},$$

which simplifies to $e^{-\lambda}(\lambda p)^n (\lambda q)^{m-n}/(n!(m-n)!)$.

The different values of X form a partition, so

$$P(Y = n) = \sum_m P(X = m, Y = n) = e^{-\lambda}(\lambda p)^n / n! \sum_{m \geq n} \frac{(\lambda q)^{m-n}}{(m-n)!}.$$

The sum is just $e^{\lambda q}$, and so the whole expression simplifies to $e^{-\lambda p}(\lambda p)^n/n!$. Y has a Poiss(λp) distribution.

Should we have guessed this answer? Recall that the Poisson distribution has been offered as a model when events happen at random, at some overall average rate. It is plausible that telephone calls to the company fit this model, and the value of λ is the average number of calls during the period. But calls specifically about house insurance will similarly arrive at random, so their distribution is expected to be Poisson. If p is the proportion of calls about houses, the average number of calls on that topic will be λp. A Poiss(λp) distribution is very plausible.

With several discrete random variables, their *joint mass function* describes their properties via a list of the possible sets of values, with the corresponding probabilities, just as in this example with two variables. To find the mass function of one of the variables, we use Theorem 2.5, noting that the different values of the other variables define a partition of the space. Because of the way a mass function is found from the joint distribution, the term *marginal distribution* is often used, but the word "marginal" is superfluous.

When we have a collection of continuous random variables, their properties are found from either their joint distribution function, or their joint density function.

Definition 4.16

Given random variables $X_1, X_2, \ldots, X_n$ on the same probability space, their *joint distribution function* is defined by

$$F(x_1, \ldots, x_n) = P(\{\omega : X_1(\omega) \leq x_1, \ldots, X_n(\omega) \leq x_n\})$$
$$= P(X_1 \leq x_1, \ldots, X_n \leq x_n).$$

If there is a function f such that

$$F(x_1, \ldots, x_n) = \int_{-\infty}^{x_1} \cdots \int_{-\infty}^{x_n} f(u_1, \ldots, u_n) du_n \ldots du_1,$$

then f is their *joint density function*.

Corollary 4.17

If $X_1, \ldots, X_n$ are independent, and have respective densities $f_1, \ldots, f_n$, then their joint density is the product of the individual densities.

Proof

By the definition of independence of random variables,

$$
\begin{aligned}
F(x_1,\ldots,x_n) &= P(X_1 \le x_1)\cdots P(X_n \le x_n) \\
&= \int_{-\infty}^{x_1} f_1(u_1)du_1 \cdots \int_{-\infty}^{x_n} f_n(u_n)du_n.
\end{aligned}
$$

Now differentiation with respect to each of $x_1,\ldots,x_n$ establishes the desired formula.   □

If X,Y are continuous, suppose the region R in which the point (x,y) falls is not the whole plane. It is then easy to see that it is impossible for X and Y to be independent if the boundary of R has any component that is *not* parallel to either of the coordinate axes – a useful sufficient condition for non-independence.

We use joint distribution functions and joint densities to find probabilities in exactly the same way as with just one variable. For a pair (X,Y) with joint density $f(x,y)$, and R a two-dimensional region, the probability that (X,Y) falls within R is $\iint_R f(x,y)dxdy$.

Example 4.18 (Buffon's Needle)

Parallel lines are ruled on a flat board, unit distance apart. A needle of length L, where $L < 1$, is spun and dropped at random on the board. What is the probability that it intersects one of the lines?

Our model of randomness means that the distance of the *centre* of the needle from the nearest line, Y, has a $U(0,1/2)$ distribution. Independently, the orientation X of the needle with the parallel lines has a $U(0,\pi)$ distribution. Thus the joint density of (X,Y) is $f(x,y) = 2/\pi$ over the rectangle $0 < x < \pi$ and $0 < y < 1/2$.

The needle will cross the line provided the angle X is steep enough, i.e. if $\alpha < X < \pi - \alpha$ where $\sin(\alpha) = 2Y/L$.

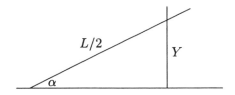

Expressing this condition directly in terms of X and Y, we require that $Y < (L/2)\sin(X)$. The probability that the needle crosses the line is the integral of the density function over this region, i.e.

$$\int_0^\pi \int_0^{(L/2)\sin(x)} (2/\pi)dy\,dx = (L/\pi)\int_0^\pi \sin(x)dx = 2L/\pi.$$

If you conduct this experiment 1000 times, and the needle crosses the line K times, then the ratio $K/1000$ should be close to $2L/\pi$. Thus $2000L/K$ is a rather convoluted "estimate" of the value of π.

The densities of the individual variables are easily deduced from the joint density. As usual, we work through distribution functions, and we will simplify the notation by taking just two variables, X and Y with joint density $f(x,y)$. Here $P(X \le x) = \int_{-\infty}^x \int_{-\infty}^\infty f(u,v)dv\,du$. Differentiate the right side with respect to x to obtain $\int_{-\infty}^\infty f(x,v)dv$, so that

$$f_X(x) = \int_{-\infty}^\infty f(x,v)dv$$

gives the so-called *marginal density* f_X of X. Informally, to obtain the density of any individual component, integrate out the unwanted components over the full range of integration – just as, in the discrete case, we *sum* over the values of the unwanted variables.

One way of keeping a check on your health is to watch your weight. There is no absolute ideal range for your weight, it depends on your build and, in particular, on your height. Suppose X is your weight in kg, and Y is your height in cm. For any specified value of Y, a dietician would be able to describe the corresponding density function for X that would represent the standard variability expected in people of normal health. The notation $f(x|y)$ seems a natural choice to indicate the density of X, conditional upon $Y = y$. Here a problem arises: height is a continuous variable, so whatever the value of y, the event $Y = y$ has zero probability, and conditioning on events of zero probability is not yet defined.

The way forward is to take an interval $(y - h/2, y + h/2)$ around the value y, for some tiny value h. There will be non-zero probability that Y falls in that interval, and wherever in the interval it does fall, we would expect the ideal weight distribution hardly to change. The distribution of X, conditional on $y - h/2 \le Y \le y + h/2$, can be specified.

Let $f(x,y)$ be the joint density of X and Y, and let $f_Y(y)$ be the density of Y alone. Then

$$P(X \leq x | y - h/2 \leq Y \leq y + h/2) = \frac{P(X \leq x \cap y - h/2 \leq Y \leq y + h/2)}{P(y - h/2 \leq Y \leq y + h/2)}$$

$$= \frac{\int_{-\infty}^{x} \int_{y-h/2}^{y+h/2} f(u,v) dv du}{\int_{y-h/2}^{y+h/2} f_Y(v) dv}.$$

Provided that f_Y is reasonably smooth, we can write the last denominator as $h f_Y(y) + O(h^2)$. Then, differentiating with respect to x, the right side becomes $\int_{y-h/2}^{y+h/2} f(x,v) dv / (h f_Y(y) + O(h^2)) = (f(x,y) + O(h))/(f_Y(y) + O(h))$, having cancelled a common factor h. (See Appendix 9.3D for the O-notation.) Let $h \to 0$; this motivates:

Definition 4.19

Given X and Y with joint density $f(x,y)$, and Y having density $f_Y(y)$, the *conditional density* of X, given $Y = y$ is $f(x|y) = f(x,y)/f_Y(y)$.

As an immediate corollary, we have

$$f(x,y) = f_Y(y) f(x|y)$$

for the joint density, useful when we know the distribution of one variable, and how the distribution of another depends on it.

For example, suppose we have a stick of unit length which we break at a random point; we then take the *larger* of the two pieces, and break that piece at a randomly chosen point. What is the probability the three pieces can form a triangle?

This problem is similar to that in Example 1.12, the difference being in the way the two points to break the stick are chosen. Before we solve it, think about how the two answers are likely to differ. Here we have deliberately selected the larger piece for the second break, which looks as though it should lead to an *increased* chance for a triangle.

Let Y denote the position of the first break, and X that of the second. Then Y is $U(0,1)$; if $Y = y < 1/2$, then X has the $U(y,1)$ distribution, whereas when $Y = y > 1/2$, X has the $U(0,y)$ distribution. Thus the joint density is

$$f(x,y) = \begin{cases} 1/(1-y) & 0 < y < 1/2 \quad y < x < 1 \\ 1/y & 1/2 < y < 1 \quad 0 < x < y \end{cases}$$

and $f(x,y)$ is zero elsewhere. The shaded area in Figure 1.2 still represents the region where a triangle can be formed; and so the probability P we seek comes from integrating $f(x, y)$ over that region. Hence

$$
\begin{aligned}
P &= \int_0^{1/2} \int_{1/2}^{y+1/2} \frac{1}{1-y} dx dy + \int_{1/2}^1 \int_{y-1/2}^{1/2} \frac{1}{y} dx dy \\
&= \int_0^{1/2} \frac{y}{1-y} dy + \int_{1/2}^1 \frac{1-y}{y} dy \\
&= 2\ln(2) - 1 \approx 0.3863
\end{aligned}
$$

which is indeed higher than our previous answer, $1/4$, as we anticipated.

If X has density $f(x)$, we might be interested in X only when its values lie in a certain range. Let $Y = X | a \le X \le b$; we seek the density of Y. Outside the interval (a, b), the density is zero, so suppose $a < y < b$, and write

$$
G(y) = P(Y \le y) = \frac{P(X \le y \cap a \le X \le b)}{F(b) - F(a)} = \frac{F(y) - F(a)}{F(b) - F(a)}.
$$

Differentiating, we see that Y has density

$$
g(y) = \frac{f(y)}{F(b) - F(a)} \quad \text{when } a < y < b.
$$

Thus the density of Y is found by restricting to the interval (a, b), and scaling up the original density of X to make the total area under that curve equal to unity.

Theorem 4.20

Suppose X and Y are independent random variables with respective densities f and g. Their sum $X + Y$ has density h, where

$$
h(t) = \int_{-\infty}^{\infty} f(x)g(t - x)dx. \tag{4.2}
$$

Proof

We follow the advice to work through distribution functions. Since X and Y are independent, their joint density is the product of f and g, so $X + Y$ has distribution function

$$
H(t) = P(X + Y \le t) = \iint_{x+y \le t} f(x)g(y)dx dy.
$$

The double integral can be written as a repeated integral, so that

$$
H(t) = \int_{-\infty}^{\infty} f(x)dx \int_{-\infty}^{t-x} g(y)dy = \int_{-\infty}^{\infty} f(x)G(t - x)dx,
$$

where G is the distribution function. Now differentiate with respect to t; the formula is proved.   □

Equation (4.2) is known as the *convolution* of the two functions f and g. Take care to get the limits of integration correct, as any formula for f or g may apply to only part of the real line, with the density zero elsewhere. Take the case when X and Y are each $U(0,1)$. Then since $f(x) = 1$ only when $0 \le x \le 1$, we have $h(t) = \int_0^1 g(t-x)dx$.

When $t \le 1$, this becomes $h(t) = \int_0^t 1dx = t$, but for $1 < t \le 2$ we obtain $h(t) = \int_{t-1}^1 1dx = 2 - t$. The density of $X + Y$ has the triangular shape

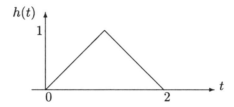

Suppose we have a sequence (X_i) of independent random variables, all with the Exponential distribution $E(\lambda)$. Write $S_n = X_1 + X_2 + \cdots + X_n$, and suppose S_n has density $f_n(t)$, clearly non-zero when $t \ge 0$. We know that $S_1 = X_1$ has density $f_1(t) = \lambda \exp(-\lambda t)$. We will prove, by induction, that

$$f_n(t) = \frac{\lambda^n t^{n-1}}{(n-1)!} \exp(-\lambda t)$$

for $n \ge 1$. For the inductive step, use Equation (4.2) to obtain

$$f_{n+1}(t) = \int_{-\infty}^{\infty} f_n(x) f_1(t-x)dx = \int_0^t \frac{\lambda^n x^{n-1}}{(n-1)!} e^{-\lambda x} \lambda e^{-\lambda(t-x)} dx,$$

which quickly simplifies to what is required. You will recognise f_n as the density of a Gamma variable.

This result allows us to show the close affinity between the Poisson and Exponential distributions, and to justify the assertion that the Poisson should be a good model for processes such as the number of radioactive particles emitted in a period. Suppose in the interval $(0, t)$, the number emitted is $N(t)$, and the time intervals between emissions are independent, all with this Exponential distribution. Then (crafty little observation)

$$N(t) \ge n \Leftrightarrow S_n \le t,$$

and

$$P(N(t) = n) = P(N(t) \ge n) - P(N(t) \ge n + 1).$$

Hence

$$P(N(t) = n) = P(S_n \le t) - P(S_{n+1} \le t) = \int_0^t (f_n(u) - f_{n+1}(u))du.$$

But using the expression we have found for $f_{n+1}(u)$ and integrating it once by parts, this last integral collapses to $(\lambda t)^n e^{-\lambda t}/n!$.

This shows that $N(t)$, the number of emissions in a period of length t, has a Poiss(λt) distribution. So long as we can convince ourselves that the events of interest – misprints, earthquakes, meeting celebrities in the street – are happening at random, at some overall average rate, the number that occur in a time interval should plausibly have a Poisson distribution.

Although I counselled against memorising any formula for the density of $Y = h(X)$ in terms of the density function of X, I concede that the corresponding formula for the transformation of a *pair* of variables X, Y into the pair U, V is worth having. But do not forget to check that the transformation is invertible.

Theorem 4.21

Suppose X and Y have joint density $f(x, y)$, and that $U = u(X, Y)$, $V = v(X, Y)$ is some one–one transform, i.e. there is an inverse $X = x(U, V)$, $Y = y(U, V)$.

Then the joint density of (U, V) is $g(u, v) = f(x(u, v), y(u, v)).J(u, v)$, where the *Jacobian* J is the absolute value of the determinant

$$\begin{vmatrix} \frac{\partial x}{\partial u} & \frac{\partial x}{\partial v} \\ \frac{\partial y}{\partial u} & \frac{\partial y}{\partial v} \end{vmatrix}.$$

Proof

By definition, for any event A,

$$P((U, V) \in A) = \iint_A g(u, v)dudv.$$

Whatever the event A, since the transformation is invertible, the event $(U, V) \in A$ will correspond to $(X, Y) \in B$ for some well-defined region B. So

$$\iint_A g(u, v)dudv = \iint_B f(x, y)dxdy;$$

in the right side of this equation, make the change of variables from (x, y) to (u, v). This leads to

$$\iint_A g(u, v)dudv = \iint_A f(x(u, v), y(u, v)).J(u, v)dudv.$$

As this holds for all events A, the two integrands are equal – the formula we have claimed. $\square$

To illustrate this, suppose X and Y are independent, each having the Exponential distribution $E(\lambda)$. They might represent the times to wait for two buses. Write $U = X + Y$ and $V = X/Y$, the sum and the ratio of these quantities. In the notation of the theorem, $f(x, y) = \lambda^2 \exp(-\lambda(x + y))$ on the positive quadrant $x > 0$, $y > 0$. Further, since $x = vy$, then $u = y(1 + v)$, and we have the inverse transformations

$$y = \frac{u}{1 + v}, \qquad x = \frac{uv}{1 + v}.$$

The determinant in the expression for the Jacobian is

$$\begin{vmatrix} \frac{v}{1+v} & \frac{u}{(1+v)^2} \\ \frac{1}{1+v} & \frac{-u}{(1+v)^2} \end{vmatrix} = \frac{-uv}{(1+v)^3} - \frac{u}{(1+v)^3} = \frac{-u}{(1+v)^2},$$

so $J = u/(1 + v)^2$. In terms of u and v, $f(x, y)$ is $\lambda^2 \exp(-\lambda u)$, so $g(u, v) = \lambda^2 u \exp(-\lambda u)/(1 + v)^2$ over $0 < u$, $0 < v$.

To find the density of either U or V on its own, we integrate out the other variable. Thus the density of U is

$$\int_0^\infty g(u, v) dv = \lambda^2 u \exp(-\lambda u) \int_0^\infty \frac{dv}{(1 + v)^2}.$$

This last integral is easily calculated to be unity, so U has density $\lambda^2 u \exp(-\lambda u)$ on $u > 0$. Similarly, the density of V is $1/(1 + v)^2$ on $v > 0$, which means that U and V are *independent*, as their joint density is just the product of their individual densities.

Intuition suggests that, in general, information about the sum of two random variables should give some sort of clue about their ratio. But we have seen that this is false in this example: U and V are independent. This is a consequence of the particular choice of the Exponential distribution for X and Y.

We have offered the distribution $E(\lambda)$ as appropriate for the waiting time for emissions of radioactive particles from a nuclear source. Label the current time as zero, let X be the time to wait for the next emission, and let Y be the time since the last one. Then X and Y will both have $E(\lambda)$ distributions, and $X + Y$ is the time between two successive emissions. But $E(X + Y) = 2/\lambda$, which seems inconsistent with our belief that the mean time between emissions is $1/\lambda$. Something seems to have gone wrong!

The explanation is that we are not comparing like with like. The diagram shows the times of emissions as *:

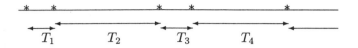

Times of emission, and inter-emission periods.

The times $T_1, T_2, T_3, \ldots$ are the successive periods to wait between emissions, and they do indeed have independent Exponential distributions with mean $1/\lambda$. If each time interval has the same chance of being chosen, its mean is $1/\lambda$. But if we select a random *time point*, and measure the length of the interval in which that time falls, it is far more likely to fall in a longer interval such as T_2 than shorter ones like T_1 or T_3. Selecting an arbitrary origin, then using it to define the time period between successive emissions, biases the choice in favour of longer time periods. Here the bias is considerable – a mean length of $2/\lambda$, not $1/\lambda$. The calculations were all fine, care was needed to see exactly how the intervals were being selected.

This phenomenon is sometimes called the *bus paradox*: if, instead of radioactive particles, we are waiting for buses that turn up at random, we appear to have the contradictory results:

- the mean time between buses arriving is $1/\lambda$;

- the mean time between the bus we missed and the one we catch is $2/\lambda$.

Just as means and variances are a quick and useful indication of some of the main properties of single random variables, there is a corresponding summary of the interaction between a pair of variables.

Definition 4.22

If X and Y have means μ_X and μ_Y respectively, their *covariance* is

$$\mathrm{Cov}(X, Y) = E((X - \mu_X)(Y - \mu_Y)).$$

When X and Y have finite variances σ_X^2 and σ_Y^2, the *correlation* between X and Y is $\mathrm{Corr}(X, Y) = \mathrm{Cov}(X, Y)/(\sigma_X \sigma_Y)$.

Expanding the definition of $\mathrm{Cov}(X, Y)$, we deduce the alternative calculation $\mathrm{Cov}(X, Y) = E(XY) - E(X)E(Y)$. When we take $Y = X$, we see that $\mathrm{Cov}(X, X) = \mathrm{Var}(X)$. The etymology of "covariance" tells exactly what a covariance is supposed to do - to show how X and Y vary together. The covariance will tend to be positive if large (small) values of X tend to be associated with large (small) values of Y, and negative if the reverse is true. A correlation is just a scaled value of a covariance, taking into account the variation in the two variables. Its value does not depend on the units in which X and Y are measured, and, as Exercise 4.20 will show, correlations always fall in the range from -1 to 1.

Theorem 4.23

If X and Y are independent, their covariance is zero (if it exists).

Proof

We deal with the continuous case only; the discrete case is proved similarly. Let the respective densities be f and g, so that

$$E(XY) = \iint xyf(x)g(y)dxdy = \int xf(x)dx \int yg(y)dy = E(X)E(Y),$$

which establishes the result.   □

Notice that the theorem is one way only, it does *not* say that if the covariance of two random variables is zero, then those variables are independent. Exercise 4.23 provides a simple counterexample.

EXERCISES

4.20. By noting that $E((X + \theta Y)^2)$ is non-negative, so this quadratic in θ cannot have distinct real roots, prove the Cauchy–Schwarz inequality, that $(E(XY))^2 \leq E(X^2)E(Y^2)$.

Replace X, Y by $X - \mu_X, Y - \mu_Y$ to show that $-1 \leq \mathrm{Corr}(X, Y) \leq 1$.

4.21. Let X and Y have joint density $2\exp(-x - y)$ over $0 < x < y < \infty$. Find their marginal densities; the density of X, given $Y = 4$; and the density of Y, given $X = 4$. Show that X and Y are not independent.

Find the joint density of $U = X + Y$ and $V = X/Y$. Are U and V independent?

4.22. Verify that discrete random variables are independent if, and only if, their joint mass function is the product of their individual mass functions.

4.23. Take R to be the triangular region in the plane with vertices at $(-1, 0), (1, 0)$ and $(0, 1)$, and suppose X, Y are uniformly distributed over R. Show that $E(XY) = 0 = E(X)E(Y)$, so $\text{Cov}(X, Y) = 0$, but that X and Y are not independent.

4.24. X and Y are independent, each $U(0, 1)$. Define $W = \min(X, Y)$ and $V = \max(X, Y)$ as the smaller and larger of these variables. Find the densities of V, W; are they independent? Show that $E(VW) = 1/4$.

4.25. On a particular examination paper, a pupil whose mean score is μ will actually score X, where X has a $U(\mu - 10, \mu + 10)$ distribution. Over the pupils taking the test, μ is assumed to have a $U(10, 90)$ distribution. Find the density of the test score of a randomly chosen pupil.

Given that Fred has scored x, what is the distribution of his underlying mean score?

4.26. Suppose the joint mass function of X and Y, the numbers of goals scored by home and away teams in a soccer league is as shown in the table.

Y/X	0	1	2	3	≥ 4
0	0.108	0.100	0.079	0.026	0.026
1	0.066	0.132	0.079	0.050	0.029
2	0.063	0.055	0.066	0.008	0.018
3	0.016	0.032	0.008	0.010	0.013
≥ 4	0.008	0.003	0.005	0.000	0.000

Find the (marginal) mass functions of X and Y. What is the distribution of X, conditional on $Y = 1$? Confirm that X and Y are not independent. Find their correlation. (Take ≥ 4 as 4.)

4.27. Suppose $\{X_i : i = 1, 2, \ldots, n\}$ are independent, with X_i having the Exponential distribution $E(\lambda_i)$. Let $W = \min\{X_1, \ldots, X_n\}$ be the smallest of these quantities. Show that W has an Exponential distribution, with parameter $\lambda_1 + \cdots + \lambda_n$.

(One way to interpret the answer is to think of the different X_i as representing the times to wait for buses running on n different routes; the customer will board the first to arrive. The different λ_i are the *rates of arrival* of the buses, so their sum is the overall rate at which some bus will arrive.)

4.28. There are N patients in a hospital ward, and the times a consultant spends with any patient are independent, each with the $E(\lambda)$ distribution. Let T be the total time the consultant spends with patients, so that (N, T) is an example of a pair of variables, one discrete and the other continuous. The properties of T are found via

$$P(T \leq t) = \sum_n P(N = n)P(T \leq t|N = n).$$

Find the density of T when N is Geometric $G_1(p)$.

4.29. We have found the density of the sum of two independent $U(0,1)$ variables. Use that result to find the density of the sum of three independent $U(0,1)$ variables. (You will need to treat the intervals $(0,1), (1,2)$ and $(2,3)$ separately, although an appeal to symmetry can reduce the work. Give yourself a gold star if you get the right answer first go - get the limits of the integrals correct.)

4.30. Another way in which a stick might be broken "at random" is to independently select random points in each of the two halves. What would be the chance of the three pieces forming a triangle in this case?

4.5 Conditional Expectation

Recall that $f(x|y) = f(x,y)/f_Y(y)$ is the density of X, conditional on $Y = y$. It is easy to see that it satisfies the two conditions to be a density function. Firstly, it is plainly non-negative and secondly

$$\int f(x|y)dx = \frac{1}{f_Y(y)} \int f(x,y)dx = \frac{f_Y(y)}{f_Y(y)} = 1.$$

Thus we can define the expectation of X, conditional on $Y = y$ as

$$E(X|Y = y) = \int x f(x|y)dx.$$

When X and Y are independent, then $f(x|y) = f(x)$, and so $E(X|Y = y) = E(X)$ whatever the value of Y, but when they are not independent, $E(X|Y = y)$ may vary with the value of Y. It thus makes sense to regard $E(X|Y)$ as a random variable $h(Y)$ that takes the value $E(X|Y = y)$ when $Y = y$. We have the elegant result:

Theorem 4.24

$E(E(X|Y)) = E(X)$. And hence $E(E(h(X)|Y)) = E(h(X))$ when $h(X)$ is a random variable.

Proof

The second result will follow from the first, simply by replacing X by $h(X)$. Now

$$
\begin{aligned}
E(E(X|Y)) &= \int E(X|Y=y)f_Y(y)dy = \int \int x\frac{f(x,y)}{f_Y(y)}dx\,f_Y(y)dy \\
&= \int \int xf(x,y)dx\,dy = \int xf_X(x)dx = E(X),
\end{aligned}
$$

as was to be established.   □

This is usually the easiest way to find the mean and variance of X, our object of interest, when it arises naturally through explicit dependence on some other variable Y. For example, the total amount X that an insurance company pays out is intimately related to Y, the number of claims, for varying amounts. And goods may have come from one of several sources, with different probabilities of being defective: the number of defectives in a sample of size n may have a $\text{Bin}(n,p)$ distribution, where p is a random variable whose distribution is described by the likelihood a given source is selected, and its characteristic proportion of defectives.

For a direct illustration of Theorem 4.24, suppose Y is $U(0,1)$, and then that X is $U(0,Y)$. To find $E(X)$, note that, conditional on $Y=y$, X is $U(0,y)$, and so has expectation $y/2$. Thus $E(X|Y)=Y/2$, and the theorem then gives $E(X)=E(E(X|Y))=E(Y/2)=1/4$ (which should accord with your intuition).

To find the variance of X, we use the theorem, with $h(X)=X^2$. Conditional on $Y=y$, since X is $U(0,y)$, so the expectation of X^2 is $y^2/3$, i.e.

$$
E(X^2|Y)=Y^2/3.
$$

Hence $E(X^2)=E(E(X^2|Y))=E(Y^2/3)=1/9$, and so $\text{Var}(X)=E(X^2)-(E(X))^2=1/9-1/16=7/144$.

As a check on this calculation, we will see how to find the mean and variance of X directly. We know that $f_Y(y)=1$ on $0<y<1$, and $f(x|y)=1/y$ on $0<x<y$. Hence $f(x,y)=1/y$ on $0<x<y<1$. Thus $f_X(x)=\int_x^1(1/y)dy=-\ln(x)$ over $0<x<1$. You should now verify the stated values of $E(X)$ and $\text{Var}(X)$.

Example 4.25

What is the mean number of draws to obtain all the different numbers at least once in a 6/49 Lottery?

Plainly, the first few draws will produce new numbers quite frequently, but as our collection gets more nearly complete, new numbers are less likely to appear. Our approach will be to *condition* on the number of new numbers each draw. Let X_n be the number of future draws to complete the collection when we have already got n of the 49 numbers, and let Y_n be the number of new numbers in the next draw. Write $\mu_n = E(X_n)$; we seek μ_0.

When n numbers have been drawn, there are $49 - n$ others left, so Y_n will have the hypergeometric distribution

$$p(n,r) = P(Y_n = r) = \binom{49-n}{r}\binom{n}{6-r} \bigg/ \binom{49}{6}.$$

In particular, $p(n,0) = (n)_6/(49)_6$. The fundamental step is using

$$E(X_n) = E(E(X_n|Y_n)).$$

For $0 \le n \le 48$, conditional on $Y_n = r$, we have $X_n = 1 + X'_{n+r}$ where X'_{n+r} is the number of further draws to complete the collection. Thus

$$E(X_n|Y_n = r) = 1 + \mu_{n+r},$$

hence

$$\mu_n = E(E(X_n|Y_n)) = \sum_{r \ge 0} p(n,r)(1 + \mu_{n+r}) = 1 + \sum_{r \ge 0} p(n,r)\mu_{n+r}$$

i.e.

$$\mu_n(1 - p(n,0)) = 1 + \sum_{r \ge 1} p(n,r)\mu_{n+r}.$$

This gives a recursive way to calculate all the values $\{\mu_0, \mu_1, \ldots, \mu_{48}\}$ by working backwards. We know that $\mu_{49} = 0$, so that formula gives μ_{48} first, and so on. The value of μ_0, the answer to the question asked, is 35.08. (It actually took only 26 draws after the UK Lottery started in 1994 for all the numbers to appear as winning numbers.)

Example 4.26

A group of n players jointly play Rock-Scissors-Paper. If all select the same strategy, or if all three strategies are each played at least once, another round is played. But if just two of the strategies, R and P say, are used, then all those who used the weaker strategy – here R – are eliminated. The game continues until just one player remains; what can we say about W_n, the number of rounds this takes, if all players independently choose a strategy at random?

This question was posed, and answered, by Maehara and Ueda (2000). There is no game at all when $n = 1$, so $W_1 = 0$, and it is easy to see that W_2 has the geometric distribution $P(W_2 = k) = 2/3^k$ for $k = 1, 2, \ldots$. When $n \geq 3$ players participate, there are $3(2^n - 2)$ ways in which exactly two of the strategies can appear, so write

$$p_n = (2^n - 2)/3^{n-1}$$

as the probability that precisely two of the strategies are played. Thus X_n, the number of rounds until someone gets eliminated has the geometric distribution $G_1(p_n)$.

Let Y_n be the number of players who remain after this elimination, and let U_n be the number of subsequent rounds. It is plain that Y_n and U_n are independent of X_n, and that the distribution of Y_n is binomial, conditional on the outcome being neither 0 nor n. Thus

$$P(Y_n = k) = \binom{n}{k} \bigg/ (2^n - 2) \quad \text{for } k = 1, 2, \ldots, (n - 1).$$

Moreover, conditional on $Y_n = k$, U_n has the distribution of W_k.

Hence $W_n = X_n + U_n$ and we shall use

$$E(W_n) = E(E(W_n|Y_n))$$

and the fact that $E(W_n|Y_n = k) = E(X_n) + E(W_k)$. This gives the recurrence relation

$$\mu_n = E(W_n) = \frac{3^{n-1} + \sum_{k=1}^{n-1} \binom{n}{k} \mu_k}{2^n - 2},$$

taking advantage of the common denominator $2^n - 2$. Take $\mu_1 = 0, \mu_2 = 1.5$, and deduce μ_n for $n \geq 3$ recursively.

EXERCISES

4.31. Suppose X is $\text{Bin}(n, p)$, where the distribution of p is (a) $U(0, 1)$; (b) Beta. Find the mean and variance of X in each case.

4.32. Find a recurrence relation for the variance of W_n in Example 4.26.

4.33. Make sure you understand the definition $\text{Var}(X|Y) = E((X - E(X|Y))^2|Y)$. Prove that

$$\text{Var}(X) = E(\text{Var}(X|Y)) + \text{Var}(E(X|Y)).$$

5
Sums of Random Variables

The command "Let X be ..." empowers the mathematical modeller. The use of a single symbol X to represent the main object of interest, and the use of good notation in an analysis, are often a long stride towards the solution. In modelling random phenomena, we may find that X is either naturally the sum of other quantities $Y_1, Y_2, \ldots Y_n$, or can be conveniently written as such a sum. Especially when the components of this sum are independent, we can take advantage of powerful general results to make meaningful deductions about X.

5.1 Discrete Variables

Theorem 5.1

Suppose $X_1, X_2, \ldots$ are independent with respective pgfs $g_1, g_2, \ldots$. Then $S_n = X_1 + X_2 + \cdots + X_n$ has pgf $g_1 g_2 \cdots g_n$.

Proof

We have seen that $g_i(z) = E(z^{X_i})$, so S_n has pgf

$$E(z^{S_n}) = E(z^{X_1 + \cdots + X_n}) = E(z^{X_1}) \cdots E(z^{X_n})$$

using the independence of the components. But this last expression is the product of the g_i, as we claimed. □

Because mass functions and pgfs determine each other, we can use this result to establish useful facts about particular discrete variables. The pgf of a Poiss(λ) variable is $\exp((z-1)\lambda)$, so if $X_1, \ldots, X_n$ are independent Poisson variables with parameters $\lambda_1, \ldots, \lambda_n$, their sum has pgf $\exp((z-1)\sum \lambda_i)$. Thus the sum of independent Poisson variables is also Poisson, with the obvious parameter.

For another demonstration, recall that a Geometric $G_1(p)$ variable has pgf $pz/(1-qz)$, so the sum of r variables, all having this distribution, has pgf $p^r z^r/(1-qz)^r$. This is the pgf of the Negative Binomial $\mathrm{NB}_r(r,p)$ distribution, so the sum of these Geometric variables has a Negative Binomial distribution. If you recall that the Geometric arises as the number of Bernoulli trials for the first Success, and the Negative Binomial as the number of trials to the rth Success, this should not be a surprise.

Example 5.2

A *permutation* of the numbers $\{1, 2, \ldots, n\}$ is any one of the $n!$ ways of writing them in some order. Given any of these permutations, let X_i be the number of integers smaller than i that precede i. Define $S_n = \sum_i X_i$: what is the distribution of S_n if the permutation is chosen at random?

We can imagine the random permutation arising inductively as follows. Having obtained $\{1, 2, \ldots, i\}$ in some order, there are $i+1$ places where the integer $i+1$ can be inserted, and our choice makes no difference at all to the values of $X_1, X_2, \ldots, X_i$. Thus X_{i+1} is independent of these, and since $P(X_{i+1} = j) = 1/(i+1)$ for $j = 0, 1, \ldots, i$, its pgf is $(1 + z + z^2 + \cdots + z^i)/(i+1)$.

By Theorem 5.1, S_n has pgf

$$\prod_{i=1}^{n} \frac{1 + z + \cdots + z^{i-1}}{i} = \frac{(1+z)(1+z+z^2)\cdots(1+z+\cdots+z^{n-1})}{n!}.$$

Knowing this pgf, we can expand it to evaluate any individual probability. The mean is easily calculated to be $n(n-1)/4$.

For an application, suppose two judges at an ice-skating contest have each ranked n skaters on "artistic impression". Label the skaters by their rank order under the first judge: the second judge will have some permutation of this order. If the judges are completely consistent, this will be the identity permutation, and the value of S_n will be zero. Provided S_n is fairly small, this is evidence they are reasonably consistent.

For example, suppose $n = 8$ and $S_n = 3$. If the two judges' opinions bore no relation to each other, this model with the random permutation would apply, and the probability of a value as low as 3 works out to be 111/8!, about 1 in 360. That is good evidence they are *not* ordering the skaters at random. (The quantity S_n is directly related to *Kendall's* τ, a statistical method of testing for correlation.)

Theorem 5.1 deals with summing a fixed number of random variables. The next theorem considers summing a *random* number of random variables. This would arise when a hospital knows the distribution of the number of road accidents per day, and also the distribution of the number of casualties per accident it will need to treat, and seeks to combine these to look at the distribution of the total number of road accident casualties it can expect.

Theorem 5.3

Suppose $X_1, X_2, \ldots$ are independent, all having the same distribution with pgf $g(z)$, and that N is a random variable independent of these X_i, having pgf $h(z)$. Write

$$S_N = \begin{cases} 0 & \text{if } N = 0; \\ X_1 + \cdots + X_n & \text{if } N = n \geq 1. \end{cases}$$

Then S_N has pgf $h(g(z))$.

Proof

The pgf of S_N is

$$m(z) = E(z^{S_N}) = \sum_n E(z^{S_N} | N = n) P(N = n).$$

But conditional on $N = n$, $z^{S_N} = z^{S_n}$, so

$$m(z) = \sum_n E(z^{S_n}) P(N = n).$$

Theorem 5.1 shows that $E(z^{S_n}) = (g(z))^n$, hence

$$m(z) = \sum_n P(N = n)(g(z)^n) = h(g(z)),$$

establishing the result.   □

Corollary 5.4

In the notation of Theorem 5.3, $E(S_N) = E(N)E(X)$, and $\mathrm{Var}(S_N) = E(N)\mathrm{Var}(X) + \mathrm{Var}(N)E(X)^2$.

Proof

We use Theorems 4.5 and 4.9, and the fact that $g(1) = 1$ when g is a pgf. The derivative of $h(g(z))$ is $h'(g(z))g'(z)$, so the mean value of S_N is $h'(g(1))g'(1) = h'(1)g'(1)$, proving the first result.

The second derivative of $h(g(z))$ is

$$h''(g(z))(g'(z))^2 + h'(g(z))g''(z).$$

Thus S_N has variance $h''(1)(g'(1))^2 + h'(1)g''(1) + h'(1)g'(1) - (h'(1)g'(1))^2$, which reduces to the stated expression.   □

No-one will be surprised at the first result. In terms of the hospital casualty example, the mean number of casualties is the product of the mean number of accidents, and the mean number of casualties per accident. But it seems unlikely that the expression for the variance of the total number of casualties would spring readily to mind.

When the N in Theorem 5.3 has the Poiss(λ) distribution, S_N is said to have a *Compound Poisson* distribution. Its exact format will vary, of course, with the distribution of X. You should verify that the corollary implies $E(S_N) = \lambda E(X)$ and $\text{Var}(S_N) = \lambda E(X^2)$ in this case.

It is an easy consequence of Theorem 5.1 that the expectation of the sum S_n is the sum of the individual expected values. In fact, this result holds whether or not the summands are independent, or have pgfs. To prove this, begin with two variables X, Y with joint mass function $p(m, n) = P(X = x_m, Y = y_n)$. Then

$$E(X + Y) = \sum_{m,n}(x_m + y_n)p(m, n) = \sum_{m,n} x_m p(m, n) + \sum_{m,n} y_n p(m, n),$$

provided the sums are absolutely convergent. The first sum can be written as $\sum_m x_m P(X = x_m) = E(X)$, since the different values taken by Y partition the space, and similarly the second sum is $E(Y)$. Thus, whatever the variables X and Y, $E(X + Y) = E(X) + E(Y)$. A straightforward induction then proves the general result:

Theorem 5.5

For any random variables $X_1, X_2, \ldots, X_n$ that have finite means,

$$E\left(\sum_{i=1}^{n} X_i\right) = \sum_{i=1}^{n} E(X_i).$$

Since we have seen that $E(aX) = aE(X)$ for any constant a, it follows that $E(\sum c_i X_i) = \sum c_i E(X_i)$. This present result and the following corollary also hold for continuous random variables.

Corollary 5.6

In the same notation,

$$\mathrm{Var}\left(\sum c_i X_i\right) = \sum c_i^2 \mathrm{Var}(X_i) + 2\sum_{i<j} c_i c_j \mathrm{Cov}(X_i, X_j).$$

The proof is left as an exercise. Of course, if the $\{X_i\}$ do happen to be independent, their covariances will be zero, and $\mathrm{Var}(\sum c_i X_i) = \sum c_i^2 \mathrm{Var}(X_i)$.

It is worth a small pause to remember why we calculate a mean and variance. Even if we can obtain the full distribution, unless it is one of the standard ones – Binomial, Poisson, Exponential etc. – it may be too complex for easy interpretation. And in any case, the values of the mean and variance can say a lot about the main properties of the distribution.

Example 5.7

Each packet of cornflakes bought contains one of N toys that children wish to collect. How many cornflakes packets do you expect to have to buy to obtain the complete collection?

For our model, we assume that all N toys are equally likely to be in any packet, and that successive packets are independent. Let X_n denote the number of extra packets we have to buy to obtain the next new toy, when now we have $n - 1$ different ones. The total number of packets will then be $S_N = X_1 + X_2 + \cdots + X_N$.

It is plain that $X_1 = 1$. Also these X_n are independent: how many packets we had to buy to get the fourth toy does not affect how many we shall need for the fifth. When we have $n - 1$ items, the chance that the next packet duplicates one of these is $(n-1)/N$, so the chance we get a new toy is $p_n = (N-n+1)/N$. Thus X_n will have the Geometric distribution $G_1(p_n)$.

The mean and variance of a $G_1(p)$ variable are $1/p$ and q/p^2 respectively. Hence we have

$$E(S_N) = \sum_{n=1}^{N} \frac{N}{N-n+1} = N\sum_{n=1}^{N} \frac{1}{n},$$

the last expression coming from reversing the order of the sum. Similarly

$$\mathrm{Var}(S_N) = \sum_{n=1}^{N} \frac{N(n-1)}{(N-n+1)^2} = N^2\sum_{n=1}^{N} \frac{1}{n^2} - N\sum_{n=1}^{N} \frac{1}{n}$$

using a similar trick. Since $\sum_{n=1}^{N} 1/n$ is approximately $\ln(N)$ for large N, and $\sum_{n=1}^{N} 1/n^2$ converges to $\pi^2/6$, we have a useful approximation. The mean number of cornflakes packets is about $N \ln(N)$, and the variance is about $N^2 \pi^2/6 - N \ln(N)$ for large N.

Of course, this neat answer depends heavily on our assumption that the company played fair and distributed the toys randomly and in equal quantities! See Exercise 5.5.

The next problem brings in the imaginative tool of an *indicator variable* to find means and variances.

Example 5.8

Suppose there are $2n$ teams in a competition, and they are paired to meet at random. Peter decides to make random guesses about these pairings. What can be said about T, the total number of correct guesses?

When n is quite small (say $n \leq 3$), it is easy enough to find the distribution of T by exhaustive counting, but there seems no obvious pattern. Just as in Example 1.9 when a secretary was scattering letters at random into envelopes, it is impossible to have exactly $n - 1$ correct guesses, so the exact distribution of T is unlikely to be simply expressed. However, the mean and variance of T can be found as follows. Label the teams that are drawn first in the real draw as $\{1, 2, \ldots, n\}$, and for $i = 1, 2, \ldots, n$ write

$$X_i = \begin{cases} 0 & \text{if guess at } i\text{'s opponent is wrong;} \\ 1 & \text{if guess at } i\text{'s opponent is correct.} \end{cases}$$

The values 1 or 0 simply *indicate* whether the guess is right or wrong, and the cunning choice of when $X_i = 1$ means that $T = \sum_{i=1}^{n} X_i$. (You should pause here to ensure you know exactly why this is so, as this is the key step.) Moreover, the product of any number of indicator variables takes only the values 0 or 1, so the expected value of this product is just the probability that all the components are equal to unity.

For any i, there are $2n - 1 = m$ (say) possible opponents for the ith team, so $P(X_i = 1) = 1/m$. Hence $E(X_i) = 1/m$, and (for later) $\text{Var}(X_i) = (m-1)/m^2$. Using Theorem 5.5, $E(T) = \sum_{i=1}^{n} E(X_i) = n/m = n/(2n - 1)$. On average, the number of correctly guessed pairings is about 0.5 when n is large.

When $i \neq j$, we identify when both X_i and X_j take the value 1. Recall that $P(X_i = 1) = 1/(2n - 1)$. When $X_i = 1$, team i and its opponent are correctly paired off, leaving $2n - 2$ other teams. There are just $2n - 3$ possible opponents for team j, so $P(X_j = 1|X_i = 1) = 1/(2n - 3)$. Hence

$$P(X_i = 1 \cap X_j = 1) = 1/((2n - 1)(2n - 3)),$$

and the covariance of X_i and X_j simplifies to $2/((2n - 1)^2(2n - 3))$, whatever the values of i and j. We now use Corollary 5.6. There are $n(n - 1)/2$ pairs i, j with $i < j$, and after simplification we have

$$\text{Var}(T) = \frac{4n(n - 1)^2}{(2n - 1)^2(2n - 3)}.$$

This, like the mean, is close to 0.5 when n is large.

One of the properties of the Poisson distribution is that its mean and variance are equal. Although the distribution of T cannot be exactly Poisson, the Poiss(0.5) distribution will be an excellent approximation for large n. A non-rigorous argument to support this is the realisation that, when n is large, the successive guesses are close to being independent. If they were independent, the Bin$(n, 1/(2n - 1))$ distribution would be appropriate – and this is close to the suggested Poisson. We follow this idea up in the next chapter.

EXERCISES

5.1. Suppose X and Y are independent Poisson variables with respective means λ and μ. Show that, conditional on $X + Y = n$, X has a Bin(n, p) distribution, where $p = \lambda/(\lambda + \mu)$.

5.2. The number of sports-related casualties has a Poisson distribution with mean 8, and, independently the number of other casualties has a Poisson distribution with mean 4. Given there are 16 casualties altogether, what is the mean number of sports-related casualties? Given there are 4 casualties in all, what is the chance they are all sports-related?

5.3. Suppose X and Y are independent with respective distributions Bin(m, p) and Bin(n, p). Find $P(X = k|X + Y = r)$. Explain, intuitively, why your answer is independent of p.

5.4. Prove the formula in Corollary 5.6.

5.5. Alter the specification of Example 5.7 so that a proportion p_0 of packets have no toys, while p_i have toy i $(i = 1, 2, \ldots, N)$, where $p_0 + p_1 + \cdots + p_N = 1$. Assuming the contents of packets remain independent, show that the mean number of purchases to obtain at least one copy of each toy is $U_1 - U_2 + U_3 - \cdots + (-1)^{N-1}U_N$, where

$$U_1 = \sum_{i=1}^{N} \frac{1}{p_i}, \quad U_2 = \sum_{0<i<j} \frac{1}{p_i + p_j}, \quad U_3 = \sum_{0<i<j<k} \frac{1}{p_i + p_j + p_k}, \quad \cdots,$$

and $U_N = 1/(p_1 + \cdots + p_N)$.

5.6. Two packs of playing cards are shuffled and set face down, side by side. The top cards of each are turned over, and compared – either they are identical (a match) or they are different. Whatever the result, the second pair are turned over and compared, then the third, and so on. Use indicator variables for packs of n cards ($n \geq 2$) to find the mean and variance of the total number of matches.

(You will note that, mathematically, this set-up is identical to that of Example 1.9, the scatter-brained secretary.)

5.7. Show that if I and J are indicator variables with the property that $P(I = 1|J = 1) = P(I = 1)$, then they are independent.

5.8. At a dinner party, n married couples are to be seated round a table. Suppose all $2n$ places are allocated at random; find the mean and variance of the number of married couples seated next to each other.

Solve the same problem, when first the women seat themselves one space apart, and then the men fill the spaces at random.

5.9. (Buffon's noodle) Recall the calculation showing that if a needle of length L is dropped on a flat board marked with parallel lines at unit distance apart, the probability is crosses a line is $2L/\pi$. Change the needle to a *noodle* of arbitrary length y, i.e. a smooth but wriggly length of spaghetti. Imagine chopping the noodle into very short lengths: use indicator functions to argue why the *mean* number of crossings should be $2y/\pi$.

5.2 General Random Variables

A central problem in Statistics is to use a sample to make inferences about a population. Ideally, the complete distribution of the population should be found, or estimated; less ambitiously, we might restrict ourselves to the mean and variance. How the sample has been selected is crucial. The most convenient assumption is that the values of the sample have been drawn at random.

Mathematically, that is expressed by saying that our sample consists of the values taken by a collection $\{X_1, X_2, \ldots\}$ of random variables that have two properties:

(i) they are independent;

(ii) they have the same distribution.

The abbreviation iidrv ("independent, identically distributed random variables") then trips off the tongue. The next result shows one way of estimating the mean and variance of this distribution.

Theorem 5.9

Suppose $\{X_1, X_2, \ldots\}$ are iidrv, with $E(X_i) = \mu$ and $\mathrm{Var}(X_i) = \sigma^2$ both finite. Write $S_n = X_1 + X_2 + \cdots + X_n$, $\bar{X} = S_n/n$ and $S^2 = \sum_{i=1}^{n}(X_i - \bar{X})^2/(n-1)$. Then $E(\bar{X}) = \mu$, $\mathrm{Var}(\bar{X}) = \sigma^2/n$ and $E(S^2) = \sigma^2$.

Proof

The fact that $E(\bar{X}) = \mu$ is an easy consequence of Theorem 5.5. Using Corollary 5.6, $\mathrm{Var}(\bar{X}) = \sum_{i=1}^{n} \sigma^2/n^2 = \sigma^2/n$, as the independence means that the covariance terms are all zero.

Note that

$$
\sum (X_i - \bar{X})^2 \quad - \quad \sum((X_i - \mu) - (\bar{X} - \mu))^2
$$
$$
= \sum (X_i - \mu)^2 - 2(\bar{X} - \mu)\sum(X_i - \mu) + n(\bar{X} - \mu)^2
$$
$$
= \sum (X_i - \mu)^2 - n(\bar{X} - \mu)^2.
$$

Take expectations. Since $E(\sum(X_i - \mu)^2) = n\sigma^2$ and $E((\bar{X} - \mu)^2) = \mathrm{Var}(\bar{X}) = \sigma^2/n$, then $E(\sum(X_i - \bar{X})^2) = n\sigma^2 - \sigma^2 = (n-1)\sigma^2$, from which the final result follows. $\qquad\square$

Let's interpret this result as a piece of Statistics. The value of $\bar{X}$ is called the *sample mean*, the arithmetic average of all the sample values. Intuitively, that value ought to be a reasonable estimate of the population mean, μ. The theorem tells us that this is so, in the sense that the average or expected value of $\bar{X}$ is this population mean. A statistician speaks of $\bar{X}$ being an *unbiased estimate* of μ. We also learn that the variance of this estimate is σ^2/n, so the standard deviation is $\sigma/\sqrt{n}$. Since the s.d. measures the variability of a quantity, in the same units, the larger the value of n, the less this variability, and the more likely is the value of $\bar{X}$ to be close to μ. That square root sign is very telling; to *halve* the s.d., we must *quadruple* the sample size - you have to work four times as hard to do twice as well.

The final result in the theorem relates to the use of the *sample variance S^2* as an estimate of the population variance σ^2. Just as with $\bar{X}$ for μ, so S^2 is an unbiased estimate of σ^2; sometimes S^2 will be too big, other times too small, on average it is just right.

We can be a little more precise about how close the value of $\bar{X}$ will be to μ using the next result.

Theorem 5.10 (Chebychev's Inequality)

Let X have mean μ and variance σ^2, assumed finite, and let $t > 0$ be given. Then $P(|X - \mu| \geq t) \leq \sigma^2/t^2$.

Proof

We give the proof when X is continuous; the proof for discrete X is on similar lines. Write $Y = |X - \mu|$, and suppose Y has density $f(y)$. Then

$$\sigma^2 = E(Y^2) = \int y^2 f(y) dy = \left(\int_{0 \leq y < t} + \int_{y \geq t} \right) y^2 f(y) dy.$$

The first integral is non-negative, and since the integrand in the second integral is at least $t^2 f(y)$, we have $\sigma^2 \geq 0 + t^2 \int_{y \geq t} f(y) dy = t^2 P(Y \geq t)$, from which Chebychev's inequality follows. $\square$

In the notation of Theorem 5.9, this result applied to $\bar{X}$ shows that, given any $t > 0$, $P(|\bar{X} - \mu| \geq t) \leq \sigma^2/(nt^2)$. In the special case when $t = \sigma$, we see that the probability that $\bar{X}$ is more than one standard deviation (of the original X_i) from its mean is less than $1/n$, whatever the common distribution of the X_i. And if t is fixed, however small, the probability that $\bar{X}$ is more than t away from its mean tends to zero as the sample size, n, increases. If a statistician has a good idea of the value of σ, then they can estimate how well $\bar{X}$ estimates μ.

To obtain corresponding results for the estimate S^2 of the variance would involve looking at the expected value of $(S^2 - \sigma^2)^2$, which involves third and fourth powers of the original variables X_i. So we make a general definition.

Definition 5.11

The nth *moment* of the random variable X is $\mu_n = E(X^n)$. The nth *central moment* is $E(X - \mu_1)^n$. The *moment generating function*, or mgf, is $M(t) = E(e^{tX})$, for those values of t for which this expectation is defined.

Thus the mean is the first moment, and the variance is the second central moment. The moment generating function does exactly as it says:

$$E(e^{tX}) = E(1 + tX + t^2 X^2/2! + t^3 X^3/3! + \cdots),$$

so the nth moment of X is the coefficient of $t^n/n!$ in this expansion. There is a close affinity between pgfs and mgfs: if X is discrete with pgf $g(z)$, then $M(t) = g(e^t)$. The proofs of the following (except for the last) are either obvious, or very similar to the corresponding result about pgfs.

1. If X has mgf $M(t)$, then $Y = aX + b$ has mgf $e^{bt} M(at)$.

2. If X and Y are independent with respective mgfs M_X and M_Y, then $M_{X+Y}(t) = M_X(t) M_Y(t)$.

3. If $\{X_i\}$ are independent, each with mgf M, and N is independent of them, and has pgf h, then $S_N = X_1 + \cdots + X_N$ has mgf $h(M(t))$.

4. If X has mgf $M(t)$, then $E(X^n) = M^{(n)}(0)$, where $M^{(n)}$ is the nth derivative of M.

5. If X and Y have the same mgf in some region $-a < t < a$ where $a > 0$, then X and Y have the same distribution. (This is essentially equivalent to Corollary 6.3 of Capiński and Kopp, 1999.)

To illustrate Definition 5.11, take the case when X has the Exponential distribution $E(\lambda)$. Its mgf is defined whenever $t < \lambda$, and simplifies to $\lambda/(\lambda - t)$. We can expand this when $|t| < \lambda$ as

$$\left(1 - \frac{t}{\lambda}\right)^{-1} = 1 + \frac{t}{\lambda} + \frac{t^2}{\lambda^2} + \cdots.$$

Reading off the coefficients, we confirm our earlier calculations $E(X) = 1/\lambda$ and $E(X^2) = 2/\lambda^2$, so $\text{Var}(X) = 1/\lambda^2$.

Moreover, the mgf of the sum of n independent random variables, all having this Exponential distribution is $\lambda^n/(\lambda - t)^n$. This is also the mgf of a $\Gamma(n, \lambda)$ distribution, so the sum of n independent Exponential random variables with the same parameter has a Gamma distribution. (We used convolutions in Chapter 4 to obtain this same result.)

Again, suppose X and Y are independent, having respective $\Gamma(a, \lambda)$ and $\Gamma(b, \lambda)$ distributions, where a and b are not necessarily integers. The mgf of X is

$$E(e^{tX}) = \int_0^\infty e^{tx} \frac{e^{-\lambda x} \lambda^a x^{a-1}}{\Gamma(a)} \, dx = \frac{\lambda^a}{\Gamma(a)} \int_0^\infty x^{a-1} e^{-(\lambda-t)x} dx.$$

There is a little trick that lets us evaluate such integrals with hardly any work. We use the fact that the integral of a density function is unity. Here, since

$$\int_0^\infty \frac{\lambda^a x^{a-1} e^{-\lambda x}}{\Gamma(a)} = 1, \quad \text{so} \quad \int_0^\infty x^{a-1} e^{-\lambda x} dx = \frac{\Gamma(a)}{\lambda^a},$$

whatever the value of $\lambda > 0$.

In the expression above for the mgf, use this result with $\lambda - t$ instead of λ. So long as $t < \lambda$, the mgf evaluates as

$$E(e^{tX}) = \frac{\lambda^a}{\Gamma(a)} \cdot \frac{\Gamma(a)}{(\lambda - t)^a} = \frac{\lambda^a}{(\lambda - t)^a}.$$

Similarly the mgf of Y is $\lambda^b/(\lambda - t)^b$ on this same interval, and $X + Y$ has mgf $\lambda^{a+b}/(\lambda - t)^{a+b}$, which implies that $X + Y$ has the $\Gamma(a + b, \lambda)$ distribution. The sum of two independent Gamma variables (with the same value of λ, the second parameter) is also Gamma.

Given any random variable X with mean μ and variance σ^2, we may *standardise* it as $Y = (X - \mu)/\sigma$ (i.e. Y is a scaled transformation of X, with mean zero and variance unity). Let $\{X_1, \ldots, X_n\}$ be independent, all with mgf $M(t)$, and write $Z_n = (X_1 + \cdots + X_n - n\mu)/(\sigma\sqrt{n})$, the standardised version of their sum. Apart from the facts that $E(Z_n) = 0$ and $\text{Var}(Z_n) = 1$, what can we say about Z_n for large values of n?

Its mgf is $H(t) = \exp(-\mu t \sqrt{n}/\sigma)(M(t/(\sigma\sqrt{n})))^n$. Noting the exponential term, and the power, it seems sensible to take logarithms. Thus

$$\ln(H(t)) = \frac{-\mu t \sqrt{n}}{\sigma} + n \ln(M\left(\frac{t}{\sigma\sqrt{n}}\right)).$$

Notice that the last term has a factor n in front, and the argument of M has a term $\sqrt{n}$ in the denominator. If we expand $M(t)$ as a power series in t, beginning

$$M(t) = 1 + \mu t + (\mu^2 + \sigma^2)t^2/2! + \cdots,$$

the first omitted term involves t^3. Thus

$$M\left(\frac{t}{\sigma\sqrt{n}}\right) = 1 + \frac{\mu t}{\sigma\sqrt{n}} + \frac{(\sigma^2 + \mu^2)t^2}{2n\sigma^2} + O\left(\frac{1}{n^{3/2}}\right).$$

Using the expansion $\ln(1 + x) = x - x^2/2 + \cdots$, this gives

$$\ln(H(t)) = \frac{-\mu t \sqrt{n}}{\sigma} + n\left(\frac{\mu t}{\sigma\sqrt{n}} + \frac{(\sigma^2 + \mu^2)t^2}{2n\sigma^2} - \frac{\mu^2 t^2}{2n\sigma^2} + O\left(\frac{1}{n^{3/2}}\right)\right).$$

The right side collapses to $t^2/2 + O(1/\sqrt{n})$. Let $n \to \infty$, so that $\ln(H(t)) \to t^2/2$. Hence $H(t) \to \exp(t^2/2)$, *whatever the distribution* of the X_i, provided only that they have a mgf defined over some region surrounding zero.

We have not met the mgf $\exp(t^2/2)$ before, nor have we seen how to reverse the argument to obtain a density function from the mgf. But if, in an inspired moment, you look at the density function that is proportional to $\exp(-x^2/2)$, you can then show that its mgf is indeed $\exp(t^2/2)$.

Definition 5.12

A random variable Z with density function $\frac{1}{\sqrt{2\pi}}\exp(-x^2/2)$ over $-\infty < x < \infty$ is said to have a *Standard Normal* distribution, with symbol $N(0,1)$. If X has density $\frac{1}{\sigma\sqrt{2\pi}}\exp(\frac{-(x-\mu)^2}{2\sigma^2})$ over the same range, then X has the Normal distribution $N(\mu,\sigma^2)$.

It is straightforward (Exercise 5.13) to show that, if X is $N(\mu,\sigma^2)$, then its mean is μ and its variance is σ^2. We have to use numerical methods to integrate the density function to find probabilities. However, it is sufficient to know the values of probabilities for a $N(0,1)$ distribution, as the transformation $Z = (X - \mu)/\sigma$ converts $N(\mu,\sigma^2)$ to $N(0,1)$. Table 5.1 shows some sample values; fuller tables can be found in most Statistics texts.

Table 5.1 Sample values of $P(Z > x)$ when Z is $N(0,1)$.

x	$P(Z > x)$	x	$P(Z > x)$	x	$P(Z > x)$
0.0	0.5000	1.2	0.1151	2.5	0.0062
0.2	0.4207	1.4	0.0808	2.75	0.0030
0.4	0.3446	1.6	0.0548	3.0	0.0013
0.6	0.2743	1.8	0.0359	3.25	0.0006
0.8	0.2119	2.0	0.0228	3.5	0.0002
1.0	0.1587	2.25	0.0122	4.0	0.00003

The density of a Normal distribution is symmetrical about its mean; as touchstones, a Normal variable is within one standard deviation of its mean about 68% of the time, and within two standard deviations just over 95% of the time.

The earlier work can be summed up as a remarkable theorem.

Theorem 5.13 (Central Limit Theorem)

Let $\{X_i\}$ be iidrv with mean μ and variance σ^2 whose mgf converges in some interval $-a < t < a$ with $a > 0$. Let $Z_n = (X_1 + \cdots + X_n - n\mu)/(\sigma\sqrt{n})$. Then the mgf of Z_n converges to $\exp(t^2/2)$ as $n \to \infty$, and so also

$$F_n(x) = P(Z_n \le x) \to F(x) = P(Z \le x) = \frac{1}{\sqrt{2\pi}} \int_{-\infty}^{x} \exp(-u^2/2)du$$

as $n \to \infty$, for any fixed z.

Proof

We have shown that the mgf of Z_n converges to the mgf of a Standard Normal variable. The extra step, showing that in this case the distribution function also converges is proved as Theorem 7.22 in Capiński and Kopp, (1999). □

This theorem is expressed in terms of the convergence of the mgf of Z_n to the mgf of a $N(0,1)$ variable, so that we expect to approximate the distribution of Z_n by a Standard Normal distribution when n is large. How fast the distribution of Z_n converges depends on many factors, especially the degree to which the distributions of the X_i are symmetrical. It is sometimes more convenient to think in terms of the sum $X_1 + \cdots + X_n$ having, approximately, the $N(n\mu, n\sigma^2)$ distribution.

The first Central Limit Theorem (CLT) proven was by de Moivre early in the eighteenth century. In our terminology, he showed how to approximate a $\text{Bin}(n, 1/2)$ distribution by a $N(n/2, n/4)$ when n is large, using Stirling's formula to deal with the factorials that arise in the binomial expressions. Laplace extended the result to $\text{Bin}(n, p)$ when $p \ne 1/2$, and Gauss used $N(\mu, \sigma^2)$ to model the distribution of errors in measurements of data. In honour of his work, the Normal is also termed the Gaussian distribution. The first rigorous proof of the CLT in the form we have stated is due to Lyapunov, about 200 years after de Moivre. We take a CLT to be an assertion that, under certain conditions, the distribution of a given family of variables is asymptotically Normal.

A nice example of a different setting for the CLT was proved by Harper (1967). Suppose the discrete random variables $\{X_n\}$ are independent, but not necessarily identically distributed. In essence, he showed that the following two conditions are together sufficient for the asymptotic normality of their sum:

C1. Each X_i takes only finitely many different values, and has pgf $g_i(z)$;

C2. If z^{R_i} is the highest power of z in $g_i(z)$, let $R(n) = \max\{R_1, \ldots, R_n\}$, and let $\sigma_n^2 = \sum_{i=1}^{n} \text{Var}(X_i)$; then $R(n)/\sigma_n \to 0$ as $n \to \infty$.

Example 5.14

In a medieval monastery, the Abbot requires the monks to make copies of a valuable manuscript. There is just one original, but when a new copy is to be made, any one of the existing manuscripts is selected at random. What can be said about the number of direct copies of the original?

To answer this, consider the position when n manuscripts, including the original, are available, and the next copy is to be made. Write

$$X_n = \begin{cases} 1 & \text{if the monk selects the original} \\ 0 & \text{otherwise} \end{cases}$$

so that $T_n = X_1 + \cdots + X_n$ is the number of direct copies when there are $n + 1$ manuscripts completed. In the notation of the two conditions, $g_i(z) = (i - 1 + z)/i$, so that $R_i = 1$, $R(n) = 1$, and

$$\sigma_n^2 = \sum_{i=1}^{n} \frac{i-1}{i^2} = \sum_{i=1}^{n} \frac{1}{i} - \sum_{i=1}^{n} \frac{1}{i^2}.$$

Since the series $\sum(1/i)$ diverges while $\sum(1/i^2)$ converges, C1 and C2 both hold.

We also have

$$E(T_n) = \sum_{i=1}^{n} \frac{1}{i} \quad \text{and} \quad \text{Var}(T_n) = \sum_{i=1}^{n} \frac{i-1}{i^2}$$

so we conclude that, asymptotically, T_n has a Normal distribution with that mean and variance. When n is large, both mean and variance are close to $\ln(n)$, so the number of direct copies has a distribution approximately $N(\ln(n), \ln(n))$.

Remark 5.15

It is easy to see that the pgf of T_n here is $\prod_{i=1}^{n}(z + i - 1)/n!$. The numerator can be expanded as

$$z(z + 1) \cdots (z + n - 1) = \sum_{k=1}^{n} s(n, k) z^k$$

where the coefficients $\{s(n, k)\}$ are known as the Stirling numbers of the First Kind. Some values are

n/k	1	2	3	4	5
1	1				
2	1	1			
3	1	3	2		
4	1	6	11	6	
5	1	10	35	50	24

The sum in row n is $n!$, and along each row, the values rise to a peak, then fall; our result can be seen as saying that, for large n, the pattern of the values follows closely that of a Normal curve.

An insurance company might desire to have so many clients that it can rely on the CLT to imply that its total liabilities over a year should have a Normal distribution, despite the highly non-Normal liabilities to any one client. Of course, it is not only useful that n should be large. If most of the clients of a single company had been near Pudding Lane in 1666, then the Great Fire of London would have upset any calculations based on sums of *independent* claims.

The CLT is a theorem about the convergence of the distribution functions of a sequence of variables to the distribution function of a Standard Normal variable. It does *not* say anything about the convergence of the corresponding density functions. Indeed, it could not, as the original variables $\{Z_n\}$ might well be *discrete*, as in this last case, and so not even have a density function!

Suppose X has the discrete distribution $\text{Bin}(n, p)$. Then X has the same distribution as the sum of n Bernoulli variables, so the CLT means that, when n is large, we can use the Normal distribution to approximate binomial probabilities. If we toss a fair coin 100 times, the number of Heads will have the $\text{Bin}(100, 1/2)$ distribution, and the CLT implies that, approximately, this is $N(50, 25)$. The range of two standard deviations either side of the mean gives the interval $(40, 60)$; the chance of getting between 40 and 60 Heads is about 95%.

When λ is large, suppose X has a $\text{Poiss}(\lambda)$ distribution. Let n be an integer close to λ, and write $\theta = \lambda/n$. Then θ is close to unity, and the sum of n independent Poisson variables, each with parameter θ, has the distribution of X. The CLT then implies that X is, approximately, $N(\lambda, \lambda)$, so we can use tables of the Normal distribution to estimate the properties of X. Many of these approximate answers are pretty good, even for values of λ as low as 20.

If you want to use the CLT to approximate a discrete distribution by a Normal distribution, you can hope to improve your accuracy by the use of the so-called continuity correction. Imagine that a multi-choice exam has 40 questions, each with four possible answers, and the Pass mark is set at 15 correct answers. What is the probability a candidate who makes random guesses at every question will pass?

Let X be $\text{Bin}(40, 1/4)$: we seek $P(X \geq 15)$. The Normal approximation to X is $N(10, 7.5)$, so we standardise and write $Z = (X - 10)/\sqrt{7.5}$. Since X takes only discrete values, and we seek the probability that X is 15 or more, the continuity correction asks us to use the value 14.5 as the Pass mark. Thus we seek $P(Z \geq (14.5 - 10)/\sqrt{7.5}) = P(Z \geq 1.643) = 0.05$ (from tables). The exact answer is the sum $\sum_{15}^{40} \binom{40}{k}(0.25)^k(0.75)^{40-k}$ which evaluates to 0.054. The approximation was swift to calculate, and does not give a misleading answer.

On this same example, let us see the continuity correction in action to find the chance that X is exactly 15. We calculate $P(14.5 \leq X \leq 15.5) = P(1.643 \leq Z \leq 2.008)$, again using $Z = (X - 10)/\sqrt{7.5}$. Normal tables give this as about 0.026, the exact answer is 0.028.

The ubiquity of the Normal as a convenient model reinforces the interest in the mean and variance of a distribution. For such a variable, its mean and variance specify the complete distribution. If we are prepared to assume that X, our variable of interest, is well approximated by a Normal whose mean and variance we can estimate fairly accurately, then we can also get a good idea of how likely X is to take extreme (worrisome?) values.

If we happened to be working with random variables whose means did not exist, or whose variances were infinite, the CLT would not apply. Take the case when the $\{X_i\}$ are iidrv, all having the Cauchy distribution with density $1/(\pi(1+x^2))$ on $-\infty < x < \infty$. The expected value of this distribution is not defined. If $\bar{X} = (X_1 + \cdots + X_n)/n$, then Exercise 5.12 asks you to show that $\bar{X}$ has this same Cauchy distribution; the sample mean has the same distribution as any of its components! However, despite such examples, the mathematical convenience of the Normal distribution, and the theoretical underpinning from the CLT, often justify assuming some quantity of interest has a $N(\mu, \sigma^2)$ distribution.

Exercise 5.13 asks you to show the following proportics of Normal variables:

P1. If X is $N(\mu, \sigma^2)$, then $aX + b$ is $N(a\mu + b, a^2\sigma^2)$ for any constants a, b, $a \neq 0$.

P2. If X and Y are independent variables with respective $N(\mu, \sigma^2)$ and $N(\lambda, \tau^2)$ distributions, then $X + Y$ is $N(\mu + \lambda, \sigma^2 + \tau^2)$.

It follows that, under the conditions of P2, the distribution of $X - Y$ is $N(\mu - \lambda, \sigma^2 + \tau^2)$ and not, as generations of students have sleepily written, $N(\mu - \lambda, \sigma^2 - \tau^2)$. A moment's reflection shows this must be wrong, as τ might exceed σ, and then we would have the impossibility of a negative variance.

Suppose Z is $N(0, 1)$, and let $Y = Z^2$. Its distribution function for $y > 0$ is

$$F_Y(y) = P(Z^2 \leq y) = \int_{-\sqrt{y}}^{\sqrt{y}} \phi(x)dx = 2\int_0^{\sqrt{y}} \phi(x)dx,$$

where $\phi(x)$ is the density of a Standard Normal. Make the substitution $x = \sqrt{u}$ to show that the density of Y is $f_Y(y) = \exp(-y/2)/\sqrt{2\pi y}$, which we can recognise as that of a $\Gamma(1/2, 1/2)$ variable. In fact, the name *chi-squared distribution on n degrees of freedom*, with symbol χ_n^2, is used for a $\Gamma(n/2, 1/2)$ variable. So we see that Z^2 has a χ_1^2 distribution. And if we add the squares of n independent Standard Normals, we get a χ_n^2 variable. Tables of these chi-squared distributions can be found in statistics texts.

Example 5.16

When Bob aims at the centre of the bull in a dartboard, his horizontal and vertical errors can be taken as independent, each with a $N(0, \sigma^2)$ distribution. We seek the distribution of the distance of the dart from the centre of the bull.

Let the horizontal and vertical errors be X and Y, and let $R^2 = X^2 + Y^2$, so that R is the distance from the bull. Write $U = X/\sigma$; then U is $N(0, 1)$, so U^2 is χ_1^2. Our previous remarks show that R^2/σ^2 has a χ_2^2 distribution. Hence

$$F_R(r) = P(R \le r) = P(R^2/\sigma^2 \le r^2/\sigma^2) = P(\chi_2^2 \le r^2/\sigma^2).$$

The density function of a χ_2^2 variable is $\exp(-x/2)/2$, the familiar Exponential density. Thus $F_R(r) = 1 - \exp(-r^2/(2\sigma^2))$, and the density function of R is $r \exp(-r^2/(2\sigma^2))/\sigma^2$ on $r > 0$.

Because there is no simple formula for the Normal distribution function, we cannot easily simulate values from it using the method based on Theorem 4.13. Plainly, it suffices to be able to simulate values of Z from a $N(0, 1)$ distribution, as then $X = \mu + \sigma Z$ is $N(\mu, \sigma^2)$. One method is based on a transformation noted by G.E.P. Box and M.E. Muller (1958).

Suppose X and Y are independent variables, each $N(0, 1)$. The point (X, Y) in Cartesian co-ordinates corresponds to (R, Θ) in polar co-ordinates, provided

$$R = \sqrt{X^2 + Y^2}$$

and

$$\cos(\Theta) = X/R, \quad \sin(\Theta) = Y/R.$$

The joint density of X and Y is

$$f(x, y) = \frac{1}{2\pi} \exp\left(-\frac{x^2 + y^2}{2}\right),$$

and we can use Theorem 4.21 to obtain the joint density $g(r, \theta)$ of R and Θ. The inverse transformation is $x = r\cos(\theta)$, $y = r\sin(\theta)$, and so

$$\left| \begin{matrix} \frac{\partial x}{\partial r} & \frac{\partial x}{\partial \theta} \\ \frac{\partial y}{\partial r} & \frac{\partial y}{\partial \theta} \end{matrix} \right| = \left| \begin{matrix} \cos(\theta) & -r\sin(\theta) \\ \sin(\theta) & r\cos(\theta) \end{matrix} \right| = r.$$

Thus

$$g(r, \theta) = \frac{1}{2\pi} \exp\left(-\frac{r^2}{2}\right).r \quad \text{for } 0 < r \text{ and } 0 < \theta < 2\pi.$$

Integrating out the variables r and θ in turn, we find $g_R(r) = r\exp(-r^2/2)$ over $r > 0$, and $g_\Theta(\theta) = 1/(2\pi)$ over $0 < \theta < 2\pi$. As the joint density is the product of these marginal densities, so R and Θ are independent.

Θ has the $U(0, 2\pi)$ distribution. Let $S = R^2$. In the previous example, we noted that S has a chi-squared distribution on two degrees of freedom, which is identical to the $E(1/2)$ distribution. Given a supply of $U(0, 1)$ values, we know how to obtain values for Θ: just multiply by 2π. And if V is $U(0, 1)$, we have seen that $-2\ln(V)$ is $E(1/2)$. Collecting all this together, we have shown

Theorem 5.17 (Box–Muller)

Suppose U and V are independent $U(0, 1)$ variables. Define

$$X = \sqrt{-2\ln(V)}\cos(2\pi U),$$
$$Y = \sqrt{-2\ln(V)}\sin(2\pi U).$$

Then X and Y are independent $N(0, 1)$ variables.

This result shows how we can transform pairs of independent $U(0, 1)$ variables into pairs of independent $N(0, 1)$ variables. It is not the most efficient method available, as calculating sines and cosines is a relatively time-consuming exercise, but the speed of modern computers has rendered this less important. The Box–Muller transform has the merit of being very easy to program. Exercise 5.25 describes another way to generate $N(0, 1)$ variables. It is just as easy to program as Box–Muller, and tends to be faster in operation.

EXERCISES

5.10. Construct an example of a random variable X with mean zero and variance $\sigma^2 > 0$ for which Chebychev's inequality is best possible, i.e. for some value of t, $P(|X| \geq t) = \sigma^2/t^2$.

5.11. Find the mgf of a $U(0, 1)$ variable, and deduce the mgf of a $U(a, b)$ variable.

5.12. Prove the assertion in the text, that $(X_1 + \cdots + X_n)/n$ has a Cauchy distribution, when the components of the sum are independent and have that Cauchy distribution.

5.13. Prove that, if X is $N(\mu, \sigma^2)$, then indeed $E(X) = \mu$ and $\mathrm{Var}(X) = \sigma^2$.

5.14. Prove the two properties P1 and P2 about Normal variables.

5.15. Suppose Z is Standard Normal. Find a formula for its nth moment.

5.16. The old term *kurtosis* is a measure of how "peaked" the distribution of a random variable is about its mean. Its formal definition is $E((X - \mu)^4)/\sigma^4$. Show that the kurtosis of a Normal variable is 3.

5.17. Let X be a random variable such that $0 \le X \le 1$, with $E(X) = \mu$, $\text{Var}(X) = \sigma^2$ and $m_4 = E((X - \mu)^4)$. Show that $m_4 + 3\sigma^4 \le \sigma^2$, with equality if, and only if, $\sigma = 0$, or X has a Bernoulli distribution.

5.18. Let X_n have a χ_n^2 distribution. Show that its mean and variance are n and $2n$ respectively.

Write $Y_n = (X_n - n)/\sqrt{2n}$ as the standardized version of X_n. Find its mgf, and, without invoking the CLT, show directly that this mgf converges to $\exp(t^2/2)$ as $n \to \infty$.

5.19. Suppose the time taken to process an application for a Library card has mean one minute with standard deviation twenty seconds. Estimate the probability that it takes over two hours to process 110 applications.

5.20. Estimate the probability that the total score on 1200 tosses of a fair die exceeds (a) 4000 (b) 4250 (c) 4400.

5.21. The score of a randomly selected student on an exam has mean 60 and variance 25.

(a) Use Chebychev's inequality to put a bound on the probability that a random score is between 50 and 70.

(b) Use the same method to put a lower bound on n, the number of students to be chosen so that the probability that their mean score is between 55 and 65 is at least 90%.

(c) Assuming that use of the CLT is justified, repeat part (b).

5.22. Use the CLT to show that

$$e^{-n} \sum_{k=0}^{n} \frac{n^k}{k!} \to 0.5$$

as $n \to \infty$. What would the sum converge to if the upper limit in the sum were changed from n to $n + \sqrt{n}$?

5.23. When estimating his monthly credit card bill, John rounds all amounts to the nearest £. Take all rounding errors as independent, $U(-0.5, 0.5)$, estimate the chance the total error exceeds £1 when his bill has twelve items.

(This indicates an approximate method of generating a $N(0,1)$ value: calculate $U_1 + \cdots + U_{12} - 6$, where the U_i are independent $U(0,1)$. Although approximate, the values generated are statistically almost indistinguishable from genuine $N(0,1)$ values, if you don't worry about very extreme quantities. This method cannot give values outside $(-6,6)$, while $P(|Z| > 6) \approx 1/500$ million, if Z is $N(0,1)$.)

5.24. In the UK 6/49 Lottery, a seventh number (the Bonus Ball) was drawn each time. Use Example 5.7 to find the mean and standard deviation of the number of draws until every number would appear at least once as the Bonus Ball. Assuming the CLT applies, is the actual result – 262 draws needed – surprising?

5.25. (Marsaglia's polar method) Suppose the pair (X,Y) has a uniform distribution over the unit disc $\{(x,y) : x^2 + y^2 \le 1\}$, and $W = X^2 + Y^2$. Let

$$U = X\sqrt{\frac{-2\ln(W)}{W}} \quad \text{and} \quad V = Y\sqrt{\frac{-2\ln(W)}{W}}.$$

Show that U and V are independent $N(0,1)$ random variables.

Given a supply of independent $U(0,1)$ variables, describe how to generate independent $N(0,1)$ variables based on this result.

5.26. Use Harper's conditions C1 and C2 to show asymptotic normality in the cases

(a) $X_n = 0$ or n, each with probability $1/2$;

(b) X_n has the discrete $U(0,n)$ distribution.

(These are useful results in non-parametric statistics. Part (a) corresponds to Wilcoxon's signed ranks statistic, part (b) to Kendall's τ, Example 5.2.)

5.3 Records

When data arise sequentially, we may be particularly interested in when new records are set. Matters are generally simpler when the data come from continuous random variables, as then ties do not occur. We restrict attention to this case.

Definition 5.18

Given a sequence of real numbers $(x_1, x_2, \ldots)$, all different, the first value x_1 is a *record*. A subsequent value x_n is an *upper record* if x_n exceeds each of $\{x_1, \ldots, x_{n-1}\}$, or a *lower record* if x_n is less than each of $\{x_1, \ldots, x_{n-1}\}$.

Since any lower record corresponds to an upper record of the sequence $(-x_1, -x_2, \ldots)$, we will look at upper records only, and refer to them simply as "records". In the rest of this section, we take $(X_1, X_2, \ldots)$ to be continuous iidrv, and also define the indicator variables

$$I_n = \begin{cases} 1 & \text{if } X_n \text{ is a record} \\ 0 & \text{otherwise.} \end{cases}$$

Theorem 5.19

$P(I_n = 1) = 1/n$ for $n \geq 1$ and, if $m \neq n$, then I_m and I_n are independent.

Proof

The event $I_n = 1$ occurs when X_n is the largest of the collection $\{X_1, \ldots, X_n\}$. Under our assumptions, all these variables are equally likely to be the largest, so $P(I_n = 1) = 1/n$.

When $m < n$, then $P(I_n = 1 | I_m = 1) = P(X_n = \max(X_1, \ldots, X_n) | X_m = \max(X_1, \ldots, X_m))$. But plainly, with $n > m$, information about which of the quantities $\{X_1, \ldots, X_m\}$ is the largest has no bearing on whether or not $X_n > X_r$ for $r = 1, 2, \ldots, n-1$, and so $P(I_n = 1 | I_m = 1) = P(I_n = 1)$. Exercise 5.7 shows that this implies independence. $\qquad\square$

A consequence of this result is that $U_n = \sum_{r=1}^{n} I_r$, the number of records among $(X_1, \ldots, X_n)$, has exactly the same distribution as the variable T_n in Example 5.14, and so $P(U_n = r)$ can be read off as the coefficient of z^r in the pgf $\Pi_{i=1}^{n}(z + i - 1)/n!$. Hence we also have $E(U_n) = \sum_{i=1}^{n}(1/i)$ and $\text{Var}(U_n) = \sum_{i=1}^{n}(1/i) - \sum_{i=1}^{n}(1/i^2)$. Records are quite likely early in the sequence, but soon become rare: for instance, the mean number among the first 100 observations is calculated as 5.19 (with a standard deviation of 1.88), while after 1000 observations, the mean has increased only to 7.49 (s.d. = 2.42). A potential use of such calculations is in statistical inference. We may suspect that, far from the observations being iid, there is an increasing trend – global warming, perhaps. If these suspicions are valid, U_n will tend to have an unusually large value, and so it can act as a test statistic for the null hypothesis that there is no trend. When Z is Standard Normal, $P(Z > 1.645) = 0.05$, so under this null hypothesis, the value of U_{100} would exceed $5.19 + 1.645 \times 1.88 = 8.28$ only 5% of the time: nine or more upper records in a series of 100 observations is prima facie evidence of a positive trend.

In the second Borel–Cantelli lemma (Theorem 2.15), take $A_n \equiv (I_n = 1)$; since $\sum P(A_n) = \sum(1/n)$ is divergent, we can be certain that infinitely many records will occur. On the other hand, let T_r be the time of the rth record. We know that $T_1 = 1$, but $T_2 = n \Leftrightarrow X_1 > \max(X_2, \ldots, X_{n-1}) \cap (X_n > X_1)$. There are $n!$ equally likely ways of ordering the values $\{X_1, \ldots, X_n\}$, and plainly $(n-2)!$ of these are such that $X_n > X_1 > (X_2, \ldots, X_{n-1})$. Thus $P(T_2 = n) = (n-2)!/n! = 1/(n(n-1))$ for $n \geq 2$. Hence $E(T_2) = \sum_{n=2}^{\infty} n \times \frac{1}{n(n-1)} = \sum_{n=2}^{\infty} \frac{1}{n-1} = \infty$; the mean time to wait, even for the second record, is infinite!

To find the distribution of T_r, note that $T_r = n \Leftrightarrow X_n$ is the rth record $\Leftrightarrow$ there are $(r-1)$ records among $(X_1, \ldots, X_{n-1})$ and X_n is a record. But the question as to whether X_n is a record is independent of how many records arose earlier, and so

$$P(T_r = n) = P(U_{n-1} = r - 1)P(I_n = 1) = P(U_{n-1} = r - 1)/n,$$

and we already know the distribution of U_{n-1}.

When considering how long to wait between records, we can argue that when $m \leq n$, then $P(T_r > n | T_{r-1} = m) = P(\text{No record at } m+1, \ldots, n | X_m$ is the $(r-1)$th record), and use the independence between the times of now records and the times of earlier ones to write down

$$P(T_r > n | T_{r-1} = m) = \frac{m}{m+1} \cdot \frac{m+1}{m+2} \cdots \cdot \frac{n-1}{n} = \frac{m}{n}. \tag{5.1}$$

This has a nice interpretation when $n = 2m$: conditional on some record (it does not matter which) occurring at time m, the next record is equally likely to occur before or after time $2m$. We also have an asymptotic result:

Theorem 5.20

For $0 < x < 1$, $P(T_{r-1}/T_r < x) \to x$ as $r \to \infty$.

Proof

Let $[y]$ denote the largest integer that does not exceed y. Then, for any $r \geq 2$

and $0 < x < 1$, we have

$$P\left(\frac{T_{r-1}}{T_r} < x\right) = \sum_{m=r-1}^{\infty} P\left(\frac{T_{r-1}}{T_r} < x | T_{r-1} = m\right) P(T_{r-1} = m)$$

$$= \sum_{m=r-1}^{\infty} P(T_r > \frac{m}{x} | T_{r-1} = m) P(T_{r-1} = m)$$

$$= \sum_{m=r-1}^{\infty} \frac{m}{[m/x]} P(T_{r-1} = m)$$

using Equation (5.1). But because

$$\frac{m}{[m/x]} - x = \frac{m - x[m/x]}{[m/x]} = \frac{x(m/x - [m/x])}{[m/x]},$$

we have

$$0 \le \frac{m}{[m/x]} - x \le \frac{x}{[m/x]} \le \frac{x}{m/x - 1} = \frac{x^2}{m-x} \le \frac{1}{m-1}.$$

Hence

$$P(T_{r-1}/T_r < x) = \sum_{m=r-1}^{\infty} (x + \epsilon_m) P(T_{r-1} = m) = x + \sum_{m=r-1}^{\infty} \epsilon_m P(T_{r-1} = m)$$

where $0 \le \epsilon_m \le 1/(m-1)$; of course, with $m \ge r-1$, we also have $\epsilon_m \le 1/(r-2)$. Thus $x \le P(T_{r-1}/T_r < x) \le x + 1/(r-2)$. Let $r \to \infty$ to prove the result.   □

In the long run, the distribution of the ratio T_{r-1}/T_r approaches $U(0,1)$.

EXERCISES

5.27. Let T_3 be the time of the third record in a sequence of continuous iidrv. Show that, if $n \ge 3$, then $P(T_3 = n) = \frac{1}{n(n-1)} \sum_{r=1}^{n-2} \frac{1}{r}$, and deduce that the median of T_3 is 7.

5.28. Show that $P(T_{r+2} > n | T_r = m) = \frac{m}{n} \sum_{k=m}^{n-2} \frac{1}{k} + \frac{m}{n-1}$ if $n \ge m+2$.

5.29. Use the previous result to show that if a new record arises at time m, when m is large, the median time to wait for the next but one record is approximately a further $4.36m$ observations.

5.30. Let J_n be the indicator function of a lower record at time n, with $V_n = \sum_{i=1}^{n} J_i$ so that $W_n = U_n + V_n$ is the total number of records, upper and lower. (Note that $W_1 = 2$.) Find $E(W_n)$ and $\text{Var}(W_n)$.

6
Convergence and Limit Theorems

The notion of convergence of a sequence of real numbers is intuitively easy to understand. We can all agree that the sequence $(\frac{1}{n}) = (1, \frac{1}{2}, \frac{1}{3}, \ldots)$ converges to zero, and that $((-1)^n) = (-1, 1, -1, 1, -1, \ldots)$ does not converge to anything. The details of the (ϵ, N) approach to convergence may not be to everyone's taste, but when a particular sequence is written down, it is usually plain whether or not it converges, and what its limit is, to a good approximation. Matters are far less clearcut with a sequence of random variables.

The Central Limit Theorem tells us about the behaviour of the *distribution functions* of a sequence of random variables. It gives sufficient conditions for the convergence, as an ordinary sequence of real numbers, of the probabilities $(P(Z_n \leq z))$ to the value of $P(Z \leq z)$. In Chapter 3, we derived the Poisson distribution by looking at the convergence of individual probabilities in the sequence of $\text{Bin}(n, \lambda/n)$ distributions. This is convergence of a sequence of mass functions. But neither of these results mention the convergence of the random variables themselves.

Before we look at the different ways in which random variables might converge, we collect together a number of inequalities that are useful in themselves, as well as steps on the way to establishing our main results.

117

6.1 Inequalities

We have already met Chebychev's inequality in Theorem 5.10, and the Cauchy–Schwartz inequality in Exercise 4.20. The first new one in our collection is named in honour of Chebychev's most famous student, A.A. Markov.

Theorem 6.1 (Markov's Inequality)

If $E(X)$ is finite, then, for any $a > 0$,

$$P(|X| \geq a) \leq \frac{E(|X|)}{a}.$$

Proof

Let Y be the random variable that takes the values a and 0, according as $|X| \geq a$ or not. Then clearly $|X| \geq Y$, and hence

$$E(|X|) \geq E(Y) = aP(|X| \geq a),$$

from which Markov's result follows.   □

Theorem 6.2 (Jensen's Inequality)

If ϕ is a convex function, then

$$E(\phi(X)) \geq \phi(E(X)).$$

Proof

(Recall that a convex function is one whose graph is always below the chord joining two points on it. This is equivalent to the tangent (where it exists) being always below the graph, or the second derivative (where it exists) being non-negative.)

A Taylor expansion of ϕ centred on $\mu = E(X)$ shows that

$$\phi(x) = \phi(\mu) + \phi'(\mu)(x - \mu) + \frac{\phi''(c)(x - \mu)^2}{2}$$

for some c between x and μ. But this last term is non-negative, so we have

$$\phi(x) \geq \phi(\mu) + \phi'(\mu)(x - \mu)$$

for all x. Hence

$$\phi(X) \geq \phi(\mu) + \phi'(\mu)(X - \mu).$$

Taking expectations now proves the result.   □

Corollary 6.3

For any $n \geq 1$, $E(|X|^n) \geq (E(|X|))^n$.

The proof is immediate, since $\phi(x) \equiv |x|^n$ is clearly convex when $n \geq 1$.

Corollary 6.4 (Lyapunov's Inequality)

If $r \geq s > 0$, then $E(|X|^r)^{1/r} \geq E(|X|^s)^{1/s}$.

Proof

To simplify the notation, let $Y = |X|$, and define $\phi(x) = \ln(E(Y^x))$. First we show that ϕ is convex. Since

$$\phi'(x) = \frac{E(Y^x.\ln(Y))}{E(Y^x)},$$

so

$$\phi''(x) = \frac{E(Y^x.(\ln(Y))^2)E(Y^x) - (E(Y^x.\ln(Y)))^2}{(E(Y^x))^2}.$$

We can write

$$(E(Y^x.\ln(Y)))^2 = (E(Y^{x/2}.(Y^{x/2}\ln(Y))))^2,$$

and use the Cauchy–Schwarz inequality to see that the right side does not exceed $E(Y^x).E(Y^x(\ln(Y))^2)$. But that means that the numerator in the expression for $\phi''(x)$ is non-negative, so ϕ is convex.

The interpretation that the chord of a convex function lies above its graph shows that, when $r \geq s > 0$,

$$\frac{\phi(r) - \phi(0)}{r} \geq \frac{\phi(s) - \phi(0)}{s}.$$

But clearly $\phi(0) = 0$, so we have

$$\frac{\ln(E(Y^r))}{r} \geq \frac{\ln(E(Y^s))}{s},$$

and Lyapunov's inequality is immediate. □

Theorem 6.5

For any $r > 0$, then $E(|X + Y|^r) \leq C_r(E(|X|^r) + E(|Y|^r))$ for some constant C_r.

Proof

For any real numbers x and y,

$$|x + y| \leq |x| + |y| \leq 2\max(|x|, |y|).$$

Hence

$$|x + y|^r \leq 2^r \max(|x|^r, |y|^r) \leq 2^r(|x|^r + |y|^r).$$

Thus $|X + Y|^r \leq C_r(|X|^r + |Y|^r)$, from which the result follows.   □

Our final inequality is a neat idea linked with the name of Herman Chernoff, but he states that Herman Rubin is the true source.

Theorem 6.6 (Chernoff's Lemma)

Suppose X has mgf $M(t)$. Then, for all $t \geq 0$,

$$P(X \geq 0) \leq M(t).$$

Proof

We have

$$M(t) = E(e^{tX}) \geq E(e^{tX}|X \geq 0)P(X \geq 0).$$

But $\exp(tX) \geq 1$ when $X \geq 0$ and $t \geq 0$, from which the result follows.   □

As an application, suppose X has the $\Gamma(n, 1)$ distribution, and we wish to bound $P(X \geq 2n)$. Since X has mgf $(1 - t)^{-n}$ for $t < 1$, the corresponding mgf of $X - 2n$ is $\exp(-2nt).(1 - t)^{-n}$. Thus

$$P(X \geq 2n) = P(X - 2n \geq 0) \leq e^{-2nt}(1 - t)^{-n}$$

for all $0 \leq t < 1$. Choose $t = 1/2$, which minimises the right side, from which we obtain

$$P(X \geq 2n) \leq (2/e)^n.$$

EXERCISES

6.1. Suppose X has the $U(-1, 1)$ distribution. Compare the exact value of $P(|X| > a)$ with what Markov's and Chebychev's inequalities give, for all $a > 0$.

6.2. Let X be Poisson with mean 25, and let $p = P(X > 30)$. Evaluate p, and compare your value with an estimate from the Central Limit Theorem, and with a bound from Markov's inequality.

6.3. There are N hazardous activities, labelled $\{1, 2, \ldots, N\}$, α_i is the chance you select the ith, and p_i is the chance you escape injury any day you try activity i. You have the choice of selecting one activity initially, and sticking to that every day, or of making a fresh choice independently each day. Find the respective probabilities of being injury-free after n days.

What comparison does the corollary to Jensen's inequality give? Evaluate the numerical answers when $N = 3$, $\alpha_1 = 0.95$, $\alpha_2 = 0.04$, $\alpha_3 = 0.01$ and $p_1 = 0.9999$, $p_2 = 0.99$ and $p_3 = 0.9$ in the cases $n = 10$, $n = 100$ and $n = 1000$.

6.4. (a) Use the Cauchy–Schwarz inequality (Exercise 4.20) to show that, if V is a non-negative random variable, then $E(1/V) \geq 1/E(V)$.

(b) In the special case where $V \geq 0$ has a distribution symmetrical about $\mu = E(V) > 0$, and $\text{Var}(V) = \sigma^2$, show that

$$E\left(\frac{1}{V}\right) \geq \frac{1}{\mu} + \frac{\sigma^2}{\mu^3}.$$

6.5. Lucy has £250 000 with which to accumulate a portfolio of shares. She has the choice of spending £1000 on each of the 250 dealing days in the year, or of making one bulk purchase, with the price of shares being their average over the year. Ignore dealing costs, interest receivable and assume share prices fluctuate randomly: use part (a) of the previous exercise to compare the mean number of shares under either scheme. (The phrase "pound cost averaging" is used in financial circles to describe the superior method.)

6.2 Convergence

Definition 6.7

Suppose X and $\{X_n : n = 1, 2, \ldots\}$ are random variables all defined on the same probability space, with respective distribution functions F and $\{F_n : n = 1, 2, \ldots\}$. We say that (X_n) converges in distribution to X, and write $X_n \xrightarrow{D} X$ if $F_n(t) \to F(t)$ at all values of t where F is continuous.

We also say that (X_n) converges *in probability* to X, and write $X_n \overset{P}{\to} X$ if $P(|X_n - X| > \epsilon) \to 0$ for all fixed $\epsilon > 0$.

When X is the trivial random variable that takes the value c with probability one, we usually write $X_n \overset{D}{\to} c$, or $X_n \overset{P}{\to} c$.

Recall how the Poisson distribution arose from a sequence of $\mathrm{Bin}(n, \lambda/n)$ distributions. In the language of this chapter, and by a slight abuse of notation, we could write $\mathrm{Bin}(n, \lambda/n) \overset{D}{\to} \mathrm{Poiss}(\lambda)$. The first version of the Central Limit Theorem is similarly expressed as stating that the sum of iidrv, scaled to have mean zero and variance unity, converges in distribution to $N(0, 1)$. We also noted that, in the CLT, the assumptions of independence and of identical distribution could be relaxed in certain ways, and still have the same conclusion. The same is true for convergence in distribution to a Poisson limit.

We first saw this in Example 1.9, where a secretary scattered letters at random into envelopes. Define the indicator variable

$$I_{jn} = \begin{cases} 1 & \text{if letter } j \text{ among } n \text{ is correctly addressed} \\ 0 & \text{otherwise} \end{cases}$$

and let $X_n = \sum_{j=1}^{n} I_{jn}$. Then these indicator variables all have the same distribution, taking the value 1 with probability $1/n$, but they are plainly not independent. Nevertheless, we saw that $P(X_n = j) \to e^{-1}/j!$, which is equivalent to $X_n \overset{D}{\to} \mathrm{Poiss}(1)$.

In general, suppose

$$I_{jn} = \begin{cases} 1 & \text{with probability } p_{jn} \\ 0 & \text{otherwise} \end{cases}$$

and $X_n = \sum_{j=1}^{n} I_{jn}$. We might hope to conclude that $X_n \overset{D}{\to} \mathrm{Poiss}(\lambda)$, provided that

(i) $\sum_{j=1}^{n} p_{jn} \to \lambda$ (which ensures that $E(X_n) \to \lambda$);

(ii) $\max p_{jn} \to 0$ (so that no single component dominates the sum);

(iii) the variables $\{I_{jn} : j = 1, 2, \ldots, n\}$ show very weak dependence.

This last condition will be the hardest to pin down but several of the examples where we used indicator functions to find the mean and/or variance of a particular quantity do, in fact, have Poisson distributions as limits. We cite Example 5.8, guessing the number of pairings in a round robin draw, and Exercise 5.8, random seatings of couples.

Example 6.8

An ordinary deck of 52 cards is shuffled, and dealt out face up one at a time. What is the chance that we ever have a Queen next to a King?

To attack this, let

$$I_j = \begin{cases} 1 & \text{if the cards in positions } j, j+1 \text{ are Q, K in either order;} \\ 0 & \text{otherwise} \end{cases}$$

with $X = \sum_{j=1}^{51} I_j$ being the number of occasions where a Queen is found next to a King. Plainly $p_j = P(I_j = 1) = 2 \times (4/52) \times (4/51)$, and $E(X) = 51P(I_j = 1) = 8/13$. The distribution of X is complex: of necessity, $0 \leq X \leq 7$, and the computation of the exact distribution means examining many cases, depending of the sizes of the gaps between consecutive Queens.

But if the Poisson is a good approximation, we would estimate $P(X = 0)$ as $\exp(-8/13) = 0.5404$, and so the chance a Queen is next to a King at least once is taken as $1 - 0.5404$, say 46%. A more precise value comes from Bonferroni's inequality, Exercise 1.6. Define $A_j \equiv \{I_j = 1\}$. Then (check these) $S_1 = 8/13 = 0.6154$, $S_2 = 32/221 = 0.1448$, $S_3 = 92/5525 = 0.0167$, so the desired probability is just less than $S_1 - S_2 + S_3 = 0.4872$, call it 48.5%. The answer based on the Poisson approximation is pretty good for most purposes, and was achieved with far less effort than is needed for the Bonferroni calculation!

The following example shows why, in the definition of convergence in distribution, we ask only that convergence occurs at points of continuity of F. Let X take the value zero with probability one. Its distribution function is then

$$F(t) = \begin{cases} 0 & \text{if } t < 0; \\ 1 & \text{if } t \geq 0. \end{cases}$$

Now let X_n take the value $-1/n$ with probability one, so that it has distribution function

$$F_n(t) = \begin{cases} 0 & \text{if } t < -1/n; \\ 1 & \text{if } t \geq -1/n. \end{cases}$$

Then whatever value of t we take, it is easy to confirm that $F_n(t) \to F(t)$, so there would be no harm here in insisting that F_n should converge to F at all points. We would write $X_n \overset{D}{\to} 0$.

But now let $Y_n = -X_n$, so that $P(Y_n = 1/n) = 1$, with distribution function

$$G_n(t) = \begin{cases} 0 & \text{if } t < 1/n; \\ 1 & \text{if } t \geq 1/n. \end{cases}$$

Then $G_n(t) \to F(t)$ whenever $t \neq 0$. However, $G_n(0) = 0$ for all n, and $F(0) = 1$, so $G_n(0) \not\to F(0)$. If we had insisted that the convergence be for all values of t, we would have had the untenable position of claiming that $X_n \overset{D}{\to} 0$, but $-X_n \overset{D}{\not\to} 0$. By asking only that convergence take place at places where F is continuous, we avoid this embarrassment.

Of course, in the CLT, the limit distribution function $F(t) = P(Z \leq t)$ is continuous everywhere, so there was no reason to make the distinction when we wrote down that result.

In the example we have just considered, not only do we have $X_n \overset{D}{\to} 0$, we also have $X_n \overset{P}{\to} 0$, since $P(|X_n - 0| > \epsilon) = 0$ whenever $n > 1/\epsilon$. To demonstrate that the two notions of convergence are different, suppose the $\{X_n\}$ are iidrv taking the values 0 or 1, each with probability $1/2$. Let X be independent of this sequence, and also have this same Bernoulli distribution. It is then obvious that $X_n \overset{D}{\to} X$. But because X and X_n are independent, then $|X_n - X|$ also takes the values 0 or 1 each 50% of the time (agreed?), so when $0 < \epsilon < 1$, $P(|X_n - X| > \epsilon) = 1/2$. Hence $X_n \overset{P}{\not\to} X$, showing that the two notions of convergence are different. However:

Theorem 6.9

Suppose $X_n \overset{P}{\to} X$. Then $X_n \overset{D}{\to} X$.

Proof

Write $F_n(t) = P(X_n \leq t)$ and $F(t) = P(X \leq t)$, and suppose $\epsilon > 0$ is given. Then, since the events $(X > t + \epsilon)$ and $(X \leq t + \epsilon)$ partition the space,

$$F_n(t) = P(X_n \leq t \cap X > t + \epsilon) + P(X_n \leq t \cap X \leq t + \epsilon).$$

But when $(X_n \leq t)$ and $(X > t + \epsilon)$, plainly $|X_n - X| > \epsilon$; and $(X_n \leq t \cap X \leq t + \epsilon) \subset (X \leq t + \epsilon)$. Thus

$$F_n(t) \leq P(|X_n - X| > \epsilon) + F(t + \epsilon).$$

Using similar reasoning,

$$\begin{aligned} F(t - \epsilon) &= P(X \leq t - \epsilon \cap X_n > t) + P(X \leq t - \epsilon \cap X_n \leq t) \\ &\leq P(|X_n - X| > \epsilon) + F_n(t). \end{aligned}$$

Putting the two inequalities in terms of $F_n(t)$, we have shown that

$$F(t - \epsilon) - P(|X_n - X| > \epsilon) \leq F_n(t) \leq P(|X_n - X| > \epsilon) + F(t + \epsilon).$$

Now if $X_n \overset{P}{\to} X$, then $P(|X_n - X| > \epsilon) \to 0$. So, asymptotically, $F_n(t)$ is sandwiched between $F(t - \epsilon)$ and $F(t + \epsilon)$. Hence, if F is continuous at t, $F_n(t) \to F(t)$, and $X_n \overset{D}{\to} X$. □

This theorem shows that convergence in probability is a stronger notion than convergence in distribution. Convergence in probability says that X_n is *very likely* to be close to X when n is large.

Definition 6.10

The sequence (X_n) converges *strongly* to X if

$$P(\omega : X_n(\omega) \to X(\omega)) = P(X_n \to X) = 1.$$

The term *almost surely* is also used for this form of convergence. We write $X_n \overset{a.s.}{\longrightarrow} X$.

This definition allows there to be some outcomes ω for which the values $X_n(\omega)$ do not converge to $X(\omega)$, but the total probability attached to these rogue values of ω is zero. Thus we can act as though the sequence of values (X_n) will definitely converge to X.

For example, suppose the experiment were to select a value U from the $U(0,1)$ distribution. I make you an offer you can't refuse: let $X_n = 1$ if $\cos(n\pi U) = 1$, and $X_n = 0$ otherwise. If the sequence $(X_n) \to 0$, you get nothing; but if $X_n \not\to 0$, I will give you £1000. How generous is my offer?

Not particularly generous! Suppose U is rational: then there will be an infinite sequence of values n for which nU is an even integer, so $\cos(n\pi U) = 1$, $X_n = 1$ along this sequence, and you get the £1000. But if U is irrational, then nU is never an integer, $\cos(n\pi U)$ never equals unity, and X_n is always zero. Of course, although there are infinitely many rational values for U, the total probability attached to them is zero. Here $X_n \overset{a.s.}{\longrightarrow} 0$, and I shall never have to pay out.

The next example is to show that convergence in probability, and strong convergence, are indeed different.

Example 6.11

Take the probability space on which all random variables are defined as that corresponding to the $U(0,1)$ distribution. Thus $\Omega = (0,1)$, and the probability of any interval in Ω is its length.

Define

$$X_n(\omega) = \begin{cases} n & 0 \le \omega < 1/n \\ 0 & 1/n \le \omega \le 1 \end{cases}$$

so that $P(X_n = n) = 1/n$ and $P(X_n = 0) = 1 - 1/n$. It is plain that $X_n \xrightarrow{P} 0$, since, when $0 < \epsilon < 1$,

$$P(|X_n| > \epsilon) = P(X_n = n) = 1/n.$$

Moreover, for any $\omega > 0$, select N so that $1/N < \omega$. If $n \ge N$, then $1/n < \omega$, so $X_n(\omega) = 0$. Hence

$$\{\omega : X_n(\omega) \to 0\} = \{\omega : 0 < \omega \le 1\},$$

which implies that $P(\omega : X_n(\omega) \to 0) = 1$, and $X_n \xrightarrow{a.s.} 0$ also.

Now we shall define a sequence (Y_n) which converges to zero in probability, but not almost surely. Indeed, there will be *no* values of ω for which $Y_n(\omega) \to 0$. Far from having $P(Y_n \to 0) = 1$, we have $P(Y_n \to 0) = 0$.

To achieve this goal, we define the variables Y_n in blocks of $1, 2, 4, 8, \ldots$. Each variable takes only the values zero and unity, so it is enough to specify the outcomes ω for which $Y_n(\omega) = 1$. Take $Y_1(\omega) = 1$ for $0 \le \omega \le 1$; for Y_2 and Y_3, let $Y_2(\omega) = 1$ when $0 \le \omega < 1/2$, and $Y_3(\omega) = 1$ on $1/2 < \omega \le 1$.

You can probably see what is coming. Y_4 will be unity on $0 \le \omega < 1/4$, the intervals for Y_5, Y_6 and Y_7 are $(1/4, 1/2], (1/2, 3/4]$ and $(3/4, 1]$. For Y_8 to Y_{15} similarly split $(0, 1)$ into eight equal intervals so that each variable has probability $1/8$ of being unity, and $7/8$ of being zero, and so on. Thus $P(Y_n \ne 0)$ decreases to zero through the values $1, 1/2, 1/4, 1/8, \ldots$ (in blocks), and $Y_n \xrightarrow{P} 0$. The (Y_n) are known as *Rademacher functions*.

Now fix ω, and look at the sequence $Y_n(\omega)$. For example, if $\omega = 1/3$, this sequence is $(1; 1, 0; 0, 1, 0, 0; 0, 0, 1, 0 \ldots)$, the semi-colons merely indicating the breaks in the blocks for the definition of (Y_n). Whatever outcome ω we had chosen, the pattern would be similar: the values are mostly zeros but, once in each block of sizes $1, 2, 4, 8, \ldots$ there is exactly one place where the value is unity. Hence the sequence $(Y_n(\omega))$ converges nowhere, as we claimed. Convergence in probability and convergence almost surely are different notions. The next result shows how they are related.

Theorem 6.12

If $X_n \xrightarrow{a.s.} X$, then $X_n \xrightarrow{P} X$.

Proof

We first show an equivalent definition of almost sure convergence. Write $C = \{\omega : X_n(\omega) \to X(\omega)\}$ and, for fixed $\epsilon > 0$, let

$$A_n = \{\omega : |X_n(\omega) - X(\omega)| < \epsilon\} \quad \text{and} \quad B_N = \bigcap_{n=N}^{\infty} A_n.$$

Suppose $\omega \in C$. Then, by the definition of convergence of real numbers, there is some N such that, for $n \geq N$, $|X_n(\omega) - X(\omega)| < \epsilon$. Hence $\omega \in A_n$ for all $n \geq N$, i.e. $\omega \in B_N$ for this value of N.

But $B_N \subset B_{N+1} \subset B_{N+2} \subset \cdots$; so $\omega \in \bigcup B_N = B$. Whenever $\omega \in C$, then $\omega \in B$, so $C \subset B$; this is true for all choices of $\epsilon > 0$.

Conversely, suppose $\omega \in B$ for all $\epsilon > 0$, and let $\epsilon > 0$ be fixed. For this ϵ, $\omega \in B_N$ for some N, i.e. $|X_n - X| < \epsilon$ for all $n \geq N$. Whatever $\epsilon > 0$ we choose, there is some N for which $|X_n - X| < \epsilon$ for all $n \geq N$, i.e. $X_n \to X$, and $\omega \in C$. Hence $B = C$.

So an equivalent definition is that, for all $\epsilon > 0$, $P(B_N) \to 1$, which is the same as $P(B_N^c) \to 0$.

Thus, when $X_n \xrightarrow{a.s.} X$, given $\epsilon > 0$, we have $P(\bigcup_{n \geq N} |X_n - X| \geq \epsilon) \to 0$ as $N \to \infty$. A fortiori, $P(|X_n - X| \geq \epsilon) \to 0$, and $X_n \xrightarrow{P} X$. $\square$

Once we have established that $X_n \to X$ in some sense, we might hope to find that $E(X_n) \to E(X)$, and similarly for higher moments. Such hopes are quickly dashed. In Example 6.11 we constructed a sequence of random variables (X_n) that converge almost surely to zero. But, since $X_n = n$ with probability $1/n$, $E(X_n) = 1$, so $E(X_n) \nrightarrow E(X) = 0$. If means do not converge, there is little hope for the higher moments – in this example, $E(X_n^2) \to \infty$. But it does suggest yet another form of convergence.

Definition 6.13

The sequence (X_n) converges *in rth mean* to X if $E(|X_n - X|^r) \to 0$. We write $X_n \xrightarrow{r} X$.

In particular, if $r = 1$, we would say that (X_n) has *limit in mean* X; and if $r = 2$, then $X_n \to X$ *in mean square*. See Exercise 6.10. It is quite possible for a sequence to converge in mean, but not in mean square: an easy example comes from modifying the sequence (X_n) we have just looked at to $Y_n = X_n/\sqrt{n}$. On the other hand, there is a one-way relationship:

Theorem 6.14

Suppose $r > s \geq 1$ and that $X_n \xrightarrow{r} X$. Then $X_n \xrightarrow{s} X$.

Proof

Write $Y_n = |X_n - X|$. Then Lyapunov's inequality, Corollary 6.4, shows that $(E(Y_n^r))^{1/r} \geq (E(Y_n^s))^{1/s}$. Plainly, if $E(Y_n^r) \to 0$, then $E(Y_n^s) \to 0$, which is what is required here.   □

Convergence in rth mean is neither stronger nor weaker than convergence almost surely. The sequence (Y_n) in Example 6.11 converges to zero in rth mean for all $r > 0$, but does not converge almost surely. In that same example, the sequence (X_n) converges almost surely, but not in mean. However, convergence in rth mean, when $r \geq 1$, is stronger than convergence in probability.

Theorem 6.15

If $r \geq 1$ and $X_n \xrightarrow{r} X$, then $X_n \xrightarrow{P} X$.

Proof

For any $\epsilon > 0$, Markov's inequality shows that

$$P(|X_n - X| > \epsilon) \leq \frac{E(|X_n - X|)}{\epsilon}.$$

But if $X_n \xrightarrow{r} X$, then $X_n \xrightarrow{1} X$, so the right side converges to zero, and $X_n \xrightarrow{P} X$.   □

As well as noting the relationships between the different forms of convergence, we would like to know what happens when we combine convergent sequences of random variables. Given only convergence in distribution, we cannot say very much. For example, suppose (X_n) are iidrv, each $N(0,1)$, and X also has that distribution. Let $Y_n = -X_n$, and let Y be $N(0,1)$, independent of X. It is immediate that $X_n \xrightarrow{D} X$ and that $Y_n \xrightarrow{D} Y$, but $X_n + Y_n$ is the trivial random variable, taking the value zero with probability one, while $X + Y$ is $N(0,2)$. The next result shows that matters are more satisfactory with the other modes of convergence.

Theorem 6.16

If $X_n \to X$ and $Y_n \to Y$, then $X_n + Y_n \to X + Y$, provided the mode of convergence is either in probability, almost surely, or in rth mean throughout.

Proof

Suppose first we have convergence in probability, and $\epsilon > 0$ is given. Then $|X_n + Y_n - X - Y| \leq |X_n - X| + |Y_n - Y|$. Hence

$$P(|X_n + Y_n - X - Y| > \epsilon) \leq P(|X_n - X| > \epsilon/2) + P(|Y_n - Y| > \epsilon/2).$$

Since both terms on the right side converge to zero as $n \to \infty$, the result follows.

Convergence almost surely is even easier: write $A = \{\omega : X_n(\omega) \to X(\omega)\}$, and let B be the corresponding set for Y_n and Y. Since $P(A) = P(B) = 1$, then $P(A \cap B) = 1$; and if $\omega \in A \cap B$, then $(X_n + Y_n)(\omega) \to (X + Y)(\omega)$.

For convergence in rth mean, we shall need Theorem 6.5. That result shows that

$$E(|X_n + Y_n - X - Y|^r) \leq C_r(E(|X_n - X|^r) + E(|Y_n - Y|^r)),$$

hence the final part of the theorem. $\qquad \square$

6.3 Limit Theorems

In this section, (X_n) are random variables on the same probability space, $S_n = X_1 + \cdots + X_n$, and $\bar{X} = S_n/n$ (we suppress the dependence of $\bar{X}$ on n) What can be said about $\bar{X}$ for large n? With very mild assumptions we have:

Theorem 6.17 (Weak Law of Large Numbers)

Suppose (X_n) are iidrv with mean μ and finite variance σ^2. Then $\bar{X} \xrightarrow{P} \mu$.

Proof

We use Chebychev's inequality. Recall also Theorem 5.9 which established that $E(\bar{X}) = \mu$ and that $\text{Var}(\bar{X}) = \sigma^2/n$. Hence, for any $\epsilon > 0$,

$$P(|\bar{X} - \mu| > \epsilon) \leq \frac{\sigma^2}{n\epsilon^2}.$$

Let $n \to \infty$; the result follows. $\qquad \square$

This theorem tells a statistician that the mean of a random sample converges (in a particular sense) to the mean of the underlying distribution. Hence (no surprise here) bigger samples ought generally to give better estimates than smaller ones.

Example 6.18

Suppose we wish to estimate the probability p that we succeed when we play Solitaire. Time for indicator variables to aid our thoughts. Let X_n take the values 0 or 1 according as the nth attempt fails or succeeds, so that $\bar{X} = r/n$ if the n attempts yield r successes. Here $\mu = P(X_n = 1) = p$, so the Weak Law tells us that $\frac{r}{n} \xrightarrow{P} p$; the empirical estimate of the probability converges to the correct value as we continue the sequence of trials.

This Weak Law asserts that, for any particular large sample size n, it is very likely that $\bar{X}$ will be close to $\mu = E(X)$. Although the Central Limit Theorem gives convergence only in distribution, it can be seen as refining this result, as it says that the distribution of $\bar{X}$ is approximately $N(\mu, \sigma^2/n)$. Thus the order of magnitude of the expected deviation of $\bar{X}$ from μ is $1/\sqrt{n}$, giving an idea of the *rate* at which $\bar{X}$ gets close to μ.

Neither result states that, eventually $\bar{X}$ gets close to μ and stays close: that, informally, is the import of a Strong Law, our next theorem.

Theorem 6.19 (Strong Law of Large Numbers)

Suppose (X_i) are iidrv with mean μ and finite fourth moment. Let $\bar{X} = (X_1 + \cdots + X_n)/n = S_n/n$. Then $\bar{X} \xrightarrow{a.s.} \mu$.

Proof

Plainly,

$$E((S_n - n\mu)^4) = E\left(\sum_{i=1}^{n}(X_i - \mu)^4\right) + E(\sum_{i<j}(X_i - \mu)^2(X_j - \mu)^2),$$

as all other terms in the expansion are zero, using the independence of the variables. Thus

$$E((S_n - n\mu)^4) = nE((X_1 - \mu)^4) + 3n(n-1)\sigma^4 \leq Kn^2$$

for some finite constant K.

Markov's inequality, Theorem 6.1, shows that

$$P\left(\left|\frac{S_n}{n} - \mu\right| > \epsilon\right) = P((S_n - n\mu)^4 > n^4\epsilon^4) \leq \frac{Kn^2}{n^4\epsilon^4} = \frac{A}{n^2}$$

where $A = K/\epsilon^4$. Since $\sum \frac{1}{n^2}$ converges, the first part of the Borel–Cantelli lemmas (Theorem 2.15) shows that, with probability one, only finitely many of the events $|S_n/n - \mu| > \epsilon$ occur. This holds whatever the value of $\epsilon > 0$. Thus $S_n/n \xrightarrow{a.s.} \mu$, as claimed. $\square$

We have stated the Strong Law in that format because the proof is short and neat. Here is an alternative proof, based on Chernoff's lemma, Theorem 6.6, but which requires the stronger condition that the (X_n) have a mgf $M(t)$.

The mgf of $X_n - \mu - \epsilon$ is then $H(t) = \exp(-(\mu + \epsilon)t)M(t)$. Also $H(0) = 1$, and the mean value of $X_n - \mu - \epsilon$ is $H'(0) = -\epsilon$, a negative quantity. Hence, for small positive values of t, $H(t) < 1$. Now

$$P\left(\left|\frac{S_n}{n} - \mu\right| \geq \epsilon\right) = P\left(\frac{S_n}{n} - \mu - \epsilon \geq 0\right) + P\left(\frac{S_n}{n} - \mu + \epsilon \leq 0\right)$$

and note that the mgf of $S_n/n - \mu - \epsilon$ is $\exp(-(\mu + \epsilon)t)M(t/n)^n$. Replace t/n by u, so that this mgf becomes $\exp(-(\mu + \epsilon)nu)M(u)^n = H(u)^n$.

Choose u so that $H(u) = \lambda < 1$, and Chernoff's lemma shows that

$$P\left(\frac{S_n}{n} - \mu - \epsilon \geq 0\right) \leq \lambda^n.$$

A similar argument produces a similar geometric bound, θ^n with $0 < \theta < 1$ on $P(S_n/n - \mu + \epsilon \leq 0)$, hence

$$P\left(\left|\frac{S_n}{n} - \mu\right| \geq \epsilon\right) \leq \lambda^n + \theta^n.$$

Now the same Borel–Cantelli argument we gave in the earlier proof concludes this proof of the Strong Law.

However, by working a great deal harder, the conclusion of this theorem can be shown to hold without even the assumption that fourth moments exist. It turns out that (see, e.g., Theorem 7.15 in Capiński and Kopp, 1999) provided the iidrv have a finite mean, the conclusion still holds. And now, knowing that almost sure convergence implies convergence in probability, we can see that the assumption about a finite variance in the Weak Law, Theorem 6.17, was also superfluous.

This Strong Law is one of the most important general theorems in the whole subject. For an application, we describe how to react when the odds are in your favour.

Example 6.20 (Kelly Strategy)

Suppose you come across a casino with a biased roulette wheel; the probability it shows Red is p, where $p > 1/2$. Suppose also the casino pays out even money on bets on Red, so that you get back double your stake when you win. How best should you take advantage of this?

Assuming each spin is independent of the rest, there is no point in betting on anything but Red. Even after ten Reds in a row, Red is more likely than Black next spin, so it would be an inferior strategy to bet on Black. Hence all our bets will be on Red.

Let X_0 be your initial fortune, and X_n be your fortune after n plays. The boldest strategy you could use is to bet your entire capital on Red every play. If you do this, then X_n will be either zero, if Red has ever failed to come up, or $2^n X_0$ if Red has come up every time. The probability that Red comes up every time is p^n, so your mean fortune after n plays is $E(X_n) = 2^n X_0 . p^n = (2p)^n X_0$, which increases without bound since $2p > 1$.

On the other hand, $P(X_n = 0) = 1 - p^n$, so $P(X_n = 0) \to 1$ as $n \to \infty$. Your mean fortune increases exponentially fast, but your actual fortune, X_n, converges to zero with probability one. This is not a strategy to use indefinitely, it is bound to bankrupt you – even though you do very well, "on average".

A different strategy is to vary your stake. You might decide to risk more when you are richer, less after a run of bad luck. One way to do this is to bet a fixed fraction, x say, of your current fortune at each play.

Given X_i, then X_{i+1} will be $(1 + x)X_i$ when you win, or $(1 - x)X_i$ when you lose. So when you win, multiply by $(1 + x)$, when you lose, multiply by $(1 - x)$. In n turns, if you win Y times and lose $n - Y$ times, then

$$X_n = (1 + x)^Y (1 - x)^{n-Y} X_0.$$

Divide both sides by X_0, take logs, divide by n. This shows that

$$\frac{1}{n} \ln \left(\frac{X_n}{X_0} \right) = \frac{1}{n}(Y \ln(1 + x) + (n - Y) \ln(1 - x)).$$

The left side is just the average growth rate of our fortune per play. Rewrite the right side as

$$\frac{Y}{n} \ln(1 + x) + \left(1 - \frac{Y}{n} \right) \ln(1 - x).$$

But Y/n, the fraction of times we win, converges almost surely to p by the Strong Law, so the right side converges almost surely to

$$p \ln(1 + x) + q \ln(1 - x)$$

where, as usual, $q = 1 - p$. We want to maximise our average growth rate, so we can deduce the optimal value of x from differentiation. We must solve

$$\frac{p}{1+x} = \frac{q}{1-x},$$

which leads to $x = p - q$.

The message is clear and simple: if you decide to bet a fixed fraction x of your fortune at every stage, then the best you can do is to make $x = p - q$, the size of your advantage. If $p = 52\%$, you will bet 4% of your fortune; if $p = 55\%$, you will bet 10%, and so on. The corresponding average growth rate is $p\ln(2p) + q\ln(2q)$; when $p = 52\%$, this is 0.08%.

That may look disappointing, but it is the best you can do – any other choice of x would give a worse long-term growth rate. Remember, 96% of your fortune is standing by in case of a run of losses. This strategy was first articulated by J.L. Kelly (1956).

Leo Breiman (1961) showed that if you want to maximise the long-term growth rate, or to minimise the mean time to reach a desired level of capital, you cannot improve on betting a constant fraction of your fortune. Of course, these are not the only criteria you could use but they are appealing. As Breiman wrote, "the conclusion of these investigations is that (this strategy) seems by all reasonable standards to be asymptotically best and that, in the finite case, it is suboptimal in the sense of providing a uniformly good approximation to the optimal results". In other words, if you don't use the Kelly formula, you might do a lot worse, and it is very hard to do better.

At the racetrack, there may be several horses, each at odds that offer a favourable bet. Kelly described how best to apportion your bets among them in these circumstances, assuming again that your goal is to maximise your long-term rate of return. But sometimes, using the recommended strategy may ask you to risk a large proportion of your capital: you don't have to do this! Although it will reduce your expected profits, it might make you feel more comfortable if you staked one half, or an even smaller proportion, of what the Kelly recipe suggests. A mathematical model is to help your decision-making, not to replace it.

Before we show how a similar strategy can help you split your investments between the safe and the risky, let us be clear what is meant by average growth rate. If the growth rate is compound interest at amount u, then $X_n = (1 + u)^n X_0$. Thus

$$\frac{1}{n} \ln\left(\frac{X_n}{X_0}\right) = \ln(1 + u).$$

The left side is what we have termed the average growth rate and, of course, when u is small, u and $\ln(1 + u)$ are almost identical.

Example 6.21

You have capital X_0 that you wish to invest. You can opt for National Savings, a guaranteed risk-free investment that will return 6% per year indefinitely, or a series of Venture Capitalists are happy to take your funds, offering you an 80% chance of increasing your investment by 50% in a year, but also with a 20% chance of losing the lot.

The safe option definitely turns £1000 into £1060. The risky option will turn the same sum into £1500 80% of the time, but otherwise into zero; on average, this is £1200. On average you do better with the risky option but, just as with the favourable bet in the casino, if you put everything into the risky option year on year, you are sure to become bankrupt.

Imagine that this choice will be offered to you every year, and you wish to maximise the long-term growth rate. The Kelly strategy puts a fixed fraction, x of your funds in the risky venture each year, and leaves the rest in the safe haven. The year-on-year scenario is

$$X_{k+1} = (1-x)X_k \times 1.06 + \begin{cases} xX_k \times 1.5 & \text{with probability } 0.8 \\ 0 & \text{with probability } 0.2. \end{cases}$$

If, over n years, the risky investment pays off Y times, then

$$X_n = X_0(1.06(1-x) + 1.5x)^Y(1.06(1-x))^{n-Y}$$

and so

$$\begin{aligned} \frac{1}{n}\ln\left(\frac{X_n}{X_0}\right) &= \frac{Y}{n}\ln(1.06 + 0.44x) + \left(1 - \frac{Y}{n}\right)\ln(1.06(1-x)) \\ &\to 0.8\ln(1.06 + 0.44x) + 0.2\ln(1.06(1-x)) \end{aligned}$$

almost surely. The optimal value of x is the solution of

$$\frac{0.8 \times 0.44}{1.06 + 0.44x} = \frac{0.2}{1-x}$$

i.e. $x = 7/22$. You should split your funds each year at 15/22 in the dull, boring, safe 6% home, and 7/22 in that risky area. Your overall growth rate will be $0.8\ln(1.2) + 0.2\ln(0.723) = 8.09\%$. Exercise 6.14 looks at the general case: improve your financial acumen by solving that problem.

After that instructive diversion into capital accumulation, we return to the general question of the long-run behaviour of sums. Suppose for simplicity of notation that the (X_n) are iidrv with mean 0 and variance 1, and write $S_n = X_1 + \cdots + X_n$. Then S_n/n will not only *get* close to zero, it will eventually *remain* close to zero. For all n, $S_n/\sqrt{n}$ has mean zero and variance unity and, for large n, it has, approximately, a $N(0, 1)$ distribution, so the probability that any particular $S_n/\sqrt{n}$ exceeds t can be estimated. That still leaves questions about the size of the fluctuations of $S_n/\sqrt{n}$, as n increases. The answer lies in the unforgettably named Law of the Iterated Logarithm (LIL).

I will not formally state this Law, whose proof is long and complex, but it is easy to explain what this remarkable result says. Look at the behaviour of the sequence of values $U_n = S_n/\sqrt{n \ln(\ln(n))}$. If $0 < c < \sqrt{2}$, then U_n will exceed c infinitely often; but if $c > \sqrt{2}$, then U_n will exceed c only finitely often, i.e. there is a last time when $U_n > c$. There is a symmetrical statement about when U_n is less than negative values of c.

The following is an interesting exercise in simulation. Given a sequence Y_n of independent $U(0, 1)$ variables, let $X_n = Y_n\sqrt{12} - \sqrt{3}$ so that (X_n) are iid $U(-\sqrt{3}, \sqrt{3})$, carefully chosen to have mean zero and variance unity. Plot the values of S_n/n, $S_n/\sqrt{n}$ and $S_n/\sqrt{n \ln(\ln(n))}$ as functions of n, on the same page.

The Strong Law asserts that S_n/n will converge to zero; hence, whatever $\epsilon > 0$ is given, there will be some N such that, for all $n \geq N$, we have $-\epsilon < S_n/n < \epsilon$.

The CLT states that $S_n/\sqrt{n}$ has a distribution close to $N(0, 1)$, so its values will slowly fluctuate (they are heavily dependent); in the long run, about 68% of them will satisfy $-1 < S_n/\sqrt{n} < 1$, and only about 5% of the time will $|S_n/\sqrt{n}|$ exceed 2. However, $S_n/\sqrt{n}$ is unbounded above and below, so, if we wait long enough, arbitrarily large positive and negative values will be attained.

The LIL looks at $U_n = S_n/\sqrt{n \ln(\ln(n))}$. The function $\sqrt{\ln(\ln(n))}$ is unbounded, but increases so slowly! When n is one million, it has reached 1.62, and does not struggle up to 2 until n has exceeded 5×10^{23}. Because of the enormous correlation between successive values, U_n changes very slowly. Nevertheless, this factor $\sqrt{\ln(\ln(n))}$ is large enough to tame the $N(0, 1)$ fluctuations. The LIL assures us that there will come a time (although we can never be sure that that time has been reached) after which U_n is confined to the interval $(-\sqrt{2}, \sqrt{2})$. Moreover, it will get arbitrarily close to both endpoints of that interval, infinitely often. What an amazing result!

EXERCISES

6.6. Suppose X and Y are independent strictly positive random variables with the same distribution, having finite mean μ. Identify the flaw in the argument:

$$E\left(\frac{X}{X + Y}\right) = \frac{E(X)}{E(X + Y)} = \frac{\mu}{2\mu} = \frac{1}{2},$$

but show that the value given is correct.

6.7. Suppose X_n has the discrete Uniform distribution $U(1, n)$, and write $Y_n = X_n/n$. Show that $Y_n \overset{D}{\to} U$, where U has the continuous $U(0, 1)$ distribution.

6.8. Suppose X_n has the Geometric distribution $G_1(p_n)$, where $p_n \to 0$, and that $\theta_n > 0$ is such that $p_n/\theta_n \to c$ with $0 < c < \infty$. Let $Y_n = \theta_n X_n$. Show that $Y_n \overset{D}{\to} E(c)$, where $E(c)$ is the Exponential distribution with parameter c.

6.9. Let X have the $U(0, 1)$ distribution, and expand X in decimal notation as $X = 0.Y_1 Y_2 Y_3 \ldots$, so that Y_i is the value in the ith place. Show that the Y_i are iidrv, and state their common distribution.

A *normal* number is one whose decimal expansion contains all ten digits with equal frequency. Show that $P(X$ is normal$) = 1$.

6.10. Suppose $r \geq 1$, and $X_n \overset{r}{\to} X$ with $E(|X_n|^r) < \infty$. Show that $E(|X|^r) < \infty$, and that $E(|X_n|^r) \to E(|X|^r)$. (You may need *Minkowski's inequality*, stated and proved on page 131 of Capiński and Kopp, 1999.)

6.11. In Theorem 6.16, we looked at the convergence of sums of random variables. Suppose (a_n) is a sequence of constants, converging to the finite limit a. What can be said about the convergence of $a_n X_n$ to aX, given the mode of convergence of X_n to X?

6.12. Show that, if $X_n \overset{D}{\to} X$, where X is a degenerate random variable, then $X_n \overset{P}{\to} X$.

6.13. The lifetimes (X_n) of bulbs used in street lamps are taken to be non-negative iidrv, each with strictly positive finite mean μ. Thus $S_n = \sum_{i=1}^{n} X_i$ is the time the nth bulb fails and is immediately replaced. Show that, if $N(t)$ is the number replaced up to time t, then $N(t)/t \overset{a.s.}{\longrightarrow} 1/\mu$ as $t \to \infty$.

6.14. Suppose that, as in Example 6.21, there is a safe haven with a compound growth rate of $\theta > 0$, while the risky investment has probability p of turning unit amount into $1 + u$ with $u > \theta$, and probability $q = 1 - p$ of turning unit amount into $1 - d$ with $0 < d \leq 1$. In each year, you will place a fraction x of your capital into the risky venture. What is the optimal choice of x, and the corresponding growth rate of your capital, for the various values of θ, u, d, p?

6.4 Summary

This chapter aims to make sense of the idea of convergence for a sequence of random variables. The first section describes a number of inequalities that are there mainly to help in the proofs of later results, but which have applications in their own right. For any sequence (X_n) of random variables, we have looked at different interpretations of the notion "$X_n \to X$".

The first idea is convergence in distribution: the distribution function F_n of X_n converges to the distribution function F of X, at all points where F is continuous. That last phrase is there to save technical embarrassment: it ensures (for instance) that if X_n converges to zero, then also $-X_n$ converges to zero. All the other modes of convergence that we discuss imply convergence in distribution, and this is often enough for our purposes in later work.

Next came convergence in probability. Loosely, this means that X_n is very likely to be close to X when n is large. And all the later modes we discuss imply this form of convergence.

Strong convergence, or *almost sure* convergence, gives the additional idea that not only will any particular X_n be close to X when n is large, but the whole sequence of values will remain close to X. You would correctly guess that convergence in mean square occurs when $E((X_n - X)^2)$ goes to zero as $n \to \infty$.

The Strong Law of Large Numbers sits alongside the Central Limit Theorem from the last chapter as the two most important central results in the subject. Both of them look at ways in which a suitably scaled version of a sum, $S_n \equiv X_1 + X_2 + \cdots + X_n$, will converge. The Strong Law looks at conditions under which S_n/n converges to a constant value, μ. A CLT, in our usage, gives sufficient conditions for S_n to be well approximated by a Normal variable. (Sometimes, the phrase "Central Limit Theorem" is used more widely. The main idea is to have convergence in distribution for a sequence of sums of random variables. It would include, for instance, our derivation of $\text{Bin}(n, \lambda/n) \overset{D}{\to} \text{Poiss}(\lambda)$.) The Law of the Iterated Logarithm pushes these notions just a little bit further.

One very useful application of the Strong Law is in Examples 6.20 and 6.21, where the Kelly strategy for balancing risk to optimise reward is described.

7

Stochastic Processes in Discrete Time

The phrase "stochastic process" is used when we have a collection of random variables, indexed by a time parameter, so that they have a natural order. Examples include the size of our capital after a series of investments in the stock market, or other casinos; the accumulated number of points of a football team during the season; a student's Grade Point Average as she progresses through college; your own weight as you strive for the target you set yourself; the temperature in your home.

In some instances, the random variables are naturally observed at discrete time intervals (not necessarily evenly spaced, although frequently they will be), in others near-continuous monitoring is the norm. This chapter looks at the former case, the next examines the latter. Formally

Definition 7.1

A *discrete stochastic process* is a collection of random variables, typically $\{X_0, X_1, X_2, \ldots\}$, indexed by an ordered time parameter, all defined on the same probability space.

As usual, that underlying probability space will be suppressed. We acknowledge that a rigorous development of the subject in this chapter and the next relies on a rather difficult piece of mathematics, the Daniell–Kolmogorov theorem. This result assures us that so long as the joint distribution functions of all finite collections of these random variables satisfy simple and obvious consistency conditions, there *is* a probability space on which all the variables can

139

be defined. It is the existence of this theorem that allows us to concentrate on the properties of the random variables – their distributions, their relationships with each other, limits – without distracting references to some outcome ω from the probability space behind the entire model.

Our aims are to describe the changes in (X_n) over time, to say what can be said about X_{m+n} $(n > 0)$ when we have information on $(X_0, \ldots, X_m)$, and to assess the limiting behaviour of the process over a long period.

7.1 Branching Processes

Suppose a man with a unique surname considers how many generations into the future that name might last. In his culture, surnames pass down the male line only, so the quantity of interest is how many sons he has and, in turn, the numbers of their sons, their sons' sons, and so on. As a mathematical model, we can let time be numbered by the generations, so that X_n, the size of the nth generation, is the number of times the name is represented n generations hence, with $X_0 = 1$ to indicate there is just one initiator.

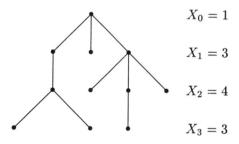

$$X_0 = 1$$
$$X_1 = 3$$
$$X_2 = 4$$
$$X_3 = 3$$

Figure 7.1 Several generations in a branching process.

The most convenient assumption is that the numbers of sons for all the men in the line are iidrv with common mass function $(p_k : k = 0, 1, \ldots)$ or, equivalently, with common pgf $g(z) = \sum p_k z^k$. This may or may not be realistic in this application, but it is the defining property of a standard branching process, whose properties we now explore.

One annoying trivial case is when $g(z) = z$, i.e. every man inevitably has exactly one son. The resulting process is very dull: X_n always takes the value unity when $n \geq 0$, and there is really nothing to add. As this case would constitute an exception to some main results, we will exclude it: the case $g(z) =$

z is not thought of as a "proper" branching process.

With that exclusion, suppose the pgf of X_n is $g_n(z)$. By definition, $g_0(z) = z$, and we know that $g_1(z) = g(z)$. Suppose we have found $g_2(z), g_3(z), \ldots, g_n(z)$, and we seek $g_{n+1}(z)$. There are two distinct ways of looking at X_{n+1}, the size of generation $n+1$: the members of that generation can be described as either

(a) the offspring of the members of generation n, so X_{n+1} is the sum of X_n independent random variables, each having pgf $g(z)$; or

(b) the nth generation offspring of the members of the first generation, so X_{n+1} is the sum of X_1 independent random variables, each with pgf $g_n(z)$.

Using Theorem 5.3, it follows that

Theorem 7.2

For all $n \geq 1$, $g_{n+1}(z) = g_n(g(z)) = g(g_n(z))$. Hence, using either expression, we find that $g_n(z) = g(g(g(\cdots(z)\cdots)))$, the nth iteration of $g(z)$.

Example 7.3

Suppose there are either no offspring or two offspring, with respective probabilities q and p, so that $g(z) = q + pz^2$. This would correspond to a model in which a cell dies with probability q, or splits into two identical daughter cells with probability p. Then

$$g_2(z) = g(g(z)) = q + p(q + pz^2)^2 = q + pq^2 + 2p^2qz^2 + p^3z^4,$$

and

$$
\begin{aligned}
g_3(z) &= q + pq^2 + 2p^2q(g(z))^2 + p^3(g(z))^4 \\
&= q(1 + pq + 2p^2q^2 + p^3q^3) + 4p^3q^2(1 + pq)z^3 \\
&\quad + 2p^4q(1 + 3pq)z^4 + 4p^6qz^6 + p^7z^8.
\end{aligned}
$$

Plainly, the exact expression for $g_n(z)$ is complex when n is large, but a computer algebra package could calculate the exact distribution of X_n for any n we cared to specify.

Theorem 7.4

If $\mu = E(X_1)$ and $\sigma^2 = \text{Var}(X_1)$ are the mean and variance of the number of offspring, then the mean and variance of the nth generation are

$$E(X_n) = \mu^n, \quad \text{Var}(X_n) = \sigma^2\mu^{n-1}(1 + \mu + \mu^2 + \cdots + \mu^{n-1}).$$

Proof

We have $\mu = g'(1)$ and $\sigma^2 = g''(1) + g'(1) - (g'(1))^2$, and the formula certainly holds when $n = 1$. We prove the general case by induction. Suppose the formula holds for some particular n. From $g_{n+1}(z) = g(g_n(z))$, we find $g'_{n+1}(z) = g'(g_n(z))g'_n(z)$ and $g''_{n+1}(z) = g''(g_n(z))(g'_n(z))^2 + g'(g_n(z))g''_n(z)$. Hence

$$E(X_{n+1}) = g'_{n+1}(1) = g'(1)g'_n(1) = \mu E(X_n) = \mu^{n+1}$$

which establishes the result for the mean. Also

$$
\begin{aligned}
\mathrm{Var}(X_{n+1}) &= g''_{n+1}(1) + \mu^{n+1} - \mu^{2n+2} \\
&= g''(1)\mu^{2n} + \mu g''_n(1) + \mu^{n+1} - \mu^{2n+2} \\
&= (\sigma^2 - \mu + \mu^2)\mu^{2n} + \mu(\mathrm{Var}(X_n) - \mu^n + \mu^{2n}) + \mu^{n+1} - \mu^{2n+2} \\
&= \sigma^2\mu^{2n} + \mu\,\mathrm{Var}(X_n)
\end{aligned}
$$

from which the inductive step follows. $\qquad\square$

Corollary 7.5

When $\mu = 1$, then $\mathrm{Var}(X_n) = n\sigma^2$, and when $\mu \neq 1$, then $\mathrm{Var}(X_n) = \sigma^2\mu^{n-1}(\mu^n - 1)/(\mu - 1)$.

The long-term behaviour is now clear when $\mu < 1$ since then both $E(X_n)$ and $\mathrm{Var}(X_n)$ converge geometrically fast to zero. When $\mu = 1$, the mean remains fixed at unity, but the variance diverges to ∞. For $\mu > 1$, both mean and variance increase geometrically fast without bound. We now establish a simple criterion for whether a branching process might continue indefinitely.

Theorem 7.6

Let x_n be the probability of extinction by generation n. Then $x_{n+1} = g(x_n)$, and $(x_n) \uparrow x$, where x is the smallest non-negative root of the equation $z = g(z)$, and is the probability of extinction. Moreover, $x = 1 \Leftrightarrow \mu \leq 1$.

Proof

The definition of a pgf shows that $x_n = P(X_n = 0) = g_n(0)$, and so $x_{n+1} = g_{n+1}(0) = g(g_n(0)) = g(x_n)$. Whenever $X_n = 0$, then $X_{n+1} = 0$, so plainly $x_n \leq x_{n+1}$; since x_n is a probability, the sequence (x_n) is bounded above by unity, so converges to some limit x with $x \leq 1$.

But $g(z) = \sum p_k z^k$ is continuous on $0 \leq z \leq 1$, so the relation $x_{n+1} = g(x_n)$ shows that $x = g(x)$. We now prove that the solution we seek is the smallest root of this equation in $[0, 1]$, using the fact that g is increasing over that interval. Let $\alpha = g(\alpha)$ be any root with $\alpha \geq 0$. Then $x_1 = g(0) \leq g(\alpha) \leq \alpha$. Inductively, suppose $x_n \leq \alpha$; then $x_{n+1} = g(x_n) \leq g(\alpha) \leq \alpha$, as required. Since $x_n \leq \alpha$ for all n, so $x = \lim(x_n) \leq \alpha$, and the solution we seek is indeed the smallest non-negative root.

Finally, geometrical considerations will show that $x = 1 \Leftrightarrow \mu \leq 1$. The case when $p_0 = 0$ is easily disposed of: here extinction is impossible, i.e. $x = 0$, and since $p_1 < 1$, it is plain that $\mu > 1$. So we may assume that $g(0) = p_0 > 0$, and we know that $g'(z) = \sum k p_k z^{k-1}$ is increasing when $z \geq 0$, with $\mu = g'(1)$.

Suppose $\mu > 1$; then since $g(1) = 1$, when z is just less than unity, we must have $g(z) < z$. As $g(0) > 0$, the Intermediate Value Theorem shows there is some x, $0 < x < 1$, with $g(x) = x$. Conversely, suppose there is some $x = g(x)$ with $0 < x < 1$. Since $g(1) = 1$, and g' is increasing over $(x, 1)$, the chord joining $(x, g(x))$ to $(1, g(1))$ lies entirely above the graph of g, hence $\mu = g'(1) > 1$. $\quad\square$

(This result illustrates why we excluded the case when $p_1 = 1$. For then $\mu = 1$, yet extinction is impossible, so the statement of this theorem would need a caveat.)

Notice that this theorem shows that $X_n \xrightarrow{a.s.} 0$ when $\mu \leq 1$, although we require $\mu < 1$ for $X_n \xrightarrow{1} 0$, as $E(X_n) = 1$ when $\mu = 1$. When $\mu < 1$, we can expect the process to die out quickly, but for $\mu = 1$, although extinction is certain, it might well take a long time: X_n could get fairly large before being fatally trapped at zero.

For $\mu > 1$, there is a dichotomy in the behaviour of X_n. If $x < 1$ is the smallest positive root of $z = g(z)$, the process either, with probability x, dies out or, with probability $1 - x$, "explodes". Let $Y_n \equiv X_n | X_n \neq 0$ be the population size, conditional on non-extinction. Then, for $r \geq 1$,

$$E(X_n^r) = E(X_n^r | X_n \neq 0)P(X_n \neq 0) + E(X_n^r | X_n = 0)P(X_n = 0)$$

so that $E(Y_n^r) = E(X_n^r | X_n \neq 0) = E(X_n^r)/P(X_n \neq 0)$. When n is large, the denominator is very close to $1 - x$, which implies that $E(Y_n) \approx E(X_n)/(1 - x)$ and $\text{Var}(Y_n) \approx \text{Var}(X_n)/(1 - x) - x(E(X_n))^2/(1 - x)^2$.

To appreciate this, we look at a numerical illustration for Example 7.3, where $g(z) = q + pz^2$. Take $p = 0.6$, so that $\mu = 1.2$, $\sigma^2 = 0.96$ and $x = 2/3$. Theorem 7.4 shows that X_{100} has a mean of about 83 million, with a standard deviation twice that. So the process will be extinct by generation 100 about $2/3$ of the time, but if it is not extinct, its mean is $E(Y_{100})$, about 250 million, and the standard deviation is about 200 million. Attempts to forecast X_{100} are subject to an enormous margin of error!

But suppose we know X_{20}, and wish to forecast X_{120}: here $E(X_{20}) \approx 38.3$, arising as a $2/3$ chance of being zero, otherwise a mean of 115. If X_{20} is not zero, it is likely to be well away from zero, so suppose $X_{20} = 100$. For the process now to die out, each of 100 independent lines must die out, which has probability $(2/3)^{100} \approx 2.5 \times 10^{-18}$. We would be on safe ground in assuming $X_{120} \neq 0$!

Moreover, as X_{120} is now taken as the sum of 100 iidrv, each having mean and variance that of the original X_{100}, so X_{120} will have mean 8.3 billion, s.d. 1.66 billion – large, but only one fifth of the mean. Attempts to forecast X_{120}, given X_{20}, will be far more successful than forecasting X_{100} ab initio.

The pattern of behaviour described in this example is typical of branching processes with $\mu > 1$.

Example 7.7 (Spread of Mutations)

An assumption sometimes made in biological models is that the number of offspring has a Poisson distribution. In equilibrium, the mean will be unity, but suppose a favourable mutation arises, so that the mean number of times it would be represented in the next generation is $\mu = 1 + \epsilon$ with $\epsilon > 0$. We seek the chance that this favourable mutation becomes established.

Here $g(z) = \exp(-(1+\epsilon)(1-z))$, and we solve $z = g(z)$. Let the solution be $x = 1 - \theta$ with $\theta > 0$ being the chance of establishment, so that taking logarithms leads to $\ln(1 - \theta) = -(1+\epsilon)\theta$. Expanding the log term in the usual fashion,

$$\theta + \theta^2/2 + \theta^3/3 + \cdots = (1+\epsilon)\theta.$$

When ϵ is small, then θ is small, hence the above equation is approximately $\theta/2 + \theta^2/3 = \epsilon$. Thus θ is just less than 2ϵ (e.g. $\epsilon = 0.05$ leads to $\theta = 0.0937$). The chance that a favourable mutation becomes established is about twice its "selective advantage". Deleterious mutations ($\epsilon < 0$) always die out; as most favourable mutations have selective advantages of at most a few percent, they also usually disappear without trace.

In the following exercises, the standard notation introduced above is taken for granted.

EXERCISES

7.1. Show that the probability that a branching process becomes extinct precisely in generation n is $p_n = x_n - x_{n-1}$, with $x_0 = 0$.

7.2. Given $g(z) = 0.4 + 0.2z + 0.4z^2$, find the pgfs of the sizes of generations one and two. Find also the chances of dying out (a) *by the* second generation; (b) *in the* second generation.

7.3. What happens in the long run when $g(z) = 0.3 + 0.5z + 0.2z^3$? Find the mean and variance of the size of generation 100, and hence the mean size conditional on non-extinction.

7.4. For what values of λ is $g(z) = 0.2 + (1.2 - \lambda)z + (\lambda - 0.4)z^3$ a pgf? When is the corresponding branching process certain to die out?

7.5. Let $S_n = X_0 + X_1 + \cdots + X_n$ be the total number in the first n generations, with pgf $h_n(z)$. Show that $h_{n+1}(z) = zg(h_n(z))$ when $n \geq 0$, with $h_0(z) \equiv z$. Assuming that $h_n(z) \to h(z)$, find $h(z)$ when $g(z) = q + pz^2$ as in Example 7.3. Under what conditions is the resulting $h(z)$ a genuine pgf? Explain the result that arises when $h(z)$ is not a pgf.

7.6. Suppose $g(z) = p/(1 - qz)$, i.e. that the offspring distribution is geometric $G_0(p)$. Show that, when $p \neq q$, then

$$g_n(z) = \frac{p((q^n - p^n) - qz(q^{n-1} - p^{n-1}))}{q^{n+1} - p^{n+1} - qz(q^n - p^n)}.$$

Find the pgf of $Y_n \equiv X_n | X_n \neq 0$. Find $g_n(z)$ when $p = q = 1/2$, and deduce that the chance of extinction in generation n is $1/(n(n+1))$. Find the distribution of Y_n defined as before, and deduce that $Y_n/n \overset{D}{\to} E(1)$.

7.7. Let $Y_n = X_n/\mu^n$ so that $E(Y_n) = 1$. Show that $\mathrm{Var}(Y_n) \to \sigma^2/(\mu^2 - \mu)$ when $\mu > 1$. Find the limiting values of $E(Y_n | X_n \neq 0)$ and $\mathrm{Var}(Y_n | X_n \neq 0)$.

7.8. Suppose that, when $n \geq 1$, Y_n immigrants arrive in addition to the offspring of generation $n - 1$, where (Y_n) are iidrv with pgf $h(z)$. Let $k_n(z)$ be the pgf of the size of generation n, taking account of the immigrants. Show that $k_{n+1}(z) = k_n(g(z))h(z)$, and write down $k_3(z)$ in terms of g, h only.

7.2 Random Walks

Definition 7.8

Let (X_n) be iidrv with $P(X_n = 1) = p$ and $P(X_n = -1) = q = 1 - p$, and let $S_0 = 0$, $S_n = X_1 + \cdots + X_n$ for $n \geq 1$. The values (S_n) are said to form a *simple random walk*; it is *symmetrical* if $p = q = 1/2$.

A simple random walk is a possible model for the fluctuations in your fortune if you gamble in a casino, winning or losing unit amounts on successive plays. It can also model the position of a drunken man on a straight road, whose steps are taken without reference to where he is, or to his past movements. The word "simple" indicates that the steps X_n can take the values ± 1 only; this restricts the applicability of the model, but allows a fairly full analysis.

The random walk has the twin properties of *time homogeneity* and *space homogeneity*. By this we mean that $P(S_{m+n} = b | S_n = a) = P(S_{m+r} = b | S_r = a)$ for all m, n, r, a, b; and similarly that $P(S_{m+n} = a | S_n = 0) = P(S_{m+n} = a + b | S_n = b)$. We use the phrase "return to the origin" to mean that S_n takes its initial value, zero, at some later time.

Theorem 7.9

When (S_n) is a simple random walk and $n \geq 0$, write $u_n = P(S_n = 0)$. Write $f_1 = P(S_1 = 0)$ and, for $n \geq 2$, write $f_n = P(S_1 \neq 0, \cdots, S_{n-1} \neq 0, S_n = 0)$ as the probability that the first return to the origin is at time n. Then their respective generating functions $U(z) = \sum_{n=0}^{\infty} u_n z^n$ and $F(z) = \sum_{n=1}^{\infty} f_n z^n$ are given by $U(z) = (1 - 4pqz^2)^{-1/2}$ and $F(z) = 1 - (1 - 4pqz^2)^{1/2}$, provided $4pqz^2 < 1$.

Proof

Since $S_0 = 0$, we have $u_0 = 1$ and $u_1 = f_1$. For $n \geq 2$,

$$u_n = P(S_n = 0) \quad = \quad \sum_{k=1}^{n} P(S_n = 0 \text{ and first return to origin is at time } k)$$

$$= \quad \sum_{k=1}^{n} P(S_n = 0, S_1 \neq 0, \ldots, S_{k-1} \neq 0, S_k = 0)$$

$$= \quad \sum_{k=1}^{n} P(S_n = 0 | S_1 \neq 0, \ldots, S_{k-1} \neq 0, S_k = 0) f_k.$$

But $P(S_n = 0 | S_1 \neq 0, \ldots, S_{k-1} \neq 0, S_k = 0) = P(S_n = 0 | S_k = 0)$ using the independence of the components (X_r). Also $P(S_n = 0 | S_k = 0) = P(S_{n-k} = 0)$ using time homogeneity. Hence, for $n \geq 1$, $u_n = \sum_{k=1}^{n} f_k u_{n-k}$. Thus

$$U(z) = \sum_{n=0}^{\infty} u_n z^n = 1 + \sum_{n=1}^{\infty} z^n \sum_{k=1}^{n} f_k u_{n-k} = 1 + \sum_{k=1}^{\infty} f_k z^k \sum_{n=k}^{\infty} u_{n-k} z^{n-k},$$

which means that

$$U(z) = 1 + F(z)U(z). \tag{7.1}$$

The event $S_n = 0$ occurs when there have been equal numbers of positive and negative steps, so $n = 2m$ is even, and the binomial distribution leads to $u_{2m} = \binom{2m}{m} p^m q^m$. Hence we can write down the series for $U(z)$. Now recall the binomial theorem for a negative index: when $|4\theta| < 1$, the expansion of $(1-4\theta)^{-1/2}$ can be seen to be $1 + \sum_{n=1}^{\infty} \binom{2m}{m} \theta^m$, leading to $U(z) = (1-4pqz^2)^{-1/2}$. Then Equation (7.1) implies the given expression for $F(z)$. □

Corollary 7.10

The simple random walk is certain to return to the origin if, and only if, $p = 1/2$. When $p = 1/2$, the time T to first return has pgf $1 - (1-z^2)^{1/2}$, and $E(T) = \infty$.

Proof

Let E denote the event that the random walk returns to the origin sometime, so that E is the union of the disjoint events $\{E_k : k = 1, 2, \ldots\}$, where $E_k \equiv S_1 \neq 0, \ldots, S_{k-1} \neq 0, S_k = 0$. Hence

$$P(E) = \sum_{k=1}^{\infty} P(E_k) = \sum_{k=1}^{\infty} f_k = F(1) = 1 - \sqrt{1 - 4pq} = 1 - |1 - 2p|.$$

Thus $P(E) = 1$ if, and only if, $p = 1/2$, otherwise $P(E) < 1$. And when $p = 1/2$, the pgf of the time to return is $F(z) = 1 - (1 - z^2)^{1/2}$. Thus $E(T) = F'(1) = \lim \frac{1}{2}(1 - z^2)^{-1/2}.2z$ as $z \uparrow 1$, i.e. $E(T) = \infty$. □

Some students have an uneasy feeling when they see the two statements
(i) return to the origin in finite time is certain
(ii) the mean time to return is infinite
in conjunction. But they are not contradictory: they simply assert that both $\sum f_k = 1$ and $\sum k f_k = \infty$.

We can use results from branching processes to find the probability that a simple random walk ever hits some particular non-zero value. Write $A_r =$ The simple random walk ever reaches $-r$, for $r = 1, 2, \ldots$. Because the steps in the walk are of unit size, in order to reach $-r$ the walk must earlier visit -1, then -2 etc. But once we have reached $-k$, the chance of ever reaching $-k - 1$ is the same as the chance of ever reaching -1, starting at zero; equivalently, the chance we ever hit zero, starting at 1. Independence over disjoint time intervals then shows that $P(A_r) = (P(A_1))^r$ for $r \geq 1$.

Now consider the branching process with pgf $q + pz^2$ of Example 7.3. Each of the k members of the current generation either has no offspring, or has two offspring. Thus, in the next generation, there is a change of -1 with probability

q and of $+1$ with probability p, independently for each member of the current generation. The size of the next generation in this branching process will have the same distribution as the value of a simple random walk that is now at position k, after k further steps. In this branching process starting with one member, any sequence leading to extinction corresponds to a simple random walk that starts at $+1$ ever visiting 0. Since the chance of extinction in the branching process is $x = \min\{1, q/p\}$, so $P(A_1) = x$ and $P(A_r) = x^r$.

To find the probability of ever visiting $+r > 0$ in the random walk, interchange the roles of p and q. Collecting these together, we have proved:

(1) if $p = q = 1/2$, it is certain that any value $\pm r$ is visited;

(2) if $p < 1/2$ and $r > 0$, $P(\text{Visit} - r) = 1$, $P(\text{Visit} + r) = (p/q)^r$;

(3) if $p > 1/2$ and $r > 0$, $P(\text{Visit} + r) = 1$, $P(\text{Visit} - r) = (q/p)^r$.

It follows from (1) that when $p = 1/2$, we are certain to visit any state at least N times, no matter how large N is, and hence we are certain to visit every state infinitely often. For, consider returns to zero. We are certain to return at least once, and when we return, time homogeneity implies we are certain to return again; and again; and again, etc. And for any $r \neq 0$, we are certain to hit it at least once; and now that we know we return to the origin infinitely often, space homogeneity shows we return to r infinitely often.

Example 7.11 (Gambler's Ruin)

The gambler's ruin problem requires a modification to a simple random walk. Given $c > 0$, let $0 \leq a \leq c$, and suppose $S_0 = a$: we seek the probability that $S_n = 0$ before $S_n = c$.

This models a gambler whose initial fortune is $a \geq 0$ playing an opponent whose initial fortune is $c - a \geq 0$. At each play of the game, the gambler wins or loses unit amount from the opponent with probabilities p and q, and the game ends when either player's fortune is zero. In the spirit of pessimism, we concentrate on the chance that it is the gambler who is ruined.

Write $p_a = P(S_n = 0 \text{ before } S_n = c | S_0 = a) = P(\text{Ruin}|S_0 = a)$. Plainly $p_0 = 1$ and $p_c = 0$, so let $0 < a < c$ and look at the position after the first play.

$$
\begin{aligned}
p_a &= P(\text{Ruin} \cap X_1 = +1 | S_0 = a) + P(\text{Ruin} \cap X_1 = -1 | S_0 = a) \\
&= pP(\text{Ruin}|X_1 = +1, S_0 = a) + qP(\text{Ruin}|X_1 = -1, S_0 = a) \\
&= pP(\text{Ruin}|S_0 = a + 1) + qP(\text{Ruin}|S_0 = a - 1) = pp_{a+1} + qp_{a-1}
\end{aligned}
$$

using time and space homogeneity.

The standard approach to solve such a *difference equation* is to write $p_a = \theta^a$, which leads to

$$\theta^a = p\theta^{a+1} + q\theta^{a-1} \quad \text{whenever } 1 \le a \le c-1,$$

i.e. $p\theta^2 - \theta + q = 0$, which has solutions $\theta = 1$ and $\theta = q/p$. When $p \ne q$, these solutions are different, and the general solution of the difference equation is $p_a = A + B(q/p)^a$; when $p = q = 1/2$, the general solution is $p_a = C + Da$. In either case, the boundary conditions $p_0 = 1$, $p_c = 0$ enable the constants A, B or C, D to be found. The answers are

$$p_a = P(\text{Ruin}|S_0 = a) = \begin{cases} 1 - a/c & \text{if } p = q = 1/2; \\ \frac{\theta^c - \theta^a}{\theta^c - 1} & \text{where } \theta = q/p, \text{ if } p \ne q. \end{cases}$$

If we interchange the roles of p and q, and also of a and $c - a$, we see the game from the opponent's perspective, and we can write down the chance she is ruined. It turns out that

$$P(\text{Gambler ruined}) + P(\text{Opponent ruined}) = 1,$$

which means that the game is certain to end sometime, and the formula for p_u lets us find $1 - p_a$, our chance of victory.

Let T_a be the time the game lasts, conditional on $S_0 = a$. Here $T_0 = T_c = 0$, so take $0 < a < c$ and again condition on the outcome of the first play. With a parallel argument, we now find

$$T_a = 1 + pT_{a+1} + qT_{a-1},$$

where T_{a+1} and T_{a-1} are the durations of games starting at $a + 1$ and $a - 1$. Write $\mu_a = E(T_a)$, so that

$$\mu_a = 1 + p\mu_{a+1} + q\mu_{a-1}, \quad 1 \le u \le c-1$$

with $\mu_0 = \mu_c = 0$. This is very similar to the recurrence relation for the (p_a), and you should verify that

$$\mu_a = E(\text{Duration of game}) = \begin{cases} a(c - a) & \text{if } p = q = 1/2; \\ \frac{c - a - cp_a}{p - q} & \text{if } p \ne q, \end{cases}$$

where p_a was found earlier.

So in a game with $p = q = 1/2$, suppose you begin with £1 and your opponent begins with £99. The chance you bankrupt her is only $1/100$ but, on average, the game lasts 99 plays. (This is plainly an example where the average behaviour is atypical: half the time, you lose the first play and the match is over, a further $1/8$ of the time you become ruined on the third play.)

The case $c = \infty$ corresponds to a gambler playing an infinitely rich opponent – effectively the position most of us are in if we visit a casino. Even if the game were fair, with $p = q = 1/2$, then $p_a \to 1$ as $c \to \infty$, i.e. ruin is certain. For $p \neq q$, then $p_a \to 1$ if $q > p$ (no surprise there), and $p_a \to (q/p)^a$ if $q < p$. This shows another way to establish the earlier result

$$P(\text{Ever visit } - r | S_0 = 0) = (q/p)^r \quad \text{if } q < p.$$

Another way to analyse the simple random walk is to trace its *path*, i.e. the set $\{(n, S_n) : n = 0, 1, 2, \ldots\}$; we refer to n as the *time* and S_n as the *value*.

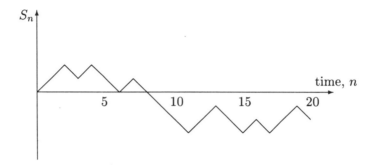

Let a, b be any integers, take $n > 0$, and write

(i) $N(a, b, n) =$ the number of possible paths from $(0, a)$ to (n, b);

(ii) $N^0(a, b, n) =$ the number of such paths that revisit the time axis, i.e. hit some $(k, 0)$ with $k > 0$.

Theorem 7.12 (Reflection Principle)

Given $a > 0$ and $b > 0$, $N^0(a, b, n) = N(-a, b, n)$.

Proof

It is enough to establish a one–one correspondence between all paths from $(0, -a)$ to (n, b), and all paths from $(0, a)$ to (n, b) that contain some value $(k, 0)$.

Plainly any path from $(0, -a)$ to (n, b) takes the value zero at least once; suppose the first such time is k. Whatever the path from $(0, -a)$ to $(k, 0)$, reflect it in the time axis to give a path from $(0, a)$ to $(k, 0)$, and leave the path from $(k, 0)$ to (n, b) alone. This establishes the correspondence. $\square$

Example 7.13 (Enough Change?)

Suppose soft drinks cost 50p each, and a vending machine will accept 50p or £1 coins. How likely is it that there is no suitable change when a customer inserts a £1 coin? Jiří Anděl (2001) noted how the Reflection Principle can help. Assume the operator pre-loads a number a of 50p coins as potential change, and that N customers buy drinks, b of them using 50p coins and the rest using £1 coins. We assume that the coins are inserted in a random order, and also that $N - b \leq a + b$, otherwise it is certain that some customer will be denied change.

Fix attention on the number of 50p coins in the machine after each drink purchased: it increases or decreases by unity according to the type of coin used. Asking that no customer is denied change is the same as asking that a random walk, starting at $(0, a)$ and ending at $(N, a + 2b - N)$ never crosses the x-axis, which is equivalent to asking that a walk starting at $(0, a + 1)$ and ending at $(N, a + 2b + 1 - N)$ never even touches the x-axis.

There are $\binom{N}{b}$ paths from $(0, a+1)$ to $(N, a+2b+1-N)$ without restrictions, and the Reflection Principle implies that the number of these paths that *do* touch the x-axis is the same as the number of unrestricted paths from $(0, -a-1)$ to $(N, a + 2b + 1 - N)$. The latter is $\binom{N}{a+b+1}$, and hence the chance that change is always available is $[\binom{N}{b} - \binom{N}{a+b+1}]/\binom{N}{b}$. To ensure this chance is 100%, we must have $a \geq N - b$, of course, but the operator may be content to ensure the chance is some reasonable value, say 95%. See Exercise 7.13.

Theorem 7.14 (Ballot Theorem)

Suppose that candidate A scores x votes, and candidate B scores y votes, with $x > y$. Then, assuming the votes are randomly mixed before counting, the chance that A is always ahead of B during the count is $(x - y)/(x + y)$.

Proof

Let $X_n = +1$ or -1 according as the nth vote counted is cast for A or B, so that S_n is A's accumulated lead. We seek the probability that the path from $(0, 0)$ to $(x + y, x - y)$ never revisits the time axis.

For this to occur, A must get the first vote, which has probability $x/(x+y)$, and we look at paths from $(1, 1)$ to $(x - y, x + y)$. There are $N(1, x - y, x + y - 1)$ of them altogether, and $N^0(1, x - y, x + y - 1)$ of them *do* visit some $(k, 0)$. The Reflection Principle implies that $N^0(1, x-y, x+y-1) = N(-1, x-y, x+y-1)$, and we have the general formula

$$N(a,b,n) = \binom{n}{(n+b-a)/2}$$

since any path from $(0,a)$ to (n,b) must have $(n+b-a)/2$ positive steps, and $(n+a-b)/2$ negative ones. Hence the number of paths from $(1,1)$ to $(x+y, x-y)$ that never visit any $(k,0)$ is

$$\binom{x+y-1}{x-1} - \binom{x+y-1}{x} = \frac{x-y}{x+y}\binom{x+y}{x}.$$

Thus the probability that A is in the lead for the whole count is

$$\frac{x}{x+y}\cdot\frac{x-y}{x+y}\cdot\binom{x+y}{x}\bigg/\binom{x+y-1}{x-1}$$

which reduces to $(x-y)/(x+y)$, as claimed.   □

The Reflection Principle also helps prove a remarkable result about the simple, symmetric random walk, relating its chance of being *at* zero after $2m$ steps to its chance of *never visiting* zero during the first $2m$ steps, or of always being on the same side of zero in that period.

Theorem 7.15

In the symmetric simple random walk, if $m \geq 1$ then

$$\begin{aligned}
P(S_{2m} = 0) &= P(S_1 \neq 0, S_2 \neq 0, \ldots, S_{2m} \neq 0) \\
&= P(S_1 \geq 0, S_2 \geq 0, \ldots, S_{2m} \geq 0).
\end{aligned}$$

Proof

Suppose $X_1 = +1$ and take $y > 0$. The number of paths from $(1,1)$ to $(2m, y)$ that never visit any $(k, 0)$ is

$$N(1, y, 2m-1) - N^0(1, y, 2m-1) = N(1, y, 2m-1) - N(-1, y, 2m-1)$$

by the Reflection Principle. Hence the total number of paths from $(1,1)$ to some $(2m, y)$ with $y > 0$ that never visit any $(k, 0)$ is

$$\sum_{y=2}^{2m}(N(1,y,2m-1)-N(-1,y,2m-1)) = \sum_{y=2}^{2m}\left(\binom{2m-1}{m-1+y/2} - \binom{2m-1}{m+y/2}\right)$$

which telescopes down to $\binom{2m-1}{m}$.

Similarly, when $X_1 = -1$, there are the same number of paths that remain always negative up to time $2m$, so there are $2\binom{2m-1}{m}$ paths that never visit any $(k, 0)$ when $1 \leq k \leq 2m$. Each such path has probability 2^{-2m}, so

$$P(S_1 \neq 0, S_2 \neq 0, \ldots, S_{2m} \neq 0) = 2\binom{2m-1}{m} \cdot 2^{-2m},$$

which is exactly the same as $\binom{2m}{m} 2^{-2m} = P(S_{2m} = 0)$.

For the other result, clearly

$$P(S_1 \neq 0, S_2 \neq 0, \ldots, S_{2m} \neq 0) = 2P(S_1 > 0, S_2 > 0, \ldots, S_{2m} > 0).$$

But $P(S_1 > 0, \ldots, S_{2m} > 0) = P(S_2 > 0, \ldots, S_{2m} > 0 | S_1 = 1)(1/2)$, and space homogeneity implies this conditional probability is also $P(S_2 > -1, \ldots, S_{2m} > -1 | S_1 = 0)$. Now time homogeneity shows this is equal to $P(S_1 \geq 0, \ldots, S_{2m-1} \geq 0)$. But whenever $S_{2m-1} \geq 0$, then $S_{2m-1} \geq 1$, hence $S_{2m} \geq 0$. The factors 2 and $(1/2)$ cancel, leaving the result we seek.   □

Corollary 7.16

In the simple symmetric random walk, again write $u_{2m} = P(S_{2m} = 0)$. Then the probability that the *last* return to zero, up to and including time $2n$ was at time $2k$ is $u_{2k}u_{2n-2k}$. And this is also the probability that S_r is positive over exactly $2k$ time intervals up to time $2n$. (S_r is said to be positive on the interval $(k, k+1)$ if either $S_k > 0$ or $S_{k+1} > 0$.)

Proof

Let $\alpha(2k, 2n)$ be the probability that the last return to zero up to and including time $2n$ was at time $2k$. Then

$$\begin{aligned} \alpha(2k, 2n) &= P(S_{2k} = 0, S_{2k+1} \neq 0, \ldots, S_{2n} \neq 0) \\ &= P(S_{2k} = 0)P(S_{2k+1} \neq 0, \ldots, S_{2n} \neq 0 | S_{2k} = 0). \end{aligned}$$

Time homogeneity shows this conditional probability is the same as $P(S_1 \neq 0, \ldots, S_{2n-2k} \neq 0 | S_0 = 0)$, and the theorem shows the latter is $P(S_{2n-2k} = 0)$. Hence $\alpha(2k, 2n) = u_{2k}u_{2n-2k}$, as stated; and plainly $\alpha(2k, 2n) = \alpha(2n-2k, 2n)$.

For the second assertion, note that S_r can be zero only when r is even, so (S_r) is positive over the interval $(2m, 2m+1)$ if, and only if, it is positive over the interval $(2m+1, 2m+2)$. Hence the number of intervals up to time $2n$ over which (S_r) is positive is necessarily even. Let $\beta(2k, 2n)$ be the probability it is positive over exactly $2k$ intervals.

Plainly $\beta(2n, 2n) = P(S_1 \geq 0, \ldots, S_{2n} \geq 0) = u_{2n}$ by the theorem, and $\beta(0, 2n) = P(S_1 \leq 0, \ldots, S_{2n} \leq 0) = u_{2n}$ by symmetry. Thus $\beta(2k, 2n) =$

$u_{2k}u_{2n-2k}$ when $2k = 0$ or $2k = 2n$; we use induction to establish this for $2 \leq 2k \leq 2n - 2$. It is easy to calculate that the given formula holds for all k when $n \leq 2$, so suppose it holds for all time periods up to and including $2n - 2$. Let $A_m \equiv$ first return to zero is at time $2m$, and B_k be that S_r is positive over $2k$ time intervals up to time $2n$. Then

$$\beta(2k, 2n) = \sum_{m \geq 1} P(B_k \cap A_m)$$

$$= (1/2) \sum_{m \geq 1} (P(B_k \cap A_m | X_1 = 1) + P(B_k \cap A_m | X_1 = -1)).$$

But $P(B_k \cap A_m | X_1 = 1) = f_{2m} P(\text{Positive } 2k - 2m \text{ times in } (2m, 2n) | S_{2m} = 0)$ which, by the inductive assumption and time homogeneity, is $f_{2m} u_{2k-2m} u_{2n-2k}$. Similarly $P(B_k \cap A_m | X_1 = -1) = f_{2m} u_{2k} u_{2n-2m-2k}$. Here f_{2m} is as in Theorem 7.9, which also shows that $\sum f_{2m} u_{2N-2m} = u_{2N}$. Thus $\beta(2k, 2n) = (1/2)(u_{2n-2k} u_{2k} + u_{2k} u_{2n-2k}) = u_{2k} u_{2n-2k}$ as claimed. $\square$

Given $n > 0$, let T be a random variable with $P(T = 2k) = P(T = 2n - 2k) = \alpha(2k, 2n) = u_{2k} u_{2n-2k}$ for $0 \leq k \leq n$. This corollary shows that T has the distribution both of the time of last return to zero up to time $2n$, and of the number of time periods up to $2n$ that the random walk is positive. Symmetry implies that $E(T) = n$, and, for many people, this fact and general intuition suggest that T is likely to be close to n. However, values near n are *much less likely* than values near the extremes of 0 or $2n$. The exercises ask you to verify this, and we give numerical calculations for $2n = 24$ as an illustration.

$2k$	0, 24	2, 22	4, 20	6, 18	8, 16	10, 14	12
$P(T = 2k)$	0.1612	0.0841	0.0661	0.0580	0.0537	0.0515	0.0509

EXERCISES

7.9. Bets on Red in a Las Vegas casino win or lose the amount of the stake with respective probabilities 18/38 and 20/38. Your initial capital is 100 units, you quit when ruined or if you reach 400 units. Compare your chances of ruin if you (a) bet one unit every play; (b) bet 100 units every play; (c) bet your fortune every play. Solve the same problem for a UK casino, taking the respective chances as 73/148 and 75/148 (which is a reasonably accurate description).

7.10. An infinitely rich casino returns six times your stake (hence a profit of five times your stake) if your bet on the numbers one to six wins, otherwise you lose your stake. Let $x = P(\text{You win})$, and let p_a be your probability of ruin, starting with amount a, betting unit amount. Show that

$$p_a = xp_{a+5} + (1 - x)p_{a-1} \quad \text{if } a \geq 1.$$

What condition on x makes the game favourable to you? Use this equation, or the branching process analogy in the text, to find p_a when (a) $x = 1/7$; (b) $x = 1/6$; (c) $x = 1/5$.

7.11. In the simple random walk, find $E(S_n)$ and $\text{Var}(S_n)$. For $p - 0.6$, use the Central Limit Theorem to estimate the probability that the random walk is more than ten positive steps from its start point after 100 steps.

7.12. Verify the assertion made in the text, that for fixed n, the sequence $\alpha(2k, 2n)$ is decreasing for $0 \leq k \leq n/2$. Show that, if T_n has the distribution $\alpha(2k, 2n)$, then $P(T_n/(2n) \leq x) \to \frac{2}{\pi} \arcsin(\sqrt{x})$ (i.e. that $T_n/(2n) \overset{D}{\to} B(1/2, 1/2)$, the Beta distribution.) (Stirling's formula for $m!$ may be useful.)

7.13. In Example 7.13 with $N = 30$, how many 50p coins should be preloaded into the machine to give at least a 95% chance that all customers obtain change, in the cases (a) $b = 15$ and (b) $b = 10$.

7.3 Markov Chains

Definition 7.17

Let $S = \{s_1, s_2, \ldots\}$ be a set of *states*, and $(X_n : n = 0, 1, \ldots)$ be random variables such that $P(X_n \in S) - 1$ for all n. Suppose also that, for any n, times $m_1 < m_2 < \cdots < m_n$, and states $s_1, s_2, \ldots, s_n$,

$$P(X_{m_n} = s_n | X_{m_1} = s_1, \ldots, X_{m_{n-1}} = s_{n-1}) = P(X_{m_n} = s_n | X_{m_{n-1}} = s_{n-1})$$

(the *Markov property*). Then $((X_n), S)$ is a *Markov chain*.

Informally, the values of the stochastic process (X_n) always belong to S and, given its values at times in the past, all but the most recent information can be disregarded for predicting a future value.

The sizes of generations in branching processes, and the positions in a random walk, are examples of Markov chains. The initial example by A.A. Markov was a two-state chain: he reduced Pushkin's tone poem Eugene Onegin to its series of vowels and consonants (e.g. this paragraph "The sizes of" would become ccvcvcvcvc). He then observed that, wishing to predict v or c for the next letter, it was (almost) enough to know only the current letter, previous information could be discarded.

Every collection of independent random variables taking values in a given set is trivially a Markov chain, and Markov chains can be thought of as one step more complicated than an independent sequence. Their applications are diverse – weather patterns, progression in hierarchies within a firm, spread of diseases, queues, and many more. Their usefulness is a consequence of them being sufficiently complex to be a reasonable model of many real phenomena, but also sufficiently simple to permit considerable mathematical analysis.

We will restrict our attention to time-homogeneous chains, i.e. where $P(X_{m+n} = j|X_m = i)$ depends only on the states i and j and the time difference n, and not on the value of m.

Notation: Write $P(X_n = j|X_0 = i) = p_{ij}^{(n)}$. We write p_{ij} instead of $p_{ij}^{(1)}$, and refer to it as the (one-step) *transition probability*. The matrix P whose entries are (p_{ij}) is the *transition matrix*. In such a matrix, $p_{ij} \geq 0$ for all i, j, and $\sum_j p_{ij} = 1$ for each i. The matrix $P^{(n)} = (p_{ij}^{(n)})$, with $P^{(0)}$ taken as the identity matrix I, is the n-step transition matrix, and we write $p_i^{(n)} = P(X_n = i)$, so that the vector of probabilities $\mathbf{p^{(n)}} = (p_1^{(n)}, p_2^{(n)}, \ldots)$ gives the distribution of X_n.

Example 7.18

With states $\{1, 2, 3\}$, let $P = \begin{pmatrix} 1/3 & 1/3 & 1/3 \\ 1/2 & 0 & 1/2 \\ 2/3 & 0 & 1/3 \end{pmatrix}$ be the transition matrix.

To interpret it, look along each row in turn: if now in state 1, we are equally likely to be in any one of the three states next time; if now in 2, we move to 1 or 3 with equal probabilities; and if now in 3, we remain there with probability $1/3$, otherwise move to 1.

Our focus is on computing $p_{ij}^{(n)}$, given P, and describing long-run behaviour. The fundamental tool is

Theorem 7.19 (Chapman–Kolmogorov)

$p_{ij}^{(m+n)} = \sum_k p_{ik}^{(m)} p_{kj}^{(n)}$ and so $P^{(n)} = P^n$, $\mathbf{p}^{(m+n)} = \mathbf{p}^{(m)} P^n$.

Proof

$p_{ij}^{(m+n)} = P(X_{m+n} = j|X_0 = i) = \sum_k P(X_{m+n} = j \cap X_m = k|X_0 = i)$. But the term in the summation is equal to $P(X_{m+n} = j|X_m = k \cap X_0 = i)P(X_m =$

$k|X_0 = i)$, and the Markov property shows that $P(X_{m+n} = j|X_m = k \cap X_0 = i) = P(X_{m+n} = j|X_m = k)$, proving the first equation.

The fact that $P^{(n)} = P^n$ is a simple induction. And

$$P(X_{m+n} = j) = \sum_i P(X_{m+n} = j|X_m = i)P(X_m = i)$$

$$= \sum_i p_i^{(m)} p_{ij}^{(n)} = (\mathbf{p}^{(m)} P^{(n)})_j$$

so $\mathbf{p}^{(m+n)} = \mathbf{p}^{(m)} P^n$ follows. □

Thus knowledge of P and the distribution of the chain at any given time lead to the distribution at any future time. If we are now in state i, we shall remain there for a random time having the Geometric distribution $G_1(1 - p_{ii})$; on leaving that state, we jump to state $j \neq i$ with probability $p_{ij}/(1 - p_{ii})$. This holds, independently, for each visit to a state.

Definition 7.20

Suppose $\mathbf{w} = (w_i)$, where $w_i \geq 0$ and $\sum_i w_i = 1$. Then if $\mathbf{w}P = \mathbf{w}$, $\mathbf{w}$ is said to be a *stationary* or an *equilibrium* distribution for P.

To justify this terminology, suppose $\mathbf{w}$ satisfies this definition. Then $\mathbf{w}P^n = (\mathbf{w}P)P^{n-1} = \mathbf{w}P^{n-1}$ and so, by induction, $\mathbf{w}P^n = \mathbf{w}$ for all $n \geq 0$. Hence, if $\mathbf{p}^{(m)} = \mathbf{w}$ for some m,

$$\mathbf{p}^{(m+n)} = \mathbf{p}^{(m)} P^n = \mathbf{w}P^n = \mathbf{w},$$

i.e. if the distribution of the state is ever given by $\mathbf{w}$, it remains fixed at that distribution thereafter.

Markov's original example had transition matrix

$$P = \begin{array}{c} \text{Vowel} \\ \text{Consonant} \end{array} \begin{pmatrix} 0.128 & 0.872 \\ 0.663 & 0.337 \end{pmatrix}.$$

The successive powers P^2, P^3, P^4 are

$$\begin{pmatrix} 0.595 & 0.405 \\ 0.308 & 0.692 \end{pmatrix}, \begin{pmatrix} 0.345 & 0.655 \\ 0.498 & 0.502 \end{pmatrix}, \begin{pmatrix} 0.478 & 0.522 \\ 0.397 & 0.603 \end{pmatrix}$$

so that, for example, $P(X_4 = \text{Vowel}|X_0 = \text{Consonant}) = p_{21}^{(4)} = 0.397$. The sequence (P^n) converges to $\begin{pmatrix} 0.432 & 0.568 \\ 0.432 & 0.568 \end{pmatrix}$ as $n \to \infty$, so, whatever the initial state, the long-run chance a letter is a vowel is 0.432. You should now verify that $(0.432, 0.568)$ is a stationary distribution.

No-one will be surprised that, for large n, the initial state is irrelevant here. This behaviour is typical of a large number of Markov chains: the sequence $(p_{ij}^{(n)}) \to \pi_j$ as $n \to \infty$, and $\pi = (\pi_j)$ satisfies $\pi P = \pi$. The quantity π_j is then the long-run probability of being in state j. The conditions under which $p_{ij}^{(n)} \to \pi_j$ are examined later. In the next example, we assume this limiting behaviour holds, and see how to apply it.

Example 7.21 (Parrondo's Paradox)

If every individual bet on the roulette wheel of a casino is unfavourable, it is not possible to construct a combination of bets that is favourable. It might appear to follow from this that if a number of separate games are all unfavourable, there is no way to combine them to give a favourable game. This example, due to Juan Parrondo and explained by Harmer and Abbott (1999) shows that such a conclusion would be incorrect.

We look at coin-tossing games. You win one unit if Heads occur, lose one unit otherwise, tosses are independent. In Game A, you use coin α for which $P(H) = 1/2 - \epsilon$ every toss. Plainly, game A is fair when $\epsilon = 0$, and unfavourable when $\epsilon > 0$.

In game B, you use either coin β, for which $P(H) = 0.1 - \epsilon$, or coin γ for which $P(H) = 0.75 - \epsilon$. The decision about which coin to use depends on your current fortune F: when $F \equiv 0 \mod 3$, i.e. $F = 3n$ for some integer n, use coin β, otherwise use γ. In this game, the successive values of F form a 3-state Markov chain. Writing $u = 0.1 - \epsilon$, $v = 0.75 - \epsilon$, the transition matrix is

$$P = \begin{pmatrix} 0 & u & 1-u \\ 1-v & 0 & v \\ v & 1-v & 0 \end{pmatrix}.$$

When $\epsilon = 0$, the solution of $\pi P = \pi$ is $\pi = \frac{1}{13}(5, 2, 6)$, and the long-run winning chance on a play is $\pi_0 \times 0.1 + (1 - \pi_0) \times 0.75$ which evaluates as 0.5. This game is fair when $\epsilon = 0$, and is plainly unfavourable when $\epsilon > 0$. (The winning chance is approximately $0.5 - 0.87\epsilon$ when ϵ is small.)

In Game C, Games A and B are played randomly: whichever is played now, either is equally likely to be played next time. This can be described via a Markov chain with the same three states as before, but different transition probabilities. The new transition matrix has the same format as before, but since Games A and B are equally likely to be played, the new values are $u = 0.3 - \epsilon$ and $v = 0.625 - \epsilon$. You should confirm that the stationary vector in the case $\epsilon = 0$ is $\pi = \frac{1}{709}(245, 180, 284)$, and then the long-run chance of winning on a play is $\pi_0 \times 0.3 + (1 - \pi_0) \times 0.625 = 727/1418 \approx 0.513$; *it exceeds 50%!* Although A and B are both fair games, and C is a random mixture of them, C favours you. Plainly, by taking $\epsilon > 0$ small, we can ensure both A and B are strictly disadvantageous, while C remains favourable.

Why has this paradox arisen? When $\epsilon = 0$, the fairness of B depends on a delicate balance in using the good coin γ 8 times to every 5 times the bad coin β is used. The proportion of times that $F \equiv 0 \mod 3$ is $5/13 \approx 0.385$. In C, $F \equiv 0 \mod 3$ a proportion $245/709 \approx 0.346$ of the time, a drop of some four percentage points. When you use the fair coin α, the value of F is immaterial, but this reduction in the overall frequency that $F \equiv 0 \mod 3$ feeds through into a reduction in the frequency that β is used, and an increase in the use of γ, for which $P(H) = 0.75$.

Definition 7.22

If, for some $n \geq 0$, $p_{ij}^{(n)} > 0$, we say that j is *accessible* from i, and write $i \rightsquigarrow j$. If $i \rightsquigarrow j$ and $j \rightsquigarrow i$, we write $i \leftrightsquigarrow j$, and say that i, j *communicate*.

Theorem 7.23

"Communicates" is an equivalence relation.
(Reminder: it is thus *reflexive, symmetric* and *transitive*).

Proof

Since $p_{ii}^{(0)} = 1 > 0$, so $i \leftrightsquigarrow i$. If $i \leftrightsquigarrow j$, there are m, n such that $p_{ij}^{(m)} > 0$ and $p_{ji}^{(n)} > 0$, hence $j \leftrightsquigarrow i$. If also $j \leftrightsquigarrow k$, suppose $p_{jk}^{(r)} > 0$ and $p_{kj}^{(s)} > 0$. The Chapman–Kolmogorov theorem shows that $p_{ik}^{(m+r)} \geq p_{ij}^{(m)} p_{jk}^{(r)} > 0$, and also that $p_{ki}^{(s+n)} \geq p_{kj}^{(s)} p_{ji}^{(n)} > 0$, so $i \leftrightsquigarrow k$. $\square$

This equivalence relation splits the states of the chain into disjoint *equivalence classes*. Each class consists of all states that communicate with each other. We say that a class is *closed* if, once entered, it cannot be left. If all states communicate, so that there is just one class, the chain is said to be *irreducible*, otherwise it is *reducible*. The importance of the idea of a class was beautifully put by Kai Lai Chung (1960): "The class is a closely knit set of states among which there is great solidarity."

Definition 7.24

Write $f_{ij}^{(1)} = P(X_1 = j | X_0 = i)$ and, for $n \geq 2$, write $f_{ij}^{(n)} = P(X_1 \neq j, \ldots, X_{n-1} \neq j, X_n = j | X_0 = i)$ as the probability that, starting in state

i, we visit j for the first time at time n, and $f_{ij} = \sum_n f_{ij}^{(n)}$ as the probability we ever visit j, starting in i. If $f_{ii} < 1$, we say that i is *transient*, while if $f_{ii} = 1$, then i is *recurrent*.

The definitions of transience and recurrence lead to another characterisation of them. For any state i, if we return to it, future behaviour is independent of the path before the return, so the chance we return to i at least N times is f_{ii}^N. If i is recurrent, this is unity for all N so, just as with symmetric random walks, we are *certain* to return to i infinitely often. But when i is transient, $f_{ii}^N \to 0$, so returning to i infinitely often has probability zero.

Theorem 7.25

i is transient $\iff \sum_n p_{ii}^{(n)}$ converges; so i is recurrent $\iff \sum_n p_{ii}^{(n)}$ diverges. If i and j are in the same class, they are recurrent or transient together.

Proof

We use the same argument as in Theorem 7.9, taking $P_{ii}(z) = \sum\limits_{n=0}^{\infty} p_{ii}^{(n)} z^n$ and $F_{ii}(z) = \sum\limits_{n=1}^{\infty} f_{ii}^{(n)} z^n$, to see that $P_{ii}(z) = 1/(1 - F_{ii}(z))$. Let $z \uparrow 1$, so that $F_{ii}(z) \to F_{ii}(1) = f_{ii}$ and $P_{ii}(z) \to \sum_n p_{ii}^{(n)}$. If i is transient, then $f_{ii} < 1$, so $\sum_n p_{ii}^{(n)}$ converges, while if i is recurrent, then $f_{ii} = 1$, and $\sum_n p_{ii}^{(n)}$ diverges.

If i and j are in the same class, suppose $p_{ij}^{(m)} > 0$ and $p_{ji}^{(n)} > 0$. Then $p_{ii}^{(m+r+n)} \geq p_{ij}^{(m)} p_{jj}^{(r)} p_{ji}^{(n)}$ so that, if $\sum_r p_{jj}^{(r)}$ diverges, so does $\sum_k p_{ii}^{(k)}$, and vice versa. The assertion about class solidarity follows. $\square$

Recall that, in the simple random walk, return to the origin is possible only after an even number of steps. Thus the sequence $(p_{ii}^{(n)})$ is alternately zero and non-zero; if this occurs in a general Markov chain, we might find the non-zero terms converging to $x > 0$, but the series as a whole does not converge. To help deal with such situations, we make

Definition 7.26

Let d_i be the greatest common divisor of $\{n : p_{ii}^{(n)} > 0\}$. If $d_i = 1$, we say that i is *aperiodic*, while if $d_i > 1$, it is *periodic* with period d_i.

Thus all states in the simple random walk have period 2. The next result is another instance of class solidarity.

Theorem 7.27

If i and j are in the same class, then $d_i = d_j$.

Proof

Take m, n so that $p_{ij}^{(m)} > 0$, $p_{ji}^{(n)} > 0$, and suppose $p_{jj}^{(r)} > 0$. We have both $p_{ii}^{(m \mid r \mid n)} \geq p_{ij}^{(m)} p_{jj}^{(r)} p_{ji}^{(n)} > 0$, and also $p_{ii}^{(m+2r+n)} \geq p_{ij}^{(m)} p_{jj}^{(r)} p_{jj}^{(r)} p_{ji}^{(n)} > 0$. Thus d_i divides both $(m + r + n)$ and $(m + 2r + n)$, and hence their difference, r. This holds for all r with $p_{jj}^{(r)} > 0$, and so d_i divides d_j. By the same argument, d_j divides d_i, so $d_i = d_j$. Neat.   □

Once we know the classes, we can focus on any state within a class to determine the period. A *sufficient* condition for aperiodicity is that some diagonal entry p_{ii} is non-zero; another is that there are paths with both 2 steps and 3 steps from state i back to itself.

We will now argue that if a chain is irreducible, we can concentrate on the aperiodic case with no real loss. Let P be the transition matrix of an irreducible chain with period $d > 1$. For any state i, write $C(i) = \{j \cdot p_{ij}^{(dn)} > 0$ for some $n \geq 0\}$. Arguments similar to those in the last proof show that two sets $C(i)$ and $C(j)$ are either identical, or disjoint. The states split up into d sets $C_1, C_2, \ldots, C_d$ that are visited *in strict rotation*.

Let $Q = P^d$; the chain with transition matrix Q is now reducible with these d sets as its classes. And if our initial state is within C_r, since one step in Q is equivalent to d steps in P, returning us to C_r, each of these classes is aperiodic. Hence we can regard the original periodic chain as d aperiodic chains operating in parallel, with a new timescale.

We shall focus on irreducible, aperiodic chains.

Theorem 7.28 (Erdös–Feller–Pollard Theorem)

For all states i and j in an irreducible, aperiodic Markov chain,

(1) if the chain is transient, $p_{ij}^{(n)} \to 0$;

(2) if the chain is recurrent, $p_{ij}^{(n)} \to \pi_j$, where either

  (a) every $\pi_j = 0$, or

  (b) every $\pi_j > 0$, $\sum \pi_j = 1$, and $\boldsymbol{\pi}$ is the unique probability vector that solves $\boldsymbol{\pi} P = \boldsymbol{\pi}$.

(3) In case (2), let T_i denote the time to return to i. Then $\mu_i = E(T_i) = 1/\pi_i$, with the interpretation that $\mu_i = \infty$ if $\pi_i = 0$.

Proof

(1) Using obvious notation in parallel to that of Theorem 7.25, it is easy to show that $P_{ij}(z) = F_{ij}(z)P_{jj}(z)$ when $i \neq j$. Thus

$$\sum_n p_{ij}^{(n)} = P_{ij}(1) = F_{ij}(1)P_{jj}(1) = f_{ij}\sum_n p_{jj}^{(n)}.$$

In a transient chain, $\sum p_{jj}^{(n)}$ converges, so $\sum p_{ij}^{(n)}$ also converges, hence $p_{ij}^{(n)} \to 0$.

(2) Suppose first that $\boldsymbol{\pi}P = \boldsymbol{\pi}$, and that $\pi_j = 0$ for some state j. Then

$$0 = \pi_j = (\boldsymbol{\pi}P)_j = (\boldsymbol{\pi}P^n)_j = \sum_i \pi_i p_{ij}^{(n)} \geq \pi_i p_{ij}^{(n)}$$

for all i, n. Since $i \leadsto j$, there is some n with $p_{ij}^{(n)} > 0$, hence $\pi_i = 0$ holds for all i. Hence, for any stationary vector, either all components are zero, or none of them are. Proofs of the other assertions here, and of (3), are given on pages 214–6 of Grimmett and Stirzaker (1992).

$\square$

Call the chain *null recurrent* if (2)(a) applies, or *positive recurrent* under (2)(b). The theorem shows these are also class properties. Since the limiting vector $\boldsymbol{\pi}$ is also a stationary vector, it is not surprising that $\pi_i = 1/\mu_i$. For, suppose the current distribution of the state is given by $\boldsymbol{\pi}$, and consider a future period of length T. During this time, the probability of being in state i remains at π_i, so we expect to visit state i about $T\pi_i$ times – which corresponds to a mean time between visits to i of $T/(T\pi_i) = 1/\pi_i$.

We can now describe the long-term behaviour of an irreducible aperiodic chain. Since $\mathbf{p}^{(n)} = \mathbf{p}^{(0)}P^n$,

(i) in a transient or null recurrent chain, $\mathbf{p}^{(n)} \to \mathbf{0}$, i.e. $P(X_n = j) \to 0$ for any given j;

(ii) in a positive recurrent chain, $\mathbf{p}^{(n)} \to \boldsymbol{\pi} > \mathbf{0}$, i.e. $P(X_n = j) \to \pi_j > 0$ for all j, where $\boldsymbol{\pi}$ is the unique probability vector solving $\boldsymbol{\pi}P = \boldsymbol{\pi}$.

The main difference between transient and null chains is in the frequency of visits to a given state i: in a transient chain, we might never make a visit, and we cannot return infinitely often, while in a null chain we are certain to make infinitely many visits (although the mean time between visits is infinite).

The next result gives criteria, whose proofs can be found on pages 208–12 of Grimmett and Stirzaker (1992), that enable us to decide between the three alternatives.

Theorem 7.29 (Foster's Criteria)

Let P be the transition matrix of an irreducible aperiodic Markov chain with state space $S = \{0, 1, 2, \ldots\}$, and let Q be P, except that the row and column corresponding to state 0 are deleted. The chain is

(1) transient $\Leftrightarrow$ the system $Q\mathbf{y} = \mathbf{y}$ has a bounded non-zero solution;

(2) positive recurrent $\Leftrightarrow$ the system $\boldsymbol{\pi}P = \boldsymbol{\pi}$ has a solution $\boldsymbol{\pi}$ that is a probability vector.

If the chain is recurrent, there is a strictly positive solution of $\mathbf{x}P = \mathbf{x}$, unique up to a multiplicative constant. This chain is positive or null according as $\sum x_i$ is finite or infinite respectively.

Note the difference between (1), where we are looking at a column vector $\mathbf{y}$, and (2), where we look at a row vector $\boldsymbol{\pi}$. We give three examples here; others are found in the exercises, and in applications in the next chapter.

The first example does not need much of this heavy machinery. Recall Example 7.18, a three-state chain that is plainly irreducible and aperiodic. The system $\boldsymbol{\pi}P = \boldsymbol{\pi}$ reads

$$
\begin{aligned}
\pi_1/3 + \pi_2/2 + 2\pi_3/3 &= \pi_1 \\
\pi_1/3 &= \pi_2 \\
\pi_1/3 + \pi_2/2 + \pi_3/3 &= \pi_3
\end{aligned}
$$

which is easily solved, giving $\boldsymbol{\pi} = \frac{1}{25}(12, 4, 9)$ as a probability vector. This chain is positive recurrent, and, in the long run, spends time in the three states in the ratios $12 : 4 : 9$.

Example 7.30

Suppose $S = \{0, 1, 2, \ldots\}$, and $p_{ij} = 1/(i+2)$ for $j = 0, 1, \ldots, i+1$, and all $i \geq 0$. To show the chain is irreducible, we show that whatever i and j are given, then $i \rightsquigarrow j$. First, if $i \geq j$, then already $p_{ij} > 0$. And if $i < j$, a possible path from i to j is via $i+1, i+2, \ldots, j-1$. Thus the chain is irreducible, and since $p_{00} > 0$, it is obviously aperiodic.

To use Foster's criteria, we look at possible solutions of $\boldsymbol{\pi}P = \boldsymbol{\pi}$. That system reads

$$
\begin{aligned}
\pi_0/2 + \pi_1/3 + \pi_2/4 + \pi_3/5 + \cdots &= \pi_0 \\
\pi_0/2 + \pi_1/3 + \pi_2/4 + \pi_3/5 + \cdots &= \pi_1 \\
\pi_1/3 + \pi_2/4 + \pi_3/5 + \cdots &= \pi_2 \\
\pi_2/4 + \pi_3/5 + \cdots &= \pi_3 \text{ etc.}
\end{aligned}
$$

Taking the differences of consecutive equations, $0 = \pi_0 - \pi_1$, $\pi_0/2 = \pi_1 - \pi_2$, $\pi_1/3 = \pi_2 - \pi_3$ etc., from which we find $\pi_1 = \pi_0$, $\pi_2 = \pi_0/2$, $\pi_3 = \pi_0/6$. It is easy to guess that $\pi_i = \pi_0/i!$, and then to verify this guess by induction. Choosing $\pi_0 = e^{-1}$ makes $\sum \pi_i = 1$, so there is a stationary probability vector, and the chain is positive recurrent. In the long run, $P(X_n = j) \to e^{-1}/j!$, the Poiss(1) distribution.

Example 7.31 (Modified Random Walk)

Consider a simple random walk with an *impenetrable barrier* at zero, i.e. state space $\{0, 1, 2, \ldots\}$, generally with $p_{i,i+1} = p$ and $p_{i,i-1} = q$, except that $p_{00} = q$. This is plainly irreducible and aperiodic when $0 < p < 1$. The system $Qy = y$ gives $py_2 = y_1$ and, for $n \geq 1$, we have $qy_n + py_{n+2} = y_{n+1}$. By induction, we see that $y_n = py_1(1 - (q/p)^n)/(p-q)$ if $p \neq q$, or $y_n = ny_1$ if $p = q$. This can be non-zero and bounded if, and only if, $p > q$: the condition for transience. The system $\pi P = \pi$ reads $(\pi_0 + \pi_1)q = \pi_0$, $p\pi_{n-1} + q\pi_{n+1} = \pi_n$ for $n \geq 1$, with unique solution $\pi_n = \pi_0(p/q)^n$. We have π a probability vector if, and only if, $p < q$: the condition for positive recurrence. When $p = q$, $\sum \pi_n = \infty$ if $\pi_0 \neq 0$, and the chain is null recurrent.

In summary, the chain is transient, and $X_n \to +\infty$ when $p > 1/2$; positive recurrent, and $P(X_n = k) \to p^k/(q^{k-1}(q-p))$ when $p < 1/2$; and null recurrent when $p = 1/2$. These broad conclusions will accord with your intuition, if you consider what happens, on average, at each step.

To estimate how often a transient state is ever visited, or a recurrent state is visited up to time T, we can use indicator functions. Define $I_{jn} = 1$ if $X_n = j$, and $I_{jn} = 0$ otherwise, so that $Y(j) = \sum\limits_{n=0}^{T} I_{jn}$ is the total number of times state j is visited up to time T. Since

$$P(I_{jn} = 1) = \sum_i P(X_n = j|X_0 = i)P(X_0 = i) = \sum_i P(X_0 = i)p_{ij}^{(n)},$$

we have

$$E(Y(j)) = \sum_{n=0}^{T} P(I_{jn} = 1) = \sum_{n=0}^{T}\sum_i P(X_0 = i)p_{ij}^{(n)} = \sum_i P(X_0 = i)\sum_{n=0}^{T} p_{ij}^{(n)}.$$

For example, if we definitely start in state i, so that $P(X_0 = i) = 1$ and $P(X_0 = k) = 0$ for $k \neq i$, the mean number of visits to j up to time T is

$$\sum_{n=0}^{T} p_{ij}^{(n)} = \left(\sum_{n=0}^{T} P^n\right)_{ij}.$$

Take Example 7.11, the gambler's ruin problem. As a Markov chain, the states are $\{0, 1, \ldots, c\}$, and the non-zero transition probabilities are $p_{00} = p_{cc} =$

1, with $p_{i,i-1} = q$ and $p_{i,i+1} = p$ for $1 \le i \le c-1$. There are three classes: $\{0\}$, $\{c\}$, both recurrent, and $\{1, 2, \ldots, c-1\}$, plainly transient.

It is convenient to write the transition matrix P with the states in a non-standard order: let

$$P = \begin{array}{c} 1 \\ 2 \\ 3 \\ \vdots \\ c-1 \\ 0 \\ c \end{array} \left(\begin{array}{cccccc} 0 & p & 0 & \cdots & q & 0 \\ q & 0 & p & \cdots & 0 & 0 \\ 0 & q & 0 & \cdots & 0 & 0 \\ \vdots & \vdots & \vdots & \cdots & \vdots & \vdots \\ 0 & 0 & 0 & \cdots & 0 & p \\ 0 & 0 & 0 & \cdots & 1 & 0 \\ 0 & 0 & 0 & \cdots & 0 & 1 \end{array} \right) = \left(\begin{array}{cc} Q & A \\ 0 & I \end{array} \right)$$

where Q is a $(c-1) \times (c-1)$ matrix.

Because of the structure of P, the powers (P^n) have a similar structure, with $P^n = \left(\begin{array}{cc} Q^n & A_n \\ 0 & I \end{array} \right)$ for some $(c-1) \times 2$ matrix A_n. Hence

$$\sum_{n=0}^{T} P^n = \left(\begin{array}{cc} \sum_{n=0}^{T} Q^n & \sum_{n=0}^{T} A_n \\ 0 & (n+1)I \end{array} \right).$$

We are interested in the expression $S_T = \sum_{n=0}^{T} Q^n$. Now $(I-Q)S_T = \sum_{n=0}^{T} Q^n - \sum_{n=0}^{T} Q^{n+1} = I - Q^{T+1}$. But $Q^n \to 0$ as $n \to \infty$, since the states $\{1, 2, \ldots, c-1\}$ are transient, so $(I-Q)S_T \to I$, and hence $S_T \to (I-Q)^{-1}$. This shows that in the gambler's ruin problem, the mean total number of visits to state j, starting from state i, is $((I-Q)^{-1})_{ij}$, if $1 \le i, j \le c-1$.

For example, if $c = 4$ then $Q = \left(\begin{array}{ccc} 0 & p & 0 \\ q & 0 & p \\ 0 & q & 0 \end{array} \right)$, and so $(I-Q)^{-1} = $

$\frac{1}{p^2+q^2} \left(\begin{array}{ccc} 1-pq & p & p^2 \\ q & 1 & p \\ q^2 & q & 1-pq \end{array} \right)$. Starting with unit amount, the mean number of times over the whole game that we possess exactly three units is the $(1, 3)$ entry $p^2/(p^2 + q^2)$.

In this fashion, we can find the mean total number of visits to any transient state of a general Markov chain. For a recurrent state, this mean number is either zero (if we cannot reach it from our start point), otherwise infinite.

A systematic way to assess the long-term behaviour of a Markov chain with transition matrix P, i.e. the fate of X_n, conditional on $X_0 = i$, might proceed as follows.

(1) Find the classes and establish which, if any, are closed. Find the period of each closed class. If the chain has no closed classes, all states are transient and $P(X_n = j) \to 0$ for all j.

(2) For each closed aperiodic class C, use Theorem 7.29 to determine whether it is transient, null, or positive. In the first two cases, $P(X_n = j) \to 0$ for all $j \in C$, otherwise $P(X_n = j | C$ ever entered$) \to \pi_j > 0$ for all $j \in C$.

(3) For each closed class C of period $d > 1$, let P_0 be that part of P that describes transitions among the states of C alone, and let $Q = P_0^d$. For the transition matrix Q, C splits into d aperiodic subclasses, each to be treated as in (2). Ascertain the order in which these subclasses are visited.

(4) Denote the closed classes by $C_1, C_2, \ldots$, and write $R = $ Rest of the states. Write $x_{ij} = P($Eventually enter $C_j | X_0 = i)$. Considering one step,

$$x_{ij} = \sum_{k \in R} p_{ik} x_{kj} + \sum_{k \in C_j} p_{ik}, \quad i \in R, j = 1, 2, \ldots$$

from which the (x_{ij}) are to be found, and then use (3), (2).

Example 7.32

Consider the eight-state chain, whose non-zero transition probabilities are $p_{11} = p_{12} = p_{13} = p_{21} = p_{24} = p_{27} = p_{63} = p_{77} = 1/3$; $p_{35} = p_{45} = p_{56} = 1$; $p_{87} = p_{88} = 1/2$; and $p_{78} = p_{64} = 2/3$. The classes are $A = \{1, 2\}$ (aperiodic, transient); $B = \{3, 4, 5, 6\}$ (recurrent, period 3); and $C = \{7, 8\}$ (recurrent, aperiodic, with stationary vector $(3/7, 4/7)$). Class B splits into subclasses $\{3, 4\}$, $\{5\}$ and $\{6\}$, visited in that order; conditional on being in $\{3, 4\}$, state 4 is twice as likely.

Starting in any of the states in B or C, future behaviour is clear, as both these classes are closed. Write x_{1B} and x_{2B} as the probabilities that, starting in state 1 or 2 respectively, the chain ever enters class B. By considering the first transition, as in step (4) of the description above, we have

$$x_{1B} = \frac{1}{3}x_{1B} + \frac{1}{3}x_{2B} + \frac{1}{3}, \quad x_{2B} = \frac{1}{3}x_{1B} + \frac{1}{3}.$$

These solve to give $x_{1B} = 4/5$ and $x_{2B} = 3/5$ as the chances of entering, and thus remaining in, class B. Plainly, the respective chances of entering C are $1/5$ and $2/5$. Hence, for example, starting in state 1, the long-term chance of being in state 7 is $(1/5) \times (3/7) = 3/35$, and the chance of being in state 8 is $4/35$. Because of the rotation among the sub-classes in B, the mean long-term chance of being in state 3 is $(4/5) \times (1/3) \times (1/3) = 4/45$; correspondingly, the chances for the other states are $8/45$, $4/15$ and $4/15$. The calculations for when we start in state 2 are similar.

EXERCISES

7.14. Given $P = \begin{pmatrix} 1/2 & 1/2 \\ 1/3 & 2/3 \end{pmatrix}$, find P^2, the values of $p_{ii}^{(3)}$, and the unique stationary probability vector. Repeat for $P = \begin{pmatrix} 0 & 1/2 & 1/2 \\ 1 & 0 & 0 \\ 1/3 & 1/3 & 1/3 \end{pmatrix}$.

7.15. Construct a transition matrix of a twelve-state Markov chain which contains a transient class with four states having period three, accessible from an aperiodic transient class; a recurrent class of period two not accessible from any other class; and an aperiodic recurrent class with three states.

7.16. A soccer player never gets two Yellow Cards (YCs) in the same match, and gets one YC with probability p, independently of other matches. When he accumulates two YCs, he is suspended for one match, but then his YC count reverts to zero. Thus, at the beginning of any match, he is in one of three states, $0, 1, 2$ according to the current number of YCs. Set this up as a Markov chain, and find the long-run proportion of matches he misses through suspension.

The manager always selects him when he is available; if there are three intervening matches before the game against Brazil, find the chance he is not suspended for that match, given his current status.

7.17. Show that an irreducible chain with finitely many states must be positive recurrent.

7.18. To look *backwards* in time, write $q_{ij} = P(X_n = j | X_{n+1} = i)$. Show that, if the chain begins in a stationary state $\pi > 0$, then $q_{ij} = \pi_j p_{ji} / \pi_i$, and that $Q = (q_{ij})$ is a transition matrix, having the same stationary vector.

7.19. Let $S = \{0, 1, \ldots\}$ be the state space, and let $p_{0j} = a_j$ where $a_j > 0$ for all j and $\sum_j a_j = 1$. For $j \geq 1$, first let $p_{jj} = \theta$ and $p_{j,j-1} = 1-\theta$ for some θ with $0 < \theta < 1$. Show that this chain is irreducible and aperiodic, and give a condition in terms of $\mu = \sum_j j a_j$ that distinguishes the null recurrent and positive recurrent cases.

Now alter the transition matrix so that, for $j \geq 1$, $p_{jj} = \theta$ and $p_{j0} = 1 - \theta$. Show that now the chain is always positive recurrent. Why does μ matter in the first case, but not the second?

7.20. Modify the simple random walk so that the steps have sizes $-1, 0$ or $+1$ with respective non-zero probabilities q, r and p, with $q + r + p =$

1. Show that this is an irreducible aperiodic Markov chain, transient when $p \neq q$, otherwise null recurrent.

7.21. A Markov chain is said to be *doubly stochastic* if the column sums, as well as the row sums, are unity. Find a stationary probability vector when the number of states is finite. Show by example that it can be null or transient when there are infinitely many states.

7.22. Set up your visit to the roulette wheel in a UK casino as a Markov chain. The state is your current fortune, your only bets are unit stakes on groups of four numbers (out of 37 equally likely), for which the payout is at $8 : 1$. Describe, using words to interpret the maths, what will happen over the short run and the long run.

7.23. A house has n rooms; there are a_{ij} two-way doors between rooms i and j, and it is possible to move from one room to any other, not necessarily directly. Goldilocks dodges from room to room, selecting any one of the $\sum_j a_{ij}$ doors in room i at random. Formulate her progress as a Markov chain. Show that, in the long run, the chance she is in any given room is proportional to the number of doors in that room, so long as the chain is aperiodic.

Baby Bear acts in the same way, independently. What is the (long-run) chance they are in the same room?

7.24. The *Ehrenfest urn model* for the diffusion of N gas molecules between two closed containers, A and B, linked by a narrow tube, supposes that, at discrete time points, one of the particles is selected at random, and switched to the other container. Let X_n be the number of molecules in A at time n. Describe the transition matrix, and find the stationary vector.

7.25. By calculating $p_{ii}^{(2n)}$ for the simple random walk, and considering the sum (over n) of these quantities, show that this random walk is transient when $p \neq 1/2$, but recurrent when $p = 1/2$. (Use Stirling's formula.) Is it null recurrent, or positive recurrent?

Look at the simple symmetric random walk in the plane, i.e. move from (x, y) to one of its four neighbours $(x-1, y)$, $(x+1, y)$, $(x, y-1)$ and $(x, y + 1)$ with equal probability. Show that the probability of being back at the start point after exactly $2n$ steps is $\binom{2n}{n}^2 / 4^{2n}$. Classify this random walk as transient, null or positive recurrent.

8

Stochastic Processes in Continuous Time

In the last chapter, we looked first at two specific models, which were later seen to be examples of discrete time Markov chains. We saw how to derive exact probabilities and limiting values for a process having the Markov property, i.e. knowing the current state, we can ignore the path to that state. Similarly, in continuous time, it is a great mathematical convenience to assume this Markov property. We begin by looking at continuous time Markov chains, with some specific applications. The general theory of such chains requires a much deeper mathematical background than is assumed for this book; the excellent texts by Chung (1960) and Norris (1997) provide fascinating reading matter, and justify the assertions we make without proof.

The following two sections illustrate the progress that can be made when making the Markov assumption would be unreasonable. We look at models of queues, and at processes where components are replaced from time to time. The final section, on Brownian motion, links two central themes, the Markov property and the Normal distribution.

8.1 Markov Chains in Continuous Time

Definition 8.1

Let $S = \{s_1, s_2, \ldots\}$ be a set of states, and $\{X_t : t \geq 0\}$ be random variables such that $P(X_t \in S) = 1$ for all $t \geq 0$. Suppose also that, for any n and times

$t_1 < t_2 < \cdots < t_n$ and states $s_1, s_2, \ldots s_n$,

$$P(X_{t_n} = s_n | X_{t_1} = s_1, \ldots, X_{t_{n-1}} = s_{n-1}) = P(X_{t_n} = s_n | X_{t_{n-1}} = s_{n-1}).$$

Then $((X_t), S)$ is a Markov chain (in continuous time).

In many applications, S will be the non-negative integers or some subset of them; we assume this holds, unless otherwise specified. As in the discrete case, we shall restrict our attention to time-homogeneous processes, and so write

$$p_{ij}(t) = P(X_t = j | X_0 = i) = P(X_{t+s} = j | X_s = i)$$

for any $s \geq 0$. The properties $0 \leq p_{ij}(t) \leq 1$ and $\sum_j p_{ij}(t) = 1$ are immediate, as is the Chapman–Kolmogorov equation

$$p_{ij}(s + t) = \sum_k p_{ik}(s) p_{kj}(t).$$

The matrix $(p_{ij}(t))$ is denoted by $P(t)$. If $\mathbf{w}$ describes the initial state distribution, then plainly $\mathbf{w}P(t)$ describes the distribution at time t. We consider only "standard" Markov chains, i.e. those in which every p_{ij} is a continuous function of t, with $P(t) \longrightarrow I$, the identity matrix, as $t \downarrow 0$. If we observe the chain at times $0, 1, 2, \ldots$ only, then $P \equiv P(1)$ can be thought of as the one-step transition matrix of a *discrete time* chain; $P^{(n)} = P^n = P(n)$. Thus all the ideas and notation from discrete time chains, such as classes, irreducibility, transience, recurrence etc., carry forward to the continuous time case. But there is one great blessing: we do not have to consider the possibility of periodicity, as our first result shows.

Theorem 8.2

For all $t \geq 0$ and all states j, $p_{jj}(t) > 0$. If $i \neq j$ and $p_{ij}(t_0) > 0$, then $p_{ij}(t) > 0$ for all $t \geq t_0$.

Proof

Let $t > 0$ be given. Since p_{jj} is continuous everywhere and $p_{jj}(0) = 1$, there is some n such that $p_{jj}(t/n) \geq 1/2$. The Chapman–Kolmogorov equation shows that $p_{jj}(t) \geq (p_{jj}(t/n))^n \geq (1/2)^n > 0$.

   Suppose $i \neq j$ and $p_{ij}(t_0) > 0$. If $t > t_0$, then $p_{ij}(t) \geq p_{ij}(t_0)p_{jj}(t - t_0) > 0$, since both terms in the product are strictly positive.   □

Questions of periodicity cannot arise, since the function $p_{jj}(t)$ is always strictly positive. The theorem also shows that, if $i \neq j$ and p_{ij} is not identically zero, then there is some $t_0 > 0$ such that, starting from state i, it is possible to be in state j at all times subsequent to t_0. Actually, a sharper result holds: if p_{ij} is not identically zero, it is strictly positive for all $t > 0$. Even if we cannot move directly from i to j, if we can get there somehow, we can get there in an arbitrarily short time.

We have assumed the functions p_{ij} are continuous. Exercise 8.1 asks you to show this continuity is uniform. It turns out that these p_{ij} are even better-behaved: they are continuously differentiable at all points $t > 0$; at $t = 0$, where their left-hand derivatives are not defined, their right-hand derivatives, denoted by $q_{ij} = p'_{ij}(0)$, also exist. Since q_{ii} is negative, it is convenient to introduce $q_i \equiv -q_{ii}$. It can be shown that every q_{ij} with $i \neq j$ is automatically finite, but if the state space S does not have finitely many states, it may happen that $q_{ii} = -\infty$. We shall not consider such models.

Rewrite the expression $\sum_j p_{ij}(t) = 1$ in the form

$$0 = \sum_j p_{ij}(t) - 1 = (p_{ii}(t) - 1) + \sum_{j \neq i} p_{ij}(t).$$

Divide both sides by t, and let $t \downarrow 0$. Then

$$0 = \lim \frac{p_{ii}(t) - 1}{t} + \lim \sum_{j \neq i} \frac{p_{ij}(t)}{t}.$$

When the sum has finitely many terms, we can always interchange the order of limit and sum, obtaining

$$0 = q_{ii} + \sum_{j \neq i} q_{ij},$$

so that $q_i = -q_{ii} = \sum_{j \neq i} q_{ij}$. Even when S is infinite, we consider only chains in which this "conservative" condition holds.

Write $Q = (q_{ij})$. This Q-matrix is a fundamental tool in finding the properties of the process. The quantity q_i is interpreted as the *rate* at which a process now in state i departs from that state in the sense that, for small t,

$$p_{ii}(t) = 1 - q_i t + o(t),$$

(see Appendix 9.3D for the notation $o(t)$), an easy consequence of the definition of a derivative. Similarly, for $i \neq j$, $p_{ij}(t) = q_{ij} t + o(t)$, and q_{ij} is the rate at which we jump to j, when we are now in i. Frequently, a model arises by specifying the values of (q_{ij}), and our task is to find the properties of $P(t)$.

For a chain currently in state i, let T denote the time it stays in that state before moving to a different one. For $s \geq 0$ and $t \geq 0$, we have

$$
\begin{aligned}
P(T \geq s + t \mid T \geq s) &= P(T \geq s + t \mid X_u = i \text{ for } 0 \leq u \leq s) \\
&= P(T \geq s + t \mid X_s = i)
\end{aligned}
$$

using the Markov property. By time homogeneity, this last expression is equal to $P(T \geq t)$, hence

$$
\begin{aligned}
P(T \geq s+t) &= P(T \geq s+t | T \geq s)P(T \geq s) \\
&= P(T \geq t)P(T \geq s),
\end{aligned}
$$

the defining property of the Exponential distribution. Thus $P(T \geq t) = \exp(-t\lambda)$ for some $\lambda > 0$. However, since then $p'_{ii}(0) = -\lambda$, we must have $\lambda = q_i$, and T has the Exponential distribution $E(q_i)$. The chain develops by remaining in its current state i for a random time having this distribution, then jumping to $j \neq i$ with probability q_{ij}/q_i. (Note the close parallel with the discrete time case, where the times we stayed in particular states had independent Geometric distributions.) In the continuous time case, if we ignore how long is spent in each state, and just look at the sequence of movements from state to state, the process we observe is called the *jump chain*.

From the Chapman–Kolmogorov equation, we have

$$
\begin{aligned}
p_{ij}(h+t) - p_{ij}(t) &= \sum_k p_{ik}(h)p_{kj}(t) - p_{ij}(t) \\
&= (p_{ii}(h) - 1)p_{ij}(t) + \sum_{k \neq i} p_{ik}(h)p_{kj}(t).
\end{aligned}
$$

Divide both sides by h, let $h \longrightarrow 0$, and assume the interchange of limit and sum is justifiable. We obtain

$$
p'_{ij}(t) = q_{ii}p_{ij}(t) + \sum_{k \neq i} q_{ik}p_{kj}(t) = \sum_k q_{ik}p_{kj}(t) \tag{8.1}
$$

the so-called Kolmogorov backward equation. ("Backward" because we look back in time to the rates of transition at time zero.) The general conditions under which this equation holds are discussed in Chung (1960); we shall assume that Equation (8.1) holds for the processes we model.

By a similar argument, the Kolmogorov forward equation

$$
p'_{ij}(t) = \sum_k p_{ik}(t)q_{kj} \tag{8.2}
$$

can be obtained. Again, its formal justification needs careful analysis. When both (8.1) and (8.2) hold, they can be combined to give the simple matrix formula

$$
P'(t) = QP(t) = P(t)Q \tag{8.3}
$$

with formal solution $P(t) = \exp(Qt)$. Although superficially attractive, this expression is as much real help in discovering $P(t)$ as the remark that, in discrete time, to find $P^{(n)}$ "all" we have to do is raise P to its nth power. But we illustrate the application of (8.3) in the simplest non-trivial case.

Example 8.3

For $S = \{1, 2\}$, suppose $Q = \begin{pmatrix} -\alpha & \alpha \\ \beta & -\beta \end{pmatrix}$, where $\alpha > 0$ and $\beta > 0$. From (8.2), we can write down

$$p'_{11}(t) = p_{11}(t)(-\alpha) + p_{12}(t)(\beta).$$

But since $p_{12}(t) = 1 - p_{11}(t)$, this equation can be written entirely in terms of $p_{11}(t)$ as

$$p'_{11}(t) = -\alpha p_{11}(t) + \beta(1 - p_{11}(t)) = \beta - (\alpha + \beta)p_{11}(t).$$

This first order differential equation is in a standard format, with solution

$$p_{11}(t) = \frac{\beta}{\alpha + \beta} + \frac{\alpha}{\alpha + \beta} \exp(-(\alpha + \beta)t)$$

when we incorporate the initial condition $p_{11}(0) = 1$.

By symmetry,

$$p_{22}(t) = \frac{\alpha}{\alpha + \beta} + \frac{\beta}{\alpha + \beta} \exp(-(\alpha + \beta)t),$$

and the other values in $P(t)$ are found using $p_{12}(t) = 1 - p_{11}(t)$, $p_{21}(t) = 1 - p_{22}(t)$.

The limiting behaviour of $P(t)$ is immediate and should be no surprise. The quantities α and β are the rates of getting out of states 1 and 2 respectively, moving to the other state, so, in the long run, the respective chances of being in 1 and 2 will be proportional to β and α. Thus $p_{i1}(t) \longrightarrow \frac{\beta}{\alpha + \beta}$ and $p_{i2}(t) \longrightarrow \frac{\alpha}{\alpha + \beta}$.

The formal solution $P(t) = \exp(Qt)$ can be expressed as

$$P(t) = I + Qt + \frac{Q^2 t^2}{2!} + \frac{Q^3 t^3}{3!} + \cdots.$$

For this 2×2 matrix, evaluating Q^n is easy, as we can write $Q = ADA^{-1}$, where $D = \begin{pmatrix} 0 & 0 \\ 0 & -\alpha - \beta \end{pmatrix}$ and $A = \begin{pmatrix} 1 & \alpha \\ 1 & -\beta \end{pmatrix}$. Thus $Q^n = AD^n A^{-1}$, and

$$P(t) = A\left(I + Dt + \frac{D^2 t^2}{2!} + \frac{D^3 t^3}{3!} + \cdots\right)A^{-1}.$$

The matrix series plainly collapses to $\begin{pmatrix} 1 & 0 \\ 0 & \exp(-(\alpha + \beta)t) \end{pmatrix}$, leading to the same answer as before.

Notice that we chose to use Equation (8.2) rather than (8.1). It is usually the case that the forward equation is more helpful.

Recalling the results of discrete time Markov chains, we might hope that in an irreducible chain, $P(t)$ converges to some limiting matrix Π, all of whose rows are the same vector π. If this is the case, we might also expect its derivative $P'(t)$ to converge to the zero matrix, in which case the first part of Equation (8.3) implies $\pi Q = 0$. Conversely, if $\pi Q = 0$, the formal expansion of $P(t) = \exp(Qt)$ as a power series shows that $\pi P(t) = \pi[I + \sum_{n \geq 1} Q^n t^n / n!] = \pi$ for all $t \geq 0$. This justifies

Definition 8.4

The probability vector π is said to be *stationary* for $P(t)$ if $\pi Q = 0$.

In the previous example, it is plain that the only probability vector such that $\pi Q = 0$ is $\pi = (\frac{\beta}{\alpha+\beta}, \frac{\alpha}{\alpha+\beta})$, the limiting distribution. In the next two examples, the state space is the positive integers, or a subset; in both cases, X_t can only increase, each state is a singleton class, and there is no non-trivial stationary vector.

Example 8.5 (Poisson Process)

Although we have seen this model already, we will look at it as a Markov chain. Let X_t be the number of emissions of radioactive particles up to time t, and $S = \{0, 1, 2, \ldots\}$. Then, assuming a constant rate of emission, $q_{i,i+1} = \lambda$ for all $i \geq 0$, and so also $q_i = -q_{ii} = \lambda$. The forward equation shows that

$$p'_{00}(t) = -\lambda p_{00}(t),$$

so $p_{00}(t) = \exp(-\lambda t)$ and, when $n \geq 1$,

$$p'_{0n}(t) = \lambda p_{0,n-1}(t) - \lambda p_{0n}(t).$$

It follows, by induction, that $p_{0n}(t) = (\lambda t)^n \exp(-\lambda t)/n!$, a familiar result.

It is not difficult to show that, if λ is allowed to vary with time, then the only change needed in the expression for $p_{0n}(t)$ is to replace λt by $\int_0^t \lambda(u)du$.

We now look at an example where the rate of change of X_t is not constant.

Example 8.6 (Pure Birth Process)

Suppose X_t is the size of some population in which births occur singly, the birth rate being λ_n when $X_t = n$, and there are no deaths. Then $q_{i,i+1} = \lambda_i$,

$q_{ii} = -\lambda_i$, all other q_{ij} are zero. For a population with initial size N, let $x_n(t) = p_{Nn}(t)$ for $n \geq N$. The forward equation gives

$$x_N'(t) = -\lambda_N x_N(t)$$

and, for $n \geq N + 1$,

$$x_n'(t) = -\lambda_n x_n(t) + \lambda_{n-1} x_{n-1}(t), \tag{8.4}$$

with initial conditions $x_N(0) = 1$, $x_n(0) = 0$ for $n \geq N + 1$.

Clearly $x_N(t) = \exp(-\lambda_N t)$, but the general expressions for x_n will be complex combinations of $\exp(-\lambda_i t)$ for $N \leq i \leq n$. The linear case $\lambda_n = n\beta$ for some $\beta > 0$ corresponds to a model in which each member of the population, independently, gives birth at rate β (the *Yule process*). For a Yule process that begins with one member, Exercise 8.7 asks you to verify the solution

$$x_n(t) = e^{-\beta t}(1 - e^{-\beta t})^{n-1} \tag{8.5}$$

for $n \geq 1$, i.e. the population size at time n has the Geometric distribution $G_1(e^{-\beta t})$.

"Birth" is just a convenient label. We might model the spread of some permanently infectious incurable disease as a pure birth process: in a population of $N + 1$ homogeneously mixing people, if n have the disease and $N + 1 - n$ do not, we could take $\lambda_n = \alpha n(N + 1 - n)$, where α depends on the infectivity and the rate of mixing. Births here occur when a new person contracts the disease. If one infected person joins a group of N who are free from the disease, the mean time for everyone to be infected is $\sum(1/\lambda_n)$, which is

$$\frac{1}{\alpha} \sum_{n=1}^{N} \frac{1}{n(N + 1 - n)} = \frac{1}{\alpha(N + 1)} \sum_{n=1}^{N} \left(\frac{1}{n} + \frac{1}{N + 1 - n} \right) = \frac{2}{\alpha(N + 1)} \sum_{n=1}^{N} \frac{1}{n}.$$

For large N, $\sum_{n=1}^{N}(1/n)$ is close to $\ln(N)$, and since $\ln(N)/(N + 1) \to 0$ as $N \to \infty$, we find that the larger the population, the sooner will everyone become infected, in this model.

To illustrate how apparently innocent assumptions can sometimes be violated, take $\lambda_n = \alpha n(n-1)/2$ in this pure birth process. This would be a natural model for a randomly mixing population that requires the interaction of exactly two individuals to produce a birth: there are $n(n - 1)/2$ pairs when the size is n, and α will relate to the probability that an interaction leads to a birth. Begin with two individuals and choose the timescale so that $\lambda_n = n(n - 1)$. Then Equation (8.4) solves to give $x_2(t) = e^{-2t}$, $x_3(t) = (e^{-2t} - e^{-6t})/2$, $x_4(t) = (3e^{-2t} - 5e^{-6t} + 2e^{-12t})/10$, and so on. The expressions are not pretty.

But, by using some ingenuity, we can show that

$$\sum_{n=2}^{\infty} x_n(t) = \sum_{r=2}^{\infty} (-1)^r (2r-1) e^{-r(r-1)t} \tag{8.6}$$

when $t > 0$. The left side is the total probability attached to the state space $S = \{2, 3, 4, \ldots\}$ at time t but, for all $t > 0$, the right side is strictly less than unity. There is positive probability that the population size is outside S, which we interpret as there being positive probability that the population size is infinite, at any finite time after the process begins!

Should we have anticipated this? In a pure birth process, the time spent in state n follows an $E(\lambda_n)$ distribution, so that $\sum_{n=2}^{\infty}(1/\lambda_n)$ is the mean time spent in S. The choice of $\lambda_n = n(n-1)$ gives a series that converges to unity, so that is the mean time to reach an infinite size. Recalling that in a continuous time Markov chain, if any phenomenon is possible, it can happen in an arbitrarily short period, we ought to expect the right side of (8.6) to be less than unity for all $t > 0$. To rescue our assumption that $\sum_j p_{ij}(t) = 1$, we can add one more state, $+\infty$, which we never leave, and take $p_{2,+\infty}(t) = x_\infty(t) = 1 - \sum_{r=2}^{\infty}(-1)^r(2r-1)e^{-r(r-1)t}$. The table shows some values.

t	0.25	0.5	0.75	1	1.5	2
$x_\infty(t)$	0.00245	0.128	0.385	0.606	0.851	0.945

In the Yule process, $\sum(1/\lambda_n) = \sum(1/(n\beta))$ is divergent, and this phenomenon does not arise.

Example 8.7 (Birth and Death Processes)

Here we model the size of a population, in which births and deaths occur singly, as a Markov chain. To avoid multiple subscripts, write $X(t)$ instead of X_t. The process is fully specified by the values of λ_n and μ_n, the rates of births and deaths when the population size is n. It makes sense to assume that $\mu_0 = 0$, and if $P_n(t) = P(X(t) = n)$, the forward equations are

$$P_0'(t) = -\lambda_0 P_0(t) + \mu_1 P_1(t)$$

with, for $n \geq 1$,

$$P_n'(t) = \lambda_{n-1} P_{n-1}(t) - (\lambda_n + \mu_n) P_n(t) + \mu_{n+1} P_{n+1}(t). \tag{8.7}$$

If $\lambda_0 = 0$, then 0 will be an absorbing state, i.e. once entered, it cannot be left. But when $\lambda_0 > 0$, the chain will usually be irreducible.

In many situations, it will be reasonable to take $\mu_n = n\mu$ for some $\mu > 0$. This corresponds to the notion that each member of the population, independently of the rest and irrespective of its age, has death probability $\mu h + o(h)$ in a small time interval of length h.

If we also take $\lambda_n = n\lambda$ for some $\lambda > 0$, we have the so-called *linear* birth and death process. Let $G(t, z) = \sum_{n=0}^{\infty} P_n(t)z^n$ be the probability generating function of this process at time t. By multiplying the nth forward equation by z^n and summing, we obtain

$$\frac{\partial G}{\partial t} = (\lambda z - \mu)(z - 1)\frac{\partial G}{\partial z}, \tag{8.8}$$

after factorisation. This partial differential equation is in a format known as Lagrange's equation, and can be solved by "the method of characteristics". The form of the solution depends on whether or not $\lambda = \mu$.

Suppose first that $\lambda = \mu$. The method of characteristics asks us to solve the system

$$\frac{dt}{1} = \frac{-dz}{\lambda(z - 1)^2} = \frac{dG}{0}$$

in the form $G =$ constant, $\lambda t = 1/(z-1)+$constant, and hence $G = f(\lambda t - \frac{1}{z-1})$ for some well-behaved function f. If there is just one member when $t = 0$, then $P_1(0) = 1$, $P_n(0) = 0$ when $n \neq 1$, and $G(0, z) = z$. Thus $z = f(1/(1-z))$, from which $f(\theta) = 1 - 1/\theta$, and so $G = 1 - 1/(\lambda t - \frac{1}{z-1})$, which can be rewritten as

$$G(t, z) = \frac{\lambda t + (1 - \lambda t)z}{\lambda t + 1 - z\lambda t}.$$

Expanding this as a power series in z and picking out the coefficients of z^n, we obtain all the values of $P_n(t)$. In particular, $P_0(t) = \lambda t/(1 + \lambda t)$. The long-term behaviour, that $P_0(t) \longrightarrow 1$, should be no surprise; as the birth and death rates are equal, the mean population size remains constant at unity, but as soon as the random fluctuations take the size to zero, it must remain there for ever. However, it would have been difficult to guess the exact form of $P_0(t)$. You should check that if we had started with a population of size $N > 1$, the pgf at time t would be $G(t, z)^N$: we can think of this population developing as the sum of N independent populations, each beginning with one member.

Exercise 8.8 asks you to show that when $\lambda \neq \mu$ and there is one member initially, the pgf is

$$\frac{\mu(1 - z) - (\mu - \lambda z)\exp(-t(\lambda - \mu))}{\lambda(1 - z) - (\mu - \lambda z)\exp(-t(\lambda - \mu))}. \tag{8.9}$$

Plainly, if $\lambda < \mu$, this expression converges to unity as $t \longrightarrow \infty$; the process inevitably dies out. But if $\lambda > \mu$, it converges to μ/λ, which we can interpret as meaning that, with this probability the process dies out, otherwise it explodes to an infinite size. This dichotomy of behaviour is the same as we found in the analysis of branching processes.

We can use Theorems 4.5 and 4.9 with Equation (8.8) to obtain the mean and variance of the population size without solving that equation. Differentiate (8.8) once and twice with respect to z, and put $z = 1$. Then

$$\frac{\partial^2 G}{\partial z \partial t}(t,1) = (\lambda - \mu)\frac{\partial G}{\partial z}(t,1)$$

and

$$\frac{\partial^3 G}{\partial z^2 \partial t}(t,1) = 2\lambda\frac{\partial G}{\partial z}(t,1) + 2(\lambda - \mu)\frac{\partial^2 G}{\partial z^2}(t,1).$$

Let $m(t) = E(X(t)) = \frac{\partial G}{\partial z}(t,1)$ and $u(t) = E(X(t)(X(t)-1)) = \frac{\partial^2 G}{\partial z^2}(t,1)$. We find $m'(t) = (\lambda - \mu)m(t)$ and $u'(t) = 2\lambda m(t) + 2(\lambda - \mu)u(t)$.

Starting with one member means that $m(0) = 1$ and $u(0) = 0$, so solving these differential equations in turn when $\lambda \neq \mu$, we obtain $m(t) = \exp((\lambda-\mu)t)$, and then $u(t) = \frac{2\lambda}{\lambda-\mu}e^{(\lambda-\mu)t}(e^{(\lambda-\mu)t} - 1)$. Thus

$$\text{Var}(X(t)) = \frac{\lambda + \mu}{\lambda - \mu}e^{(\lambda-\mu)t}(e^{(\lambda-\mu)t} - 1),$$

after simplifying. When the birth and death rates are equal, the expressions are $m(t) = 1$ and $\text{Var}(X(t)) = 2\lambda t$.

For general birth and death rates, for small values of $h > 0$ we have

$$X(t+h) = X(t) + \begin{cases} 1 & \text{with probability } \lambda_{X(t)}h + o(h); \\ 0 & \text{with probability } 1 - (\lambda_{X(t)} + \mu_{X(t)})h + o(h); \\ -1 & \text{with probability } \mu_{X(t)}h + o(h). \end{cases}$$

From the definition of a derivative, this implies

$$\frac{dm}{dt} = E(\lambda_{X(t)} - \mu_{X(t)}).$$

The linear case is $m'(t) = (\lambda - \mu)m(t)$, as we have just seen.

One way of taking account of possible competition as the population increases in size is to have $\lambda_n = n\lambda$, but $\mu_n = n^2\mu$. If $\text{Var}(X(t)$ can be neglected, then $m' = m(\lambda - m\mu)$ which, if $X(0) = 1$, solves to $m(t) = \lambda e^{\lambda t}/(\lambda + \mu(e^{\lambda t} - 1))$. As $t \longrightarrow \infty$, so $m(t) \longrightarrow \lambda/\mu$.

In the general equation (8.7), suppose $P_n(t)$ converges to some limit π_n as $t \longrightarrow \infty$, and the derivatives $P_n'(t)$ converge to zero. Then, if no death rates are zero, solving the system by recursion leads to

$$\pi_n = \frac{\lambda_0.\lambda_1.\cdots.\lambda_{n-1}}{\mu_1.\mu_2.\cdots.\mu_n}\pi_0. \tag{8.10}$$

Whether or not this corresponds to a genuine probability distribution depends entirely on whether or not the series $\sum \frac{\lambda_0 . \lambda_1 . \cdots \lambda_{n-1}}{\mu_1 . \mu_2 . \cdots \mu_n}$ converges. A convergent series means the chain is positive recurrent. If it is divergent, it could be either transient, or null recurrent. The queue $M/M/k$ in the next section illustrates these alternatives; in either the transient or the null case, the long-term probability of finding that the population size is N or fewer will be zero.

One example of a positive recurrent chain is a model for the number of dust particles in a room. "Births" here correspond to particles entering the room from outside, and it seems plausible that this should be at some constant rate α, however many particles are currently in the room, while the death rate might be taken as $\mu_n = n\mu$. Here the series converges, and the distribution of $\{\pi_n\}$ is Poiss(α/μ). Exercise 8.10 asks you to solve this "immigration–death process" in general, and verify this limit.

Example 8.8 (Spread of Epidemics)

Earlier, we modelled the spread of an incurable disease as a pure birth process. We now allow the possibility of recovery. Consider a closed, isolated, homogeneously mixing population, such as a boarding school, into which an infectious disease is introduced. We suppose that someone with the disease ceases to be infectious after a period, and that catching the disease confers lifelong future immunity. Then the N members of the population can be split into three mutually exclusive and exhaustive classes:

- $X(t)$ who are Susceptible to the disease at time t;

- $Y(t)$ who are Infectious at time t;

- $Z(t) = N - X(t) - Y(t)$ who we refer to as Removed cases – they may have recovered, or died, been isolated or, in some cases, vaccinated. They are not infectious, and cannot catch the disease.

A model of this as a Markov chain can be constructed. The states are $S = \{(x,y) : 0 \leq x, 0 \leq y, 0 \leq x + y \leq N\}$, the possible values of the pair $(X(t), Y(t))$. If the disease is transmitted by contact and no vaccination occurs, the transition rates between states will be given by

$$(x,y) \longrightarrow \begin{cases} (x-1, y+1) & \text{at rate } \beta xy \\ (x, y-1) & \text{at rate } \gamma y \end{cases}$$

for some parameters β and γ. The value of β will depend on the rate at which the population is mixing and the infectivity of the disease; γ describes the rate of recovery (or death/isolation; in each case, the effect on the future development of the disease is the same). Write $\rho = \gamma/\beta$, the *relative removal rate*, and normalise the timescale so that the general forward equation becomes

$$\frac{dP_{x,y}}{dt}(t) = (x+1)(y-1)P_{x+1,y-1}(t) - y(x+\rho)P_{x,y}(t) + \rho(y+1)P_{x,y+1}(t).$$

Ignoring the time factor, we can regard this process as a two-dimensional random walk on S. The diagrams show the possibilities for one transition, and a typical path in this random walk.

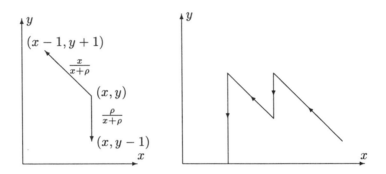

As an illustration, suppose a family of four people holiday in a remote cottage, and are joined by a fifth person with an infectious disease. They all mix homogeneously, meeting each other about ten times per day, and any encounter between an infected and a susceptible has a 5% chance of transmitting the infection; the average period of infectiousness is two days. Taking our model as appropriate, and measuring time in days, we find that $\gamma = 1/2$ and $\beta = 10 \times 5\% = 1/2$, so that $\rho = \gamma/\beta = 1$.

The initial state is $(4,1)$, and the respective chances of moving to $(4,0)$ and $(3,2)$ are $1/5$ and $4/5$. Thus, one fifth of the time the epidemic dies out with no new infections. The only way exactly one new infection occurs is when the path moves from $(3,2)$ via $(3,1)$ to $(3,0)$, and the total probability is thus $(4/5) \times (1/4) \times (1/4) = 1/20$. With such small numbers, it is plainly feasible to track all possible paths, and find their respective probabilities. Exercise 8.12 asks you to complete the calculation. This analysis is an example of using the "jump chain" to find the distribution of the number infected; it does not address the question of the duration of the epidemic. When, as here, transitions are to neighbouring states only, the phrase "embedded random walk" is used.

In general, it is possible to solve the forward equations, but the resulting expression for the general $P_{x,y}(t)$ is a complex mixture of terms. Just as we saw in the previous example, we can use the transition rates to obtain equations for the *mean* values of $X(t)$ and $Y(t)$ and deduce what happens "on average".

Let these means be denoted by $x(t)$ and $y(t)$; we find

$$\frac{dx}{dt} = -\beta xy, \qquad \frac{dy}{dt} = \beta xy - \gamma y.$$

Divide the second by the first, obtaining

$$\frac{dy}{dx} = \frac{\rho - x}{x}.$$

With the initial condition $x(0) = n$, $y(0) = \epsilon$, this has solution

$$y(t) = n - x(t) + \epsilon + \rho \ln \frac{x(t)}{n}.$$

The epidemic comes to an end when $Y(t) = 0$; and when $y = 0$, the value of x satisfies the equation

$$x = n + \epsilon + \rho \ln \frac{x}{n}.$$

Let $E = n - x$ be the mean number who catch the epidemic. Rearrange the equation for x into the form

$$1 - \frac{E}{n} = \exp\left(-\left(\frac{E + \epsilon}{\rho}\right)\right) = C \exp\left(\frac{-E}{\rho}\right),$$

where $C = \exp(-\epsilon/\rho)$. Assuming that ϵ/ρ is tiny, this is almost the same as $1 - E/n = \exp(-E/\rho)$. The nature of the solution for E here depends on the relative sizes of n and ρ. When $n > \rho$, the line $1 - E/n$ meets the two curves $\exp(-E/\rho)$ and $C \exp(-E/\rho)$ at approximately the same point, well away from zero. For example, if $n = 200$ and $\rho - 100$, the solution of $1 - E/n = \exp(-E/\rho)$ is $E \approx 160$; there has been a major epidemic.

But when $n < \rho$, the only solution of $1 - E/n = \exp(-E/\rho)$ is $E = 0$, and the solution of $1 - E/n = C \exp(-E/\rho)$ is very close to zero: the epidemic dies out before infecting much of the population. The value of ρ is called the *threshold* of the system. So far as average behaviour is concerned, so long as the size of the susceptible population is below the threshold, an epidemic will not arise, even if disease is introduced into a population. Since $\rho = \gamma/\beta$, this suggests practical steps to reduce the chance of an epidemic spreading:

- Increase γ, i.e. remove infectious individuals more quickly;

- Decrease β, i.e. reduce the rate of contact between susceptibles or, if possible, reduce the infectivity of the disease;

- Decrease n, by a process of vaccination (or isolation) that reduces the size of the susceptible population.

These are quite obvious measures: but our theory enables us to predict *how much* we can reduce the severity of an epidemic by taking them. This theory probably chimes with your experience: when some epidemic affecting children is over, the epidemic itself will have conferred immunity on many of them, and the size of the susceptible population has been reduced below the threshold. If there is no programme of vaccinating younger children, the susceptible population gradually rises until it exceeds the threshold, and a fresh epidemic then arises several years later.

This last analysis dealt with averages only. We must return to the process $(X(t), Y(t))$ to take account of random effects. Fix attention on $Y(t)$ only, the number infectious, and assume that $Y(0)$ is small. If the initial number of susceptibles is some large number x_0 then, in the early stages of an epidemic, Y behaves rather like a (linear) birth and death process with birth rate $\beta x_0 Y$ and death rate γY. If such a process begins with one member, it is certain to die out (and very likely to die out quickly) if $\gamma Y \geq \beta x_0 Y$, i.e. if $\rho \geq x_0$; it dies out with probability ρ/x_0 if $\rho < x_0$. If it does not die out, it explodes.

This suggests how the process $(X(t), Y(t))$ will progress. If $X(0) \leq \rho$, then $Y(t) \longrightarrow 0$ quite quickly, and no epidemic arises. Suppose $Y(0) > 0$ infectious individuals enter a population of size $X(0) > \rho$. Then, with probability $(\rho/X(0))^{Y(0)}$ the process dies out, but otherwise it follows near to the path described for the mean process, and we have an epidemic of size approximately E, the solution of $1 - E/X(0) = \exp(-E/\rho)$.

This can be put to the test by experiment. I used a computer to simulate 10 000 stochastic epidemics for various combinations of $X(0)$ and ρ, with one infectious individual introduced. With $X(0) = 50$ and $\rho = 100$, so that the initial population is well below the threshold, the epidemics did indeed die out quickly: two-thirds of the time, no new cases arose; 96% of the time, at most five infections occurred, and 99% ended with at most ten cases.

The intermediate case $X(0) = \rho$ is more difficult to predict. 50% of the time no new cases at all will arise, and the analogy with a birth and death process suggests that most epidemics will die out fairly quickly, although some will linger. I simulated three instances, with populations of sizes 100, 400 and 1000. Each time, six or fewer infections arose 80% of the time. If the simulations are to be believed, we could be 95% confident that the total sizes of the three epidemics would not exceed 32, 63 or 86 respectively but, in the largest population, epidemics sporadically exceeded 300.

With $X(0) = 200$ and $\rho = 100$, the simulations confirmed our speculation about a dichotomy of behaviour. At most five new cases arose in 48% of the simulations, another 2% had ten or fewer, and only 3% had between ten and 120 infections. The histogram shows the distribution of the number of cases in the remaining 47% of simulations. It has a mean of 159.4, and a standard deviation of 12.3. The mean is consistent with our estimate of some 160 cases if a genuine epidemic arises.

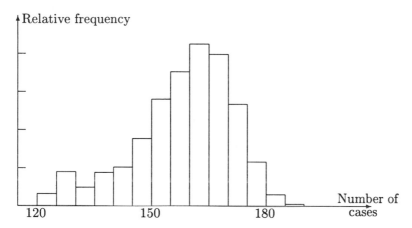

Our model plainly has weaknesses as a description of how a real epidemic may behave. The most doubtful assumptions are that the rate of losing infectivity, and the strength of infectivity, are constant during the time an individual is infectious. For many diseases, the distribution of the time of infectiousness is far removed from Exponential, and infectivity declines with time. However, since the model simply uses $\gamma Y(t)$ as an overall measure of the rate of recovery from the disease, that may well still be a reasonable approximation to reality.

Example 8.9 (War)

Who will win, when armies of sizes X_t and Y_t are in conflict? One possibility is to assume that (X_t, Y_t) is a Markov chain, with possible transitions

$$(i,j) \to \begin{cases} (i, j-1) & \text{at rate } \alpha i \\ (i-1, j) & \text{at rate } \beta j \end{cases}$$

where α and β relate to their respective fire powers. The battle ends when one army is wiped out, or is reduced to some non-viable size. Plainly the exact distribution of this chain will be very complicated (for the details, see Goldie, 1977), but the corresponding jump chain is a two-dimensional random walk, as in our model for epidemics. The forward equation is

$$\frac{dP}{dt}(m,n) = \alpha m P(m, n+1) - (\alpha m + \beta n)P(m,n) + \beta n P(m+1,n)$$

for $0 \le m \le X_0$, $0 \le n \le Y_0$, and the mean behaviour, with $x(t) = E(X_t)$ and $y(t) = E(Y_t)$ is

$$x'(t) = -\beta y, \qquad y'(t) = -\alpha x.$$

Dividing one by the other, we find $\frac{dy}{dx} = (\alpha x)/(\beta y)$, which solves to give *Lanchester's square law*

$$\alpha(X_0^2 - x^2) = \beta(Y_0^2 - y^2).$$

If fighting continues until one side is eliminated, the X-army wins whenever $\alpha X_0^2 > \beta Y_0^2$, and its final size is $\sqrt{X_0^2 - (\beta/\alpha)Y_0^2}$. The expression for the final size of the Y-army when $\beta Y_0^2 > \alpha X_0^2$ is then obvious. Military tactics have been influenced by this Law. For example, suppose the armies have the same initial size and equal fire power, but the X-army has managed to split the Y-army into two halves, intending to fight them sequentially. Take $\alpha = \beta = 1$, $X_0 = 2K$ and $Y_0 = K$ for the first battle. The square law gives a win for the X-army, with a final size of $\sqrt{4K^2 - K^2} = K\sqrt{3}$. This slightly depleted force takes on the remaining K members of the Y-army, and wins with a final size of $\sqrt{3K^2 - K^2} = K\sqrt{2}$. The X-army has annihilated its opponents, losing only $K(2 - \sqrt{2})$ men. Divide and conquer indeed!

Simulations of the stochastic random walk with $K = 100$ along the lines of those carried out for epidemics are again consistent with these estimates.

EXERCISES

8.1. Prove that the functions $p_{ij}(t)$ are *uniformly* continuous by showing that, for all $h > 0$,

$$|p_{ij}(t + h) - p_{ij}(t)| \leq 1 - p_{ii}(h).$$

8.2. Show that $p_{ii}(t) \geq \exp(-q_i t) \geq 1 - q_i t$ for all $t \geq 0$.

8.3. Given the Q-matrix $\begin{pmatrix} -1 & 1 & 0 \\ 0 & -1 & 1 \\ 1 & 0 & -1 \end{pmatrix}$, show that $p_{ii}(t) = (1 + 2\exp(-3t/2)\cos(t\sqrt{3}/2))/3$, and hence find $P(t)$.

8.4. Consider the Q-matrix on $\{1, 2, 3\}$ given by $\begin{pmatrix} -\alpha & \alpha & 0 \\ 0 & -\beta & \beta \\ \gamma & 0 & -\gamma \end{pmatrix}$.
Show that each $p_{ii}(t)$ is oscillatory if $(\alpha + \beta + \gamma)^2 < 4(\beta\gamma + \gamma\alpha + \alpha\beta)$. Show also that $p_{11}(t) \longrightarrow \beta\gamma/(\beta\gamma + \gamma\alpha + \alpha\beta)$.

8.5. In Australian Rules Football, a team may score a "goal", worth six points, or a "behind", worth one point. Let X_t denote a score difference Home $-$ Away. Assuming X_t is Markov, describe the states, and the form of the Q-matrix.

8.6. Given a Poisson process with constant rate $\lambda > 0$, alter it (separately) in the following ways. In each case, decide whether or not the process that results is Poisson.

(a) Delete every alternate event.

(b) Insert events at times $2t_0$, whenever the given process has an event at t_0.

(c) Delete any events that arise in the intervals $(3k - 1, 3k)$.

(d) Insert events at the times of an independent Poisson process, of rate $\mu > 0$.

(e) Delete any events that occur within the period $h > 0$ after the previous event.

(f) Delete each event independently, with constant probability $p > 0$.

8.7. Take $X_0 = 1$, and let X_t be a Yule process with birth rate $\lambda_n = n\beta$ for $n \geq 1$. Verify the expression (8.5) for the distribution of X_t. Find $E(X_t)$ and $\mathrm{Var}(X_t)$ directly from the differential equation for the pgf, without deriving the distribution of X_t. Find also the time t_n that maximises the value of $P(X_t = n)$ for $n \geq 2$.

8.8. Derive the expression (8.9) for the pgf of a birth and death process in which $\lambda \neq \mu$. Let $Y_t \equiv X_t | X_t > 0$. Find the pgf of Y_t, and hence its mean.

8.9. On Paradise Island, birds are immortal. Immigration occurs at rate α, birds on the island each give birth at rate β. Conditions are so good that no birds ever leave. Model this *immigration–birth* process, and find the pgf of the population size at time t, assuming the island is empty initially.

8.10. In an *immigration–death* process, the immigration rate is α, the death rate per individual is μ. Show that, if such a process begins in state 0, the pgf of the size at time t satisfies

$$\frac{\partial G}{\partial t} + \mu(z - 1)\frac{\partial G}{\partial z} = \alpha(z - 1)G.$$

Solve this equation, and verify that X_t has a Poisson distribution with mean $\rho(1 - \exp(-\mu t))$, where $\rho = \alpha/\mu$.

8.11. Microbe populations U and V each begin with one member, and develop independently as linear pure birth processes with respective birth rates θ and ϕ. Let T denote the time of the first new birth in the U-population (so that T has the $E(\theta)$ distribution); by conditioning on the value of T, find the mean size of the V-population when the first new birth occurs in U. When $\theta = 2\phi$, complete the sentence: "Although the birth rate in U is twice that in V, the mean size of the V-population when U reaches size 2 is . . . ." What other curious result do you obtain when $\theta = \phi$?

8.12. Complete the calculation started in Example 8.8, illustrating the spread of an epidemic among five people mixing in a remote cottage, by finding the distribution of the number who catch the disease.

8.13. (Spread of rumours) The Daley–Kendall (1965) model for how a rumour might spread in a closed, homogeneously mixing population splits the members into three classes:

- Ignorants, who do not know the rumour;

- Spreaders, who know the rumour and are actively spreading it;

- Stiflers, who know the rumour but have stopped spreading it.

When a Spreader A meets any person B, there is a chance she will start to tell the rumour. If B is an Ignorant, B learns the rumour, and becomes a Spreader, A continues as a Spreader. But if B knows the rumour, either as a Spreader or as a Stifler, B tells A that the rumour is stale news, and *both* become Stiflers.

Set up this process as a Markov chain, and analyse it in parallel with the epidemic model as far as you can. Is there a threshold phenomenon?

8.2 Queues

Queues abound in everyday life: in shops, at bus stops, traffic jams, visiting the doctor. Incoming planes stack at congested airports until a runway is free, manufacturers supply components that queue to be used. The next time you are placed on "hold" as you try to telephone a complaint, reflect that the modern theory of queues began when A.K. Erlang, a Danish telephone engineer, studied problems in congestion.

Because of the wide variety of types of queue, it is useful to introduce some vocabulary, and to break the modelling process down. A *customer* is someone or something requiring some sort of service, a *server* is there to provide it. Customers may arrive singly (an aircraft to a runway), or in groups (the passengers disembark and present themselves en bloc at Customs). There may be one server (your local newsagent) or dozens (the supermarket checkouts). The pattern of customer arrivals is of fundamental importance: in many models, it is reasonable to assume that customers arrive essentially at random, at some average rate, but where an appointments system operates, arrivals may be at more or less constant intervals. The nature of the service will determine the variability of service times: buying 60 minutes worth of advice from a financial adviser is one thing, but the time taken by the Benefits Agency to establish your entitlements will be very unpredictable.

Many queues operate the rule FCFS (First Come First Served), but this is far from universal. Some customers may have priority, and be served before earlier arrivals; a pop star signing autograph books thrust towards him by a crowd of fans may use SIRO (Service In Random Order). A possible example of LCFS (Last Come First Served) is when new milk bottles are put at the front of the shelf in a fridge, pushing existing bottles further in, and your flatmate lazily uses the nearest bottle. At my local railway station, with up to six staff selling tickets, new arrivals join a single central queue, the front member taking the next server to become available. At the supermarket, I must choose which of many short queues to join (and I might switch to another one if a hold-up occurs). All these factors, the "queue discipline", will affect the operation of the queue.

The main questions that Queueing Theory tries to answer are: how long does a typical customer wait for service; how busy is any server? Ideally, we would know the joint distributions of all variables at any point of time, or failing that, their means, variances and covariances. Since the exact behaviour of a queue will depend on its initial conditions, we might have to settle for a description of long-term or limiting behaviour, if that exists.

For simplicity, we shall make the assumptions that the times between arrivals, and the service times, are independent; and that the distribution of service time does not vary with the length of the queue. (It is easy to think of examples where servers will speed up if the queue gets longer, or slow down at slack times.) We also assume that all servers are equally proficient, the distributions of their service times being identical, and that a server can serve just one customer at a time. Thus our model would not apply to a car ferry, for example, where one ship simultaneously serves perhaps hundreds of customers.

David Kendall (1951) introduced the notation $A/B/c$ to give a shorthand description of many common queues. Here A defines the arrival process, B relates to the service process, and c is the number of servers. Unless otherwise specified, the queue discipline is taken to be FCFS, customers arrive singly and, if $c \geq 2$, new arrivals join a central feeder queue, as described above for the railway ticket office. We also assume there is no restriction on the length of a queue.

The simplest possible queue is $D/D/1$, where D means "deterministic", i.e. both inter-arrival time and service time are fixed quantities, with no variability at all. It is easy to see what will happen: if the inter-arrival time is shorter than the service time, the queue builds up indefinitely, the server is permanently occupied; otherwise, every new customer is served immediately on arrival, and the fraction of time the server is free can be found. End of story. The same holds for any $D/D/c$ queue. Just as we did when we looked at branching processes, we exclude these cases that have no variability at all.

Example 8.10 ($M/M/k$)

Here M means Markov, in the sense that both the inter-arrival times and the service times have the Markov property of the future being independent of the past and present. Arrivals form a Poisson process whose inter-arrival times have the $E(\lambda)$ distribution, and the service times have the $E(\mu)$ distribution. If X_t denotes the length of the queue at time t, taken to include those being served, then X_t is a birth and death process in which the birth rate is λ, and the death rate is $k\mu$ so long as $X_t \geq k$, or $n\mu$ if $X_t = n \leq k$. To emphasise the rate parameters, we write $M(\lambda)/M(\mu)/k$.

We looked at such processes in Example 8.7. The special case $k = \infty$ is the immigration–death process. For finite values of k, Equation (8.10) shows how to find any equilibrium distribution. For $n \leq k$, we obtain $\pi_n = \lambda^n \pi_0/(n!\mu^n)$, while for $n \geq k$, $\pi_n = \pi_k(\lambda/(k\mu))^{n-k}$. Provided $\sum_{n \geq k}(\lambda/(k\mu))^{n-k}$ is finite, i.e. when $\lambda < k\mu$, there is an equilibrium distribution. Normalisation shows that

$$\pi_0 = 1 \left/ \left(1 + \rho + \cdots + \frac{\rho^{k-1}}{(k-1)!} + \frac{\rho^k k\mu}{k!(k\mu - \lambda)} \right) \right., \tag{8.11}$$

where $\rho = \lambda/\mu$. As examples, in the stable $M/M/1$ queue, $\pi_n = (1 - \rho)\rho^n$, the Geometric distribution $G_0(1 - \rho)$ with mean $\rho/(1 - \rho)$; for $M/M/2$, $\pi_0 = (2 - \rho)/(2 + \rho)$, $\pi_1 = \pi_0\rho$, and $\pi_n = \pi_0\rho^n/2^{n-1}$ for $n \geq 2$.

If $\lambda \geq k\mu$, there is no equilibrium distribution. This is exactly what we would expect: $\lambda > k\mu$ means that new customers are arriving faster than all k servers can collectively dispose of them and, if $\lambda = k\mu$, there is no systematic pressure to reduce long queues.

When $\lambda < k\mu$, the values of $\{\pi_n\}$, the long-term probabilities of being in the various states, have useful interpretations. π_0 is the chance all servers are unoccupied; $\sum_{n=0}^{k-1} \pi_n$ is the chance that a newly arriving customer moves directly to service with no wait; $k\pi_0 + (k - 1)\pi_1 + \cdots + \pi_{k-1}$ is the mean number of servers who are free. We can deduce the mean time a newly arrived customer spends in the system. After his service begins, the mean time is $1/\mu$; before then, he has probability $\sum_{n=0}^{k-1} \pi_n$ of zero waiting time, and for $n \geq k$, he has probability π_n of waiting until $n - k + 1$ customers complete their service. When all servers are busy, the mean time between customers departing is $1/(k\mu)$, hence the mean total time is

$$\frac{1}{\mu} + \sum_{n=k}^{\infty} \pi_n \frac{n - k + 1}{k\mu},$$

which simplifies to

$$\frac{1}{\mu} + \frac{\mu\rho^k \pi_0}{(\mu k - \lambda)^2 (k - 1)!}.$$

Some numerical work may be illuminating: see Exercises 8.16 and 8.17.

Theorem 8.11

Suppose this $M/M/k$ queue is in equilibrium. Then the sequence of *departures* forms a Poisson process of rate λ.

Proof

The proof presented here is due to Paul Burke (1956). Let X_t denote the number in the (equilibrium) system at time t, and let L be the interval between departures. For $u \geq 0$, write

$$G_n(u) = P(L > u \cap X_t = n), \quad G(u) = \sum_n G_n(u) = P(L > u).$$

Then X_t is Markov, $G_n(0) = \pi_n$ and, just as in the derivation of the forward equations (8.2), we have

$$
\begin{aligned}
G_0'(u) &= -\lambda G_0(u) \\
G_n'(u) &= \lambda G_{n-1}(u) - (\lambda + n\mu)G_n(u) \quad 1 \leq n \leq k \\
G_n'(u) &= \lambda G_{n-1}(u) - (\lambda + k\mu)G_n(u) \quad n \geq k.
\end{aligned}
$$

Solving these by induction leads to

$$G_n(u) = \pi_n e^{-\lambda u} \quad (n \geq 0)$$

showing that $G(u) = \sum_n G_n(u) = \exp(-\lambda u)$ as required. (Notice that we have also established the independence of X_t and L.) $\square$

This result may be applicable to a system consisting of a sequence of queues, each one feeding into the next, as in a shop where a sales assistant helps you

choose goods, which you then take to a checkout to pay. Provided that customers arrive at the initial stage according to a Poisson process, that the service time at each intermediate stage has some Exponential distribution, and
the whole set-up is stable, then this result shows that *each* intermediate stage
is some $M(\lambda)/M/k$ queue. See Exercise 8.26.

In the queue $M/M/1$ with $\rho < 1$, Exercise 8.15 asks you to show that
$\pi_n = (1 - \rho)\rho^n$ when $n \geq 0$. In particular, $\pi_0 = 1 - \rho$ is the probability
the server is unoccupied ("idle") at any given time, and her "busy periods"
alternate with these idle times. An idle period begins when some customer
completes service, leaving behind an empty queue, and ends when the next
customer arrives. Thus the durations of idle periods are independent, $E(\lambda)$.
Over a long period T, the total idle time of the server will be about $T(1 - \rho)$,
and the number of idle periods will be about $T(1 - \rho)/(1/\lambda) = T\lambda(1 - \rho)$. This
must also (within unity) be the number of busy periods; the total time the
server is busy is $T - T(1 - \rho) = T\rho$, so we estimate the mean length of a busy
period as $T\rho/[T\lambda(1 - \rho)] = 1/(\mu - \lambda)$. We will present a rigorous argument to
justify this claim later.

Often, the most unrealistic assumption in taking $M/M/k$ as a model is
that any service time is Exponential. As has frequently been acknowledged,
this corresponds to the notion that the distribution of the residual length of
service is the same, no matter how long service has already taken. If we drop this
assumption, the notation $M/G/k$ means that service times have some General
distribution.

Example 8.12 ($M/G/1$)

Let X_t be the queue length at time t. You will rightly imagine that the exact
distribution of X_t will be complicated, but this queue has a neat trick associated
with it. Instead of attempting to describe the queue at all times, let Y_n denote
its length immediately after the nth customer has completed service.

It is easy to see that $\{Y_n : n \geq 0\}$ is a Markov chain, as if we know
that $Y_n = k$, then any knowledge about $Y_{n-1}, Y_{n-2}, \ldots$ is irrelevant to the
distribution of Y_{n+1}. The state space is the set of non-negative integers. Suppose
$Y_n = 0$: then when the nth customer has left, there is no-one waiting for service;
the $(n + 1)$th customer will arrive sometime, begin service immediately, and
Y_{n+1} will be the number who arrive during his service period. Conditioning on
the length of this service period leads to

$$P(Y_{n+1} = j) = p_j = \int_0^\infty f(t)e^{-\lambda t}\frac{(\lambda t)^j}{j!}\,dt$$

where f is the density function of the service time, and λ is the arrival rate.

When $Y_n = i \geq 1$, the queue diminishes by one as the $(n+1)$th customer enters service on the departure of customer n. Hence $Y_{n+1} = i - 1 + U$, where $P(U = j) = p_j$, this same expression. Thus the transition matrix is given by

$$P = \begin{pmatrix} p_0 & p_1 & p_2 & p_3 & \cdot & \cdot \\ p_0 & p_1 & p_2 & p_3 & \cdot & \cdot \\ 0 & p_0 & p_1 & p_2 & \cdot & \cdot \\ 0 & 0 & p_0 & p_1 & \cdot & \cdot \\ \cdot & \cdot & \cdot & \cdot & \cdot & \cdot \end{pmatrix}.$$

The system $\pi P = \pi$ reads:

$$\pi_0 p_0 + \pi_1 p_0 = \pi_0$$

$$\pi_0 p_1 + \pi_1 p_1 + \pi_2 p_0 = \pi_1$$

$$\pi_0 p_2 + \pi_1 p_2 + \pi_2 p_1 + \pi_3 p_0 = \pi_2 \text{ etc.}$$

Let $\Pi(z) = \sum_{n=0}^{\infty} \pi_n z^n$ and $P(z) = \sum_{n=0}^{\infty} p_n z^n$. Multiply the equation for π_n by z^n and sum to obtain

$$\Pi(z) = (\pi_0 + \pi_1)P(z) + \pi_2 z P(z) + \pi_3 z^2 P(z) + \cdots$$

which reduces to $z\Pi(z) = P(z)\{\Pi(z) + \pi_0 z - \pi_0\}$.

The definition of p_j shows that $P(z) = \int_0^{\infty} f(t) \sum_j e^{-\lambda t} \frac{(\lambda t)^j}{j!} z^j dt$ which simplifies to $\int_0^{\infty} f(t) \exp(-\lambda t(1 - z)) dt$. The relation between $\Pi(z)$ and $P(z)$ can be rewritten as

$$\Pi(z) = \pi_0 P(z) \bigg/ \left(1 - \frac{1 - P(z)}{1 - z}\right). \tag{8.12}$$

Write $R(z) = (1 - P(z))/(1 - z) = \sum_{n=0}^{\infty} z^n \sum_{j=n+1}^{\infty} p_j$, so that $R(1) = \sum_{n=0}^{\infty} \sum_{j-n+1}^{\infty} p_j = \sum_{j=1}^{\infty} j p_j = \rho$, say. This quantity ρ is just the mean number of customers who arrive during a service period. Equation (8.12) shows that Π is a genuine pgf if, and only if, $\rho < 1$. This makes intuitive sense.

We could use Foster's criteria for transience or null recurrence, Theorem 7.29, to see that the system is transient when $\rho > 1$ and null recurrent when $\rho = 1$. Again, these have clear interpretations. When $\rho > 1$, customers are arriving faster than they are being served, the queue builds up indefinitely and there is no steady state. When $\rho = 1$, the queue size returns to zero infinitely often, but the mean time between these visits is infinite: the long-term probability of finding a queue size of at most N, for any finite N, is zero.

We can use equation (8.12) in the stable case, $\rho < 1$, to find the mean equilibrium queue length at the times customers depart the system. Since $P(1) = 1$, we have $\Pi(1) = 1$ when $1 = \pi_0/(1 - R(1))$, i.e. $\pi_0 = 1 - \rho$. Then

$$\Pi'(1) = (1 - \rho) \left[\frac{P'(z)}{1 - R(z)} + \frac{P(z)R'(z)}{(1 - R(z))^2} \right]. \tag{8.13}$$

But $P'(1) = \int_0^\infty \lambda t f(t)dt = \lambda E(S) = \rho$, S being a typical service time. Also

$$R'(1) = \sum_{n=0}^\infty n \sum_{j=n+1}^\infty p_j = \sum_{j=1}^\infty p_j \sum_{n=0}^{j-1} n = \sum_{j=1}^\infty p_j \cdot \frac{j(j-1)}{2} = \lambda^2 E(S(S-1))/2.$$

Equation (8.13) shows that $\Pi'(1)$, the mean queue size, is given by

$$\rho + \frac{\rho^2 + \lambda^2 \text{Var}(S)}{2(1 - \rho)}. \tag{8.14}$$

This is known as the Pollaczek–Khinchine formula. We can also use it to find the steady state mean waiting time for a customer. Let W be the time from his arrival to when his service begins, S be the time his service takes, and Q be the number in the queue when he completes service. Then Q will be the number who arrive during time $W + S$, so $E(Q) = \lambda E(W + S)$. Using (8.14),

$$\lambda(E(W) + E(S)) = \rho + \frac{\lambda^2 \text{Var}(S) + \rho^2}{2(1 - \rho)}.$$

Since $E(S) = \rho/\lambda$, rearrangement leads to the expression

$$E(W) = \frac{\rho[E(S) + \text{Var}(S)/E(S)]}{2(1 - \rho)}$$

for the equilibrium mean waiting time until a customer begins service.

It is very striking how quickly the mean queue size and waiting time increase as ρ approaches unity: the change from $\rho = 0.99$ to $\rho = 0.999$ multiplies them tenfold. Attempting to run the queue close to its capacity, in the interests of minimising server idle time, will lead to huge (unacceptable?) increases in queue lengths. Moreover, for a given value of ρ, an increase in $\text{Var}(S)$ leads to an increase in mean waiting times and queue length. Reducing the variability of service time, even if its mean is unchanged, will improve the throughput.

We now find the distribution of B, the busy period of a server in this $M/G/1$ queue. (Unlike a customer's waiting time, the server's busy period is the same for all queue disciplines.) A busy period begins with the arrival of a customer after the server has been unoccupied; label the arrival time of this customer as $t = 0$, let x be the service time of the first customer, and suppose k other customers arrive during time x. We claim that

$$B = x + (B_1 + B_2 + \cdots + B_k)$$

where these $\{B_i\}$ are independent, all having the same distribution as B.

To justify this claim, note that it is obviously correct when $k = 0$, so we suppose that $k \geq 1$. Let B_i be the busy period that would have arisen if customer i had been the one who arrived at time zero. These B_i are plainly independent of each other, and have the same distribution as B. Since these k customers are in the system at time x, there is no gap before they begin service, and the total busy period is the expression shown.

Write $M_B(t) = E(e^{tB})$ as the moment generating function of B. Let S be the first service time, having density f and mgf $M_S(t)$, and let N denote the number of arrivals during S. Then

$$E(e^{tB}|S = x, N = k) = E(e^{t(x+B_1+\cdots+B_k)})$$
$$= e^{tx}(M_B(t))^k.$$

Conditional on $S = x$, $P(N = k) = e^{-\lambda x}(\lambda x)^k/k!$, and so

$$M_B(t) = E(E(e^{tB}|S, N)) = \int_0^\infty f(x) \sum_{k=0}^\infty e^{-\lambda x} \frac{(\lambda x)^k}{k!} e^{tx}(M_B(t))^k dx$$

$$- \int_0^\infty f(x)\exp(tx - \lambda x + \lambda x M_B(t))dx$$

$$= M_S(t - \lambda + \lambda M_B(t)).$$

This establishes

Theorem 8.13

The mgf $M_B(t)$ of the busy period in a $M/G/1$ queue satisfies the equation $M_B(t) = M_S(t - \lambda + \lambda M_B(t))$, where $M_S(t)$ is the mgf of the service time, and λ is the arrival rate.

For example, consider $M/M/1$, where $M_S(t) = \mu/(\mu - t)$ for $t < \mu$. The theorem shows that $M_B = \mu/(\mu - t + \lambda - \lambda M_B)$, which can be written as a quadratic in M_B. In order that $M_B(0) = 1$, the solution we require is easily seen to be

$$M_B(t) = \frac{\mu + \lambda - t - \sqrt{(\mu + \lambda - t)^2 - 4\mu\lambda}}{2\lambda}.$$

If you have your book of integral transforms handy, you can look this up, and find that the corresponding density function of B, when $\lambda < \mu$, is

$$f(t) = \frac{\exp(-t(\lambda + \mu))}{t\sqrt{\rho}} I_1(2t\sqrt{\lambda\mu})$$

where I_1 is a Bessel function. This should act as a warning not to expect simple answers in more general cases.

Theorem 8.13 can be used to find the mean and variance of the busy period. Differentiate, put $t = 0$ and obtain $E(B) = E(S)(1 + \lambda E(B))$, giving $E(B) = E(S)/(1-\lambda E(S))$. In $M/M/1$, where $E(S) = 1/\mu$, this shows $E(B) = 1/(\mu-\lambda)$, confirming the result we obtained earlier by a non-rigorous argument. Exercise 8.20 asks you to work out $\mathrm{Var}(B)$.

Let E_r denote the Gamma density $\Gamma(r, \alpha)$ when r is an integer. There is a nice trick, the *method of stages* that can be used on $M/E_r/1$. The service time has the same distribution as the sum of r independent $E(\alpha)$ variables, so imagine the service broken down into a sequence of r independent stages, each lasting for a period with distribution $E(\alpha)$. The state Y_t of the queue at time t can be taken as the total number of stages that those now in the queue must pass through to complete service. If a person currently being served has i ($1 \le i \le r$) more stages to complete, and j customers are waiting to begin service, $Y_t = i+rj$. The independence of the stages shows that $\{Y_t\}$ is a Markov chain on $\{0, 1, 2, \ldots\}$, so it is enough to specify its Q-matrix: each arrival adds r stages, any completion of a stage reduces Y_t by unity, so

$$q_{i,i+r} = \lambda, \quad q_{i,i-1} = \alpha, \quad q_i = \sum_{j \ne i} q_{ij}$$

for $i \ge 0$, with the obvious exception that $q_{0,-1} = 0$.

The process Y_t can be thought of as a birth and death process in which deaths occur singly at rate α, while births come in batches of size r at rate λ. The long-term probabilities, if they exist, will satisfy

$$\lambda \pi_0 = \alpha \pi_1$$
$$(\lambda + \alpha)\pi_n = \lambda \pi_{n-r} + \alpha \pi_{n+1} \quad (n \ge 1)$$

with the convention that $\pi_j = 0$ when $j < 0$. Taking $\Pi(z) = \sum_{n=0}^{\infty} \pi_n z^n$ as the pgf of this distribution, we obtain

$$(\lambda + \alpha)(\Pi(z) - \pi_0) = \lambda z^r \sum_{n=r}^{\infty} \pi_{n-r} z^{n-r} + \frac{\alpha}{z} \sum_{n=1}^{\infty} \pi_{n+1} z^{n+1}.$$

This simplifies to

$$\Pi(z)\left(\lambda + \alpha - \lambda z^r - \frac{\alpha}{z}\right) = \pi_0(\lambda + \alpha) - \frac{\alpha}{z}(\pi_0 + \pi_1 z).$$

Using the fact that $\alpha \pi_1 = \lambda \pi_0$, we can now express $\Pi(z)$ as a ratio, in which the only unknown quantity is π_0. To ensure $\Pi(1) = 1$, use L'Hôpital's rule as $z \to 1$ to find $\pi_0 = (\alpha - r\lambda)/\alpha$. Provided that $\alpha > r\lambda$, $\Pi(z)$ is a genuine pgf. This condition is equivalent to insisting that $(1/\lambda) > (r/\alpha)$, i.e. that the mean inter-arrival time exceeds the mean service time. In these cases,

$$\Pi(z) = \frac{\alpha - r\lambda}{\alpha - \lambda z \frac{(1-z^r)}{(1-z)}}$$

is the pgf of the equilibrium distribution of Y_t.

Example 8.14 $(G/G/1)$

For a single server queue with general inter-arrival and service times, we finally part company with the Markov property. Label the customers as $C_0, C_1, \ldots$ arriving at respective times $t_0 < t_1 < t_2 < \ldots$. The inter-arrival time between the $(n-1)$th and nth customers is $T_n = t_n - t_{n-1}$; these quantities $\{T_n\}$ are taken to be iid, with distribution function A, say.

Suppose that C_n waits a time W_n (≥ 0) after his arrival until he begins service, that his service time is S_n, where these $\{S_n\}$ are iid with distribution function B. We are interested in the long-term behaviour of $\{W_n\}$, and will prove

Theorem 8.15

In the $G/G/1$ queue, let S and T be independent random variables having the respective distributions of service time and inter-arrival time, and suppose $U = S - T$ has density f. Suppose also that $E(S) < E(T)$. Then this queue has an equilibrium distribution. When the queue is in equilibrium, let W denote the time that a newly arriving customer waits until he commences service. Then $P(W < \infty) - 1$, and for any $w \geq 0$, its distribution function $G(w)$ satisfies Lindley's (1952) equation

$$G(w) = \int_{-\infty}^{w} f(u)G(w-u)du = \int_{0}^{\infty} G(y)f(w-y)dy.$$

Proof

The time that C_{n+1} enters service is $t_{n+1} + W_{n+1}$; if the system is empty when he arrives, then $W_{n+1} = 0$, otherwise C_{n+1} begins service when his predecessor departs at time $t_n + W_n + S_n$. Hence

$$t_{n+1} + W_{n+1} = \max\{t_{n+1}, t_n + W_n + S_n\}.$$

Subtract t_{n+1} from both sides to give

$$W_{n+1} = \max\{0, W_n + S_n - T_{n+1}\} = \max\{0, W_n + U_n\}$$

where $U_n = S_n - T_{n+1}$. These quantities $\{U_n\}$ are iid, and their density function f can be found from knowledge of A and B.

Iterating this expression for W_{n+1}, we find

$$
\begin{aligned}
W_{n+1} &= \max\{W_n + U_n, 0\} \\
&= \max\{\max\{W_{n-1} + U_{n-1}, 0\} + U_n, 0\} \\
&= \max\{\max\{W_{n-1} + U_{n-1} + U_n, U_n\}, 0\} \\
&= \max\{W_{n-1} + U_{n-1} + U_n, U_n, 0\}.
\end{aligned}
$$

Continue this process. Three lines later, we find

$$
W_{n+1} = \max\{W_{n-2} + U_{n-2} + U_{n-1} + U_n, U_{n-1} + U_n, U_n, 0\}
$$

and so on, eventually giving

$$
W_{n+1} = \max\{U_0 + U_1 + \cdots + U_n, U_1 + \cdots + U_n, \ldots, U_{n-1} + U_n, U_n, 0\}.
$$

However, since the $\{U_i\}$ are iid, W_{n+1} has the *same distribution* as

$$
\begin{aligned}
V_{n+1} &= \max\{U_0 + U_1 + \cdots + U_n, U_0 + U_1 + \cdots + U_{n-1}, \ldots, U_0, 0\} \\
&= \max\{\Sigma_n, \Sigma_{n-1}, \ldots, \Sigma_0, 0\} = \max\{0, \Sigma_0, \Sigma_1, \ldots, \Sigma_n\},
\end{aligned}
$$

where $\Sigma_j = U_0 + U_1 + \cdots + U_j$.

It is immediate from its definition that $V_{n+1} \geq V_n \geq 0$, so the sequence $\{V_n\}$ will converge strongly to some random variable V (which might, at this stage, be infinite). But since W_n and V_n have the same distribution, if $V_n \xrightarrow{D} V$, so also $W_n \xrightarrow{D} V$; the limiting distribution of our sequence $\{W_n\}$ is the same as that of the monotone sequence $\{V_n\}$.

When $E(S) < E(T)$, then $E(U) < 0$ and the Strong Law of Large Numbers shows that $\Sigma_n/n \xrightarrow{a.s.} E(U)$. Thus there is some finite N so that $\Sigma_n/n < 0$ for all $n \geq N$. Hence $V_n = V_N$ when $n \geq N$, and V is certainly finite, being the maximum of the finite collection $\{0, \Sigma_0, \ldots, \Sigma_N\}$. Write $\Psi_n = \sum_{i=1}^{n} U_i$ so that

$$
\begin{aligned}
V &= \max\{0, \Sigma_0, \Sigma_1, \ldots\} \\
&= \max\{0, U_0, U_0 + \Psi_1, U_0 + \Psi_2, \ldots\} \\
&= \max\{0, U_0 + \max\{0, \Psi_1, \Psi_2, \ldots\}\}.
\end{aligned}
$$

But $V' = \max\{0, \Psi_1, \Psi_2, \ldots\}$ has the same distribution as V, so

$$
V = \max\{0, U_0 + V'\}
$$

where V' is independent of U_0. Thus, for any $w \geq 0$,

$$
\begin{aligned}
G(w) &= \lim_{n \to \infty} P(W_n \leq w) = \lim_{n \to \infty} P(V_n \leq w) \\
&= P(V \leq w) = P(\max\{0, U_0 + V'\} \leq w) \\
&= \int_{-\infty}^{w} f(u)G(w - u)\,du = \int_{0}^{\infty} G(y)f(w - y)\,dy
\end{aligned}
$$

as the theorem claims.   □

Pleasing though this result is, it is only in special cases that we can use it to obtain an exact expression for the form of G. As an illustration, consider the stable $M/M/1$ queue. We can use Theorem 4.20 to see that $f(u) = \frac{\lambda\mu}{\lambda+\mu}e^{\alpha u}$, where $\alpha = \lambda$ when $u < 0$ and $\alpha = -\mu$ when $u > 0$. Then you should check that $G(w) = 1 - \rho\exp(-(1-\rho)\mu w)$ for $w \geq 0$ satisfies Lindley's equation. (Exercise 8.15 asks you to derive this result more directly.)

As a digression, we note that the sequence $\{V_n\}$ and its possible limit V arise in quite diverse contexts, and this theorem can help us find the properties of V. In insurance, U_n could be the difference between the claims paid out and the income from premiums in the nth time period. Then V would be the maximum excess liability of payouts over income, and so inform the level of reserves that is prudent. Considering your health, let U_n be the difference, in calories, between your food intake and energy expended, so that the level of V, the maximum excess of consumption over expenditure, may give warnings about your need to diet, or take more exercise. Closer to the spirit of queues, there is a theory for the operation of dams and reservoirs: U_n is the difference between the demand for water and the inflow from rainfall. Our worry is that V should not get too high, or the reservoir will empty, but a practical complication arises at the other extreme: should heavy rainfall cause the U_n to be too negative, the reservoir will overflow its capacity, and successive values of U_n will not have identical distributions.

Returning to the $G/G/1$ queue, consider the possibility that $E(S) \geq E(T)$. Since $\Sigma_n/n \xrightarrow{a.s.} E(U)$, matters are plain when $E(U) > 0$: there will be some finite N so that $\Sigma_n/n > E(U)/2$ when $n \geq N$, so $\Sigma_n \xrightarrow{a.s.} \infty$, and $P(V = \infty) = 1$. There is no equilibrium distribution, it is certain that the queue builds up indefinitely.

When $E(U) = 0$, so that the mean service time and inter-arrival times are exactly equal, matters are more complex. We can appeal to the Law of the Iterated Logarithm to guess intuitively what will happen: that Law implies that, so long as $\mathrm{Var}(U) < \infty$, then $\Sigma_n/\sqrt{n \log\log(n)}$ will get close to $\sqrt{2}$ infinitely often. Thus Σ_n will exceed $\sqrt{n}$ infinitely often, so again $P(V = \infty) = 1$. This queue has no equilibrium distribution either.

Example 8.16 (Modifications)

In practical applications, realism demands that other factors are considered. The waiting area might have finite capacity: for example, a petrol filling station with c pumps (servers) may be adjacent to the road, so that if too many cars try to use it simultaneously, the highway would be blocked. The notation $A/B/c/K$ is used to indicate that when K customers are in the system, any new arrivals are simply lost (the motorist uses a different filling station.) The special case

$M/M/c/K$ is a birth and death process, with birthrate $\lambda_n = 0$ for $n \geq K$.

If a long queue simply *discourages* potential customers, the arrival process could be modified by giving new arrivals a probability, depending on the length of the queue, of "balking", i.e. again being lost to that queue for ever. Exercise 8.26 considers the case when $\lambda_n = \lambda/(n+1)$ for $n \geq 0$.

With $c \geq 2$ servers, they may have different distributions of service time. Modify $M/M/2$ so that Peter's service rate is μ_1, Paul's is μ_2, with $\mu_1 > \mu_2 > 0$. Assuming customers know which is the faster server, a customer with a choice will always prefer Peter, so if Paul is busy when Peter becomes free, the front customer is served by Peter. But suppose Peter is busy when Paul becomes free: a customer served by Paul has a mean wait $1/\mu_2$ to complete service, but if the front customer waits until Peter is free, her mean time to completion would be $2/\mu_1$. Hence, if $\mu_1 > 2\mu_2$, the front customer waits less time, on average, by ignoring Paul. If only the leading customer can enter service, individual rational choice implies that if $\mu_1 > 2\mu_2$, then Paul might just as well not be there! This queue would then operate as $M/M(\mu_1)/1$, and the whole system operates slower than it would with no selfish behaviour. See Exercise 8.24.

EXERCISES

8.14. Let $G(z,t) = \sum_{n=0}^{\infty} p_n(t)z^n$ be the pgf of the length of a $M/M/1$ queue at time t. Show that G satisfies

$$\frac{\partial G}{\partial t} = (1-z)\left[\left(\frac{\mu}{z} - \lambda\right) - \frac{\mu}{z}p_0(t)\right].$$

8.15. Show that when the queue $M/M/1$ has a stable equilibrium, it is given by $\pi_n = (1-\rho)\rho^n$ for $n \geq 0$. Suppose Jane arrives at a random time when this queue is in equilibrium; show directly, without appeal to Lindley's equation, that, if W is the time she waits to commence service, $P(W \leq w) = 1 - \rho\exp(-w(\mu-\lambda))$ for $w \geq 0$. By considering a long time period, conclude that the mean number served during a busy period is $1/(1-\rho)$.

8.16. Find the pgf of the length of a stable $M/M/2$ queue, and deduce its mean and variance. Two adjacent banks have one and two tellers respectively, all service times are $E(\mu)$. If the respective arrival rates are λ and 2λ ($\lambda < \mu$), compare (a) the chances of immediate service and (b) the mean times spent in the two systems.

8.17. For the queue $M/M/4$, take $\lambda = 4$ and suppose $\mu > 1$ so that the queue is stable. Compare the equilibrium values in the cases $\mu = 2$

and $\mu = 4/3$ for (a) the probability an arrival moves straight to service; (b) the mean number of idle servers; (c) the mean total time a customer spends in the system.

8.18. A petrol filling station has four pumps, cars arrive at random at an average rate of three per minute, service times are Exponential with mean two minutes. Give the parameters for the queue to which this corresponds; what happens in the long run?

Now suppose the forecourt has room for six cars only, in addition to any at the pumps. Find the equilibrium distribution, and estimate how many cars per hour drive on, finding the forecourt full.

Each car that drives on represents lost profit. Consider the relative merits of (a) extending the forecourt to take two more cars; (b) installing one more pump, which reduces the forecourt capacity to five waiting cars.

8.19. In an $M/G/1$ queue, 18 customers per hour arrive on average. Use the Pollaczek–Khinchine formula to obtain the mean queue size in the cases when the service time is (a) Deterministic, 3 minutes; (b) Exponential, mean 3 minutes; (c) $U[0,6]$ minutes.

8.20. Use Theorem 8.13 to find the variance of the busy period in a stable $M/G/1$ queue in terms of the mean and variance of the service time.

8.21. A single operator looks after N machines on a production line. Any machine breaks down at rate λ, and repair times are iidrv; the operator can repair only one machine at a time. Modify $M/G/1$ to describe this queue. When $G \equiv E(\mu)$, find the equilibrium distribution for the corresponding birth and death process.

8.22. A telephone exchange has K lines, and no way of placing customers on hold if all K lines are in use (an extreme case of a finite waiting room). What proportion of calls that would otherwise be made to a $M/M/K$ queue will be lost?

8.23. Explain why the method of stages does not apply directly to a $M/E_r/c$ queue when $c \geq 2$. How could you modify the definition of Y_t to extend the method to this case?

8.24. Modify a $M/M/2$ queue so that the servers have different service rates $\mu_1 > \mu_2 > 0$. The queue discipline is that when a server becomes free, customers in a waiting line are allowed, sequentially from the front, to choose to begin service (or wait). If customers act selfishly, seeking to minimise their individual mean time in the system, what will happen?

8.25. (After Burke, 1956) Customers enter a clothes shop according to a Poisson process of mean rate 120/hour. K sales assistants hover to help them choose clothes, each transaction having an Exponential distribution with mean 90 seconds. There are n cashiers who take the money when the clothes have been selected: payment time is modelled as Exponential, mean one minute. What is the minimum value of K so that the chance a customer sees at most three others waiting for an assistant is at least 90%? For this value of K, how large must n be so that the chance is at least 1/3 that a customer arriving at the pay desk sees no others waiting for a free cashier?

8.26. Modify $M/M/1$ so that the arrival rate is $\lambda/(n+1)$ when there are already n customers in the system ("discouraged arrivals"). Show that the equilibrium number in the queue is Poisson.

8.3 Renewal Theory

Suppose the lifetimes of lightbulbs are iidrv, and when a bulb fails, it is replaced immediately by a new one ("renewal"). What can be said about the times of these renewals, and the number of renewals in a given period?

That is the prototype problem of Renewal Theory, but we have met the notion already in other contexts. For example, in the queue $M/G/1$, a renewal may be thought as occurring at the times successive busy periods for the server end; in discrete time recurrent Markov chains, the times a named state is re-visited are times of renewal. We shall concentrate on the continuous time case, and use the lightbulb example as a peg for the notation:

- $\{T_n : n = 1, 2, \ldots\}$ are the successive lifetimes, all having

- density function $f(t)$ with $f(t) = 0$ when $t < 0$

- distribution function $F(t) = P(T_n \leq t)$

- survivor function $\mathcal{F}(t) = 1 - F(t) = P(T_n > t)$, and hence

- hazard rate $\phi(t) = f(t)/\mathcal{F}(t)$.

The function ϕ arises via

$$\lim_{h \downarrow 0} \frac{P(t < T_n \leq t + h | T_n > t)}{h} = \lim_{h \downarrow 0} \frac{F(t+h) - F(t)}{h\mathcal{F}(t)} = \frac{f(t)}{\mathcal{F}(t)}.$$

It is sometimes called the age-specific failure rate, and its derivation shows that it corresponds to the rate at which bulbs aged t burn out.

Since $-\phi(u) = \mathcal{F}'(u)/\mathcal{F}(u)$, we have

$$\int_0^t (-\phi(u))du = [\ln \mathcal{F}(u)]_0^t = \ln \mathcal{F}(t).$$

Thus any one of f, F, $\mathcal{F}$ and ϕ determines all the rest. In particular, if ϕ is constant, lifetimes inevitably have Exponential distributions, and vice versa.

Example 8.17

One possible model takes T to have the so-called *Pareto* density, Par(β, c), with $f(t) = c\beta/(1 + \beta t)^{c+1}$ on $t \geq 0$. The survivor function is

$$\mathcal{F}(t) = \int_t^\infty \frac{c\beta}{(1 + \beta u)^{c+1}} du = \left[-\frac{1}{(1 + \beta u)^c} \right]_t^\infty = \frac{1}{(1 + \beta t)^c},$$

so that $\phi(t)$ evaluates to $c\beta/(1 + \beta t)$.

Thus $\phi(t)$ *decreases* as t increases – the older the component, the less prone it is to immediate failure. The acronym NWU = New is Worse than Used is used for any density with $\phi(0) \leq \phi(t)$ for all $t \geq 0$ – this example shows a very strong version of this notion: reliability steadily increases with use.

Another model used is the Weibull distribution, defined for $t \geq 0$ by

$$f(t) = \alpha\rho(\rho t)^{\alpha-1} \exp(-(\rho t)^\alpha)$$

with parameters $\alpha > 0$, $\rho > 0$. In this case, $\mathcal{F}(t) = \exp(-(\rho t)^\alpha)$, so that $\phi(t) = \alpha\rho(\rho t)^{\alpha-1}$. The special case $\alpha = 1$ is the familiar Exponential $E(\rho)$ distribution which has a constant hazard rate, and the case $\alpha < 1$ is also NWU. But when $\alpha > 1$, $\phi(t)$ is an increasing function with, in particular, $\phi(t) > \phi(0)$ when $t > 0$; this is described as NBU = New is Better than Used.

Make the convention that $F(0) = 0$, so that renewals occur singly, and write X_t or $X(t)$ as the number of renewals up to time t, taking $X_0 = 0$, and refer to $(X_t : t \geq 0)$ as a *renewal process*. Renewals occur at times $W_1 = T_1$, $W_2 = W_1 + T_2$, $W_3 = W_2 + T_3, \ldots$, and the key observation that sets up the theoretical development is that

$$W_k \leq t \Leftrightarrow X_t \geq k.$$

(Do pause and ensure that you are clear why this is so: the figure may help. We met a special case of this correspondence in Chapter 4, when we explored the relationship between the Poisson and Exponential distributions.)

Illustration of $X_t = 3$, with $W_3 \leq t < W_4$.

Theorem 8.18

$E(X_t) = \sum_{k=1}^{\infty} F_k(t)$, where F_k is the distribution function of W_k.

Proof

Since X_t takes non-negative integer values only, we have

$$E(X_t) = \sum_{k=1}^{\infty} P(X_t \geq k) = \sum_{k=1}^{\infty} P(W_k \leq t) = \sum_{k=1}^{\infty} F_k(t),$$

using Exercise 4.4.   □

Definition 8.19

Write $H(t) = E(X_t)$ as the *renewal function*, with $h(t) = H'(t)$ as the *renewal density function*.

   $H(t)$, being the mean number of renewals up to time t, is a prime object of interest. As an example, suppose each T_n has the $\Gamma(2, \lambda)$ distribution. Then W_k is $\Gamma(2k, \lambda)$, and the last theorem shows that

$$H(t) = \sum_{k=1}^{\infty} \int_0^t \frac{\lambda^{2k} x^{2k-1} e^{-\lambda x}}{(2k-1)!} dx = \int_0^t \lambda e^{-\lambda x} \sum_{k=1}^{\infty} \frac{(\lambda x)^{2k-1}}{(2k-1)!} dx.$$

But the sum in the integral is $(e^{\lambda x} - e^{-\lambda x})/2$, so the integral evaluates as

$$H(t) = \frac{\lambda t}{2} - \frac{1 - e^{-2\lambda t}}{4}.$$

Hence, as $t \to \infty$, we see that $H(t)/t \to \lambda/2 = 1/E(T)$. Recalling the definition of $H(t)$, this limit will be no surprise.

Theorem 8.20

$H(t)$ is finite for all $t > 0$.

Proof

We first give a short proof that applies when $F(t) < 1$ for all finite t. Here

$$
\begin{aligned}
F_n(t) &= P(T_1 + \cdots + T_n \leq t) \leq P(\max(T_1, \ldots, T_n) \leq t) \\
&= P(T_1 \leq t, \ldots, T_n \leq t) = (P(T \leq t))^n
\end{aligned}
$$

by independence, and so, when $F(t) < 1$,

$$
H(t) = \sum_{n=1}^{\infty} F_n(t) \leq \sum_{n=1}^{\infty} (F(t))^n = F(t)/(1 - F(t)),
$$

so plainly $H(t)$ is finite if $F(t) < 1$. (Note the useful inequality $F(t) \leq H(t) \leq F(t)/(1 - F(t))$.)

If $F(t) = 1$ for some finite t, write $t_0 = \inf\{t > 0 : F(t) = 1\}$, the least such value. Given $t > 0$, let r be some integer that exceeds t/t_0, and write

$$
\mathcal{F}_r(t) = P(T_1 + \ldots + T_r > t) \geq P(T_1 > t/r, \ldots, T_r > t/r) = (\mathcal{F}(t/r))^r.
$$

Now $r > t/t_0$, so $t/r < t_0$, hence $\mathcal{F}(t/r) = 1 - \delta$ for some $\delta > 0$. Thus

$$
\begin{aligned}
F_{nr}(t) &= P(T_1 + \cdots + T_{nr} \leq t) \\
&= P((T_1 + \cdots + T_r) + \cdots + (T_{nr-r+1} + \cdots + T_{nr}) \leq t) \\
&\leq (F_r(t))^n = \delta^n
\end{aligned}
$$

using the same argument involving the maximum of n quantities as used above. Also, since $F_i(t) \geq F_{i+1}(t)$ for all i and all t, we have

$$
F_{nr}(t) + F_{nr+1}(t) + \cdots + F_{nr+r-1}(t) \leq r F_{nr}(t),
$$

and so

$$
\begin{aligned}
H(t) &= \sum_{i=1}^{r-1} F_i(t) + \sum_{n=1}^{\infty} (F_{nr}(t) + \cdots + F_{nr+r-1}(t)) \\
&\leq (r-1) + r \sum_{n=1}^{\infty} F_{nr}(t) \leq (r-1) + r \sum_{n=1}^{\infty} \delta^n
\end{aligned}
$$

which shows that $H(t) \leq (r-1) + r\delta/(1 - \delta)$, establishing our claim. $\square$

Example 8.21

Given a sequence of iid $U(0, 1)$ variables, how many are needed until the sum is at least t, for some $t > 0$? This problem plainly sits within our renewal theory context, and the mean number required is $1 + H(t) = V(t)$, say. We use Theorem 8.18, and so require $F_k(t)$, the distribution function of $U_1 + \cdots + U_k$ when these are iid $U(0, 1)$. Integrating Equation (4.2), with $f(x) = 1$ for $0 < x < 1$, we see that

$$F_k(t) = \int_0^1 F_{k-1}(t - x)dx.$$

Suppose first that $0 < t \le 1$. It is straightforward to use induction to show that $F_k(t) = t^k/k!$, so that

$$V(t) = 1 + \sum_{k=1}^{\infty} F_k(t) = \sum_{k=0}^{\infty} t^k/k! = \exp(t).$$

This is a pleasing result: the mean number of $U(0, 1)$ variables that need be chosen so that their sum exceeds t is $\exp(t)$ when $0 \le t \le 1$. In particular, when $t = 1$, we see that it takes $\exp(1) = e \approx 2.72$ such variables to reach the value unity, "on average", even though $E(U_1 + U_2) = 1$.

When $1 < t \le 2$, the expression for $F_k(t - x)$ differs according as $t - x \ge 1$, or $t - x < 1$, but induction establishes that, for $k \ge 1$,

$$F_k(t) = \frac{t^k}{k!} - \frac{(t-1)^k}{(k-1)!}.$$

This leads on to $V(t) = e^t - (t - 1)e^{t-1}$. When $t = 1$, the two expressions for $V(t)$ merge continuously, but not smoothly; the left derivative is e, the right derivative is $e - 1$.

You should now verify that, if $r \ge 2$ and $r \le t \le r + 1$ then $F_k(t) = \sum_{i=0}^{r} \frac{(-1)^i(t-i)^k}{i!(k-i)!}$, leading to the expression

$$V(t) = e^t \sum_{i=0}^{[t]} \frac{(-1)^i(t-i)^i e^{-i}}{i!}$$

for any $t \ge 0$, where $[t]$ is the largest integer that does not exceed t. Elegant though this result is, the format of $V(t)$ is not suited to numerical computation when t is large, as it is the sum of terms large in absolute value, but with alternating signs. Theorem 8.25 indicates that, for large t, we can expect this $V(t)$ to be close to $2t + 1$, since $\mu = 1/2$.

Theorem 8.22

$H(t)$ satisfies the *renewal equation*

$$H(t) = \int_0^t H(t - x)f(x)dx + F(t),$$

and hence $h(t) = \int_0^t h(t - x)f(x)dx + f(t)$.

Proof

We condition on the value of T, the time of the first renewal, i.e. $H(t) = E(X_t) = E(E(X_t|T))$. When $x > t$, it is plain that $(X_t|T = x) = 0$, while when $x \le t$, then $E(X_t|T = x) = 1 + H(t - x)$, since the renewals up to time t consist of the first one at time x, plus those in the period from x to t. Hence

$$H(t) = \int_0^t f(x)[1 + H(t - x)]dx = \int_0^t f(x)H(t - x)dx + F(t).$$

Differentiation leads to the corresponding equation for h.   □

Just prior to Theorem 8.20, we looked at an example where $E(X_t)/t \to 1/E(T)$ as $t \to \infty$. We now sharpen this, by examining the general behaviour of X_t/t for large t.

Theorem 8.23

(a) If $\mu = E(T) < \infty$, then $X(t)/t \xrightarrow{a.s.} 1/\mu$ as $t \to \infty$.

(b) If also $\sigma^2 = \text{Var}(T) < \infty$, then $\frac{X(t) - t/\mu}{\sqrt{t\sigma^2/\mu^3}} \xrightarrow{D} N(0, 1)$ as $t \to \infty$.

Proof

For any $t \ge 0$, $\quad T_{X(t)} \le t < T_{X(t)+1} \quad$ and so, when $X(t) > 0$,

$$\frac{T_{X(t)}}{X(t)} \le \frac{t}{X(t)} < \frac{T_{X(t)+1}}{X(t) + 1} \cdot \frac{X(t) + 1}{X(t)}. \tag{8.15}$$

As $t \to \infty$, so $X(t) \xrightarrow{a.s.} \infty$, hence the Strong Law of Large Numbers shows that $T_{X(t)}/X(t) \xrightarrow{a.s.} \mu$; also $(X(t) + 1)/X(t) \to 1$, and so (8.15) establishes (a).

For (b), note that

$$\frac{X(t) - t/\mu}{\sqrt{t\sigma^2/\mu^3}} \ge y \Leftrightarrow X(t) \ge \frac{t}{\mu} + y\frac{\sigma}{\mu}\sqrt{\frac{t}{\mu}}.$$

Write $\frac{t}{\mu} + y\frac{\sigma}{\mu}\sqrt{\frac{t}{\mu}} = K - \epsilon_t$, where K is an integer and $0 \le \epsilon_t < 1$. Because $X(t)$ is integral, the condition that $X(t) \ge \frac{t}{\mu} + y\frac{\sigma}{\mu}\sqrt{\frac{t}{\mu}}$ is the same as $X(t) \ge K$, which occurs exactly when $W_K \le t$. We now use the Central Limit Theorem for the sum W_K.

$$W_K \leq t \Leftrightarrow \frac{W_K - K\mu}{\sigma\sqrt{K}} \leq \frac{t - K\mu}{\sigma\sqrt{K}}$$

$$= [t - (t + y\sigma\sqrt{\tfrac{t}{\mu}} + \mu\epsilon_t)]/(\sigma\sqrt{K})$$

$$= -(y\sigma\sqrt{\tfrac{t}{\mu}} + \mu\epsilon_t) \Big/ \left(\sigma\sqrt{\tfrac{t}{\mu}} + y\tfrac{\sigma}{\mu}\sqrt{\tfrac{t}{\mu}} + \epsilon_t\right)$$

$$= -(y + \tfrac{\mu}{\sigma}\sqrt{\tfrac{\mu}{t}}\epsilon_t) \Big/ \left((1 + y\tfrac{\sigma}{\mu}\sqrt{\tfrac{\mu}{t}} + \tfrac{\mu\epsilon_t}{t})^{1/2}\right).$$

Hence

$$P\left(\frac{X(t) - t/\mu}{\sqrt{t\sigma^2/\mu^3}} \geq y\right) = P\left(\frac{W_K - K\mu}{\sigma\sqrt{K}} \leq \frac{-(y + \tfrac{\mu}{\sigma}\sqrt{\tfrac{\mu}{t}}\epsilon_t)}{(1 + y\tfrac{\sigma}{\mu}\sqrt{\tfrac{\mu}{t}} + \tfrac{\mu\epsilon_t}{t})^{1/2}}\right)$$

$$= P\left(\frac{W_K - K\mu}{\sigma\sqrt{K}} \leq -y + O\left(\frac{1}{\sqrt{t}}\right)\right)$$

when t is large. By the symmetry about zero of the Standard Normal distribution, the Central Limit Theorem gives the result we seek. $\qquad\square$

Reference to statistical tables tells us that, for large t, there is approximately a 95% chance that $X(t)$ will fall in the interval $t/\mu \pm \frac{2\sigma}{\mu}\sqrt{\frac{t}{\mu}}$; similarly, we can say that $X(t)$ will be more than $t/\mu + \frac{3\sigma}{\mu}\sqrt{\frac{t}{\mu}}$ only about one time in 700.

Consider a renewal process X_t that has been running a long time, and fix attention on the component currently in use. Recall the bus paradox of Chapter 4: selecting an interval between bus arrivals by first picking a time t, and thus the interval in which t falls, biases the choice towards longer intervals. The same occurs in renewal theory: selecting the component in use at an arbitrary time biases the choice in favour of long-lived components. Two factors are at work: this bias, which makes a component whose lifetime is kw k times as likely to be chosen as one with lifetime w; and $f(w)$, the relative frequency with which lifetimes of length w occur. Overall, the relative frequency of selecting a component with lifetime w is $wf(w)$, and since $\int_0^\infty wf(w)dw = \mu$, the density function of the lifetime of a component selected in this manner is $wf(w)/\mu$ on $w > 0$.

Plainly, the mean lifetime of such components is $\int_0^\infty (w^2 f(w)/\mu)dw = E(T^2)/\mu = \mu + \sigma^2/\mu$. In the bus paradox, with $f(w) = \lambda\exp(-\lambda w)$, then $\mu = 1/\lambda$, $\sigma^2 = 1/\lambda^2$, and the mean lifetime of $2/\lambda$ is a special case of this result.

We now seek the properties of V, the remaining lifetime of the component in current use. We have argued that its total lifetime, W, has density $wf(w)/\mu$. Conditional on $W = w$, the current time is chosen uniformly over the interval $(0, w)$, hence V has density $1/w$ on $0 < v < w$. The joint density of V and W is thus

$$g(v, w) = \frac{1}{w} \cdot \frac{wf(w)}{\mu} = \frac{f(w)}{\mu}, \qquad 0 < v < w < \infty,$$

so V has density $\int g(v, w)dw = \int_v^\infty \frac{f(w)}{\mu}dw = \frac{\mathcal{F}(v)}{\mu}$. Exactly the same argument shows that the *age* of the current component has this same density (but not independently, in general). This derivation of the density of V motivates the next definition.

Definition 8.24

An *equilibrium renewal process* differs from an ordinary renewal process only in that the lifetime of the first component has density $\mathcal{F}(t)/\mu$.

Theorem 8.25

Let $X_t^{(e)}$ be an equilibrium renewal process, and write $H^{(e)}(t) = E(X_t^{(e)})$. Then $H^{(e)}(t) = t/\mu$ for all $t \geq 0$, and so its derivative is $h^{(e)}(t) = 1/\mu$.

Proof

As in Theorem 8.18, we have $H^{(e)}(t) = \sum_{k=1}^\infty F_k^{(e)}(t)$, where $F_k^{(e)}(t)$ is the distribution function of the sum $T_1 + \cdots + T_k$. The density of T_1 is $\mathcal{F}(t)/\mu$, the rest have density $f(t)$. We will prove that the corresponding density function is $f_k^{(e)}(t) = (F_{k-1}(t) - F_k(t))/\mu$ when $k \geq 1$, with the convention that $F_0(t) - 1$.

This is true when $k = 1$ by the definition of the density of T_1. Suppose it also holds for $n = k \geq 1$. By the convolution formula (4.2),

$$\begin{aligned}
f_{n+1}^{(e)}(t) &= \int_0^t \frac{\mathcal{F}(u)}{\mu} f_n(t - u)du \\
&= \frac{1}{\mu}\left([-F_n(t-u)\mathcal{F}(u)]_0^t - \int_0^t F_n(t-u)f(u)du\right) \\
&= \frac{1}{\mu}\left(F_n(t) - \int_0^t f(u)F_n(t-u)du\right) = (F_n(t) - F_{n+1}(t))/\mu
\end{aligned}$$

which establishes the induction. Hence

$$h^{(e)}(t) = \sum_{k=1}^\infty f_k^{(e)}(t) = \sum_{k=1}^\infty \frac{1}{\mu}(F_{k-1}(t) - F_k(t)) = \frac{1}{\mu}.$$

Integration gives the expression for $H^{(e)}(t)$, since $H^{(e)}(0) = 0$.   □

In a room full of lightbulbs, replacing each of them is modelled by a renewal process. A caretaker charged with replacing all bulbs on failure is interested only in the pooled output of these processes, taken to be independent. Let $X_t^{(i)}$ denote the ith of N such processes running in parallel, and seek the properties of

$$Y_t = X_t^{(1)} + \cdots + X_t^{(N)}.$$

It is clear, even in the case $N = 2$, that Y_t is not, in general, a renewal process. But if $G_t(z)$ is the pgf of any $X_t^{(i)}$, all assumed to have the same distribution, then the pgf of Y_t is $(G_t(z))^N$. And if the system has been running for a long time, so that each is an equilibrium renewal process, we can obtain some of the properties of Y_t quite easily.

Let V denote the time until the next renewal anywhere. If V_i is the time to next renewal in the ith process, then $V = \min\{V_1, \ldots, V_n\}$. By the iid assumption, $P(V > v) = \Pi_{i=1}^{N} P(V_i > v) = (\int_v^\infty \mathcal{F}(x)dx/\mu)^N$. The density of V is

$$\frac{N\mathcal{F}(v)}{\mu} \left(\int_v^\infty \frac{\mathcal{F}(x)dx}{\mu} \right)^{N-1}. \tag{8.16}$$

For the Pareto density of Example 8.17 when $c > 1$, where $1/\mu = \beta(c-1)$, the density of V is seen to be $N\beta(c-1)/(1+\beta v)^{Nc-N+1}$, which is $\text{Par}(\beta, Nc-N)$. Hence the mean time to the next renewal is $(\beta(Nc-N+1))^{-1}$.

Over a long period T, each process has about T/μ renewals, so there will be about NT/μ in total. Hence the mean interval between renewals is estimated as $T/(NT/\mu) = \mu/N$, leading to the notation

$$W_N = \frac{N}{\mu} \min\{V_1, \ldots, V_N\}$$

as a suitably scaled time to the next renewal. We expect $E(W_N)$ to be close to unity. Using Equation (8.16), the density of W_N is

$$\frac{\mu}{N} \cdot \frac{N}{\mu} \mathcal{F}\left(\frac{\mu w}{N} \right) \left(\int_{\mu w/N}^\infty \frac{\mathcal{F}(x)dx}{\mu} \right)^{N-1}.$$

As $N \to \infty$, the lower limit in the integral decreases towards zero, so we rewrite this density in the form

$$\mathcal{F}\left(\frac{\mu w}{N} \right) \left(1 - \int_0^{\mu w/N} \frac{\mathcal{F}(x)dx}{\mu} \right)^{N-1}.$$

Use the Intermediate Value Theorem to write the integral as $\frac{\mu w}{N} \cdot \frac{\mathcal{F}(\xi)}{\mu}$ for some ξ, $0 \le \xi \le \mu w/N$. As $N \to \infty$, so $\mathcal{F}(\xi) \to 1$, and the limiting density of W_N is that of the limit of $(1 - w/N)^N$, i.e. $\exp(-w)$.

Thus, whatever the common lifetime distribution for a large number of similar renewal processes running independently in parallel, the time to wait for the next failure somewhere should have a distribution close to Exponential.

Example 8.26 (Discrete Time Renewal Processes)

This is a digression in a chapter that focuses on processes in continuous time. We retain the notation used earlier, so that X_n is the number of renewals after time zero up to and including time n. The times between renewals, T_n, are taken as iid with $P(T = k) = f_k$ and $\sum_{k=1}^{\infty} f_k = 1$. Thus simultaneous renewals do not occur. To avoid the problems associated with periodicity we found with Markov chains, we assume that the greatest common divisor of those times k with $f_k > 0$ is unity. Write $F(z) = \sum_{k=1}^{\infty} f_k z^k$ as the pgf of T, and let $u_n = P(\text{Renewal at time } n)$ with $U(z) = \sum_{n=0}^{\infty} u_n z^n$, taking $u_0 = 1$ as we begin with a new component.

By conditioning on the time of the first renewal, we have $u_n = \sum_{k=1}^{n} f_k u_{n-k}$ for $n \ge 1$ and hence, with exactly the same argument as in Theorem 7.9 which looked at returns to the origin in a random walk, $U(z) = 1 + F(z)U(z)$. Either of U or F determines the other. The pgf of $W_k = T_1 + \cdots + T_k$, the time of the kth renewal is $(F(z))^k$, and the relationship

$$W_k \le n \Leftrightarrow X_n \ge k$$

then yields the distribution of X_n.

For a simple example, take $f_1 = f_2 = 1/2$ so that $F(z) = (z + z^2)/2$, and $U(z) = (1 - z)^{-1}(1 + z/2)^{-1}$. Expand these by the binomial theorem, collect like powers of z, and find $u_n = 1 - \frac{1}{2} + \frac{1}{4} - \cdots \pm \frac{1}{2^n} = \frac{2}{3} + \frac{1}{3}(-\frac{1}{2})^n$. Thus $u_n \to 2/3 = 1/E(T)$; just as with returns to a recurrent state in discrete time Markov chains, this is a general result, and is proved in an identical fashion.

If lifetimes have a Geometric $G_1(p)$ distribution, $F(z) = pz/(1 - qz)$, so then $U(z) = (1 - qz)(1 - z)^{-1}$ and $u_n = p$ if $n \ge 1$. Such a simple result must have a simple explanation; Exercise 8.35 asks you to provide it.

EXERCISES

8.27. Find F, $\mathcal{F}$ and ϕ when T has the densities (a) $U(0,1)$ (b) $2/t^3$ for $t \ge 1$. Find $H(t)$ and $h(t)$ when each T_i is $E(\lambda)$.

8.28. An electronic counter records emissions from a radioactive source. At the beginning, and after each recording, the counter is "dead" for a random period, during which any emissions are missed. Set up a renewal process model for the instants when emissions are recorded.

8.29. Model lifetimes as having some $N(\mu, \sigma^2)$ distribution (taking μ/σ large enough to ignore the possibility of negative lifetimes). What then is the distribution of W_k? Deduce an expression for $P(X_t \geq k)$.

8.30. By considering $P(T > s + t \mid T > s)$, explain why the acronyms NBU and NWU might also be used to describe properties of lifetime distributions where the product $\mathcal{F}(s)\mathcal{F}(t)$ exceeds, or is less than, $\mathcal{F}(s + t)$ for all $s > 0$, $t > 0$.

8.31. Show that, when a lifetime T has the Pareto density $c\beta(1 + \beta t)^{-c-1}$ on $t > 0$, then $E(T_z) = E(T)(1 + \beta z)$, where T_z denotes the residual lifetime of a component aged z.

8.32. A manager can replace a component when it fails, at cost c_1, or when it has reached age ξ and is still working, at cost $c_2 < c_1$. Let K_n be the cost of the first n replacements. Show that

(a) $K_n/n \xrightarrow{a.s.} c_1 F(\xi) + c_2(1 - F(\xi)) = C$ (say), as $n \to \infty$.

(b) $K_{X(t)}/t \xrightarrow{a.s.} C/\int_0^\xi \mathcal{F}(u)du$ as $t \to \infty$.

Express this last result as an English sentence.

8.33. A different policy from the previous question is to replace on failure at cost c_1, and at the fixed times τ, 2τ, $3\tau, \ldots$, at cost c_3, irrespective of the age of the component. Show that the mean cost per unit time over a long period is $(c_1 H(\tau) + c_3)/\tau$.

8.34. In a discrete time renewal process, suppose $T = 1$, 2 or 3 with respective probabilities 1/2, 7/16 and 1/16. Find u_n explicitly, and verify that $u_n \to 16/25 = 1/E(T)$.

8.35. Provide the simple explanation for why $u_n = p$ for $n \geq 1$ if lifetimes are Geometric.

8.4 Brownian Motion: The Wiener Process

This topic has a prominent place in mathematical modelling. The name "Brownian motion" honours the botanist Robert Brown (1773–1851) who used his observations of the movements of pollen particles in a container of water as support for the hypothesis that all material is made up from molecules. In recent years, fluctuations in the prices of financial assets traded on stock markets have been modelled in a similar fashion. Large fortunes have been made, and lost, through actions taken on the basis of these mathematical models.

To motivate the formal definition, consider the simple symmetric random walk of Section 7.2, modified in two ways:

- the size of the step is changed from ± 1 to $\pm h$;

- the time interval between steps is δ, not unity.

We shall let $h \downarrow 0$ and $\delta \downarrow 0$ in such a way that a limit can be expected. Given $t > 0$, the number of steps taken before t differs from t/δ by at most unity; the mean and variance of step sizes are 0 and h^2, so the mean and variance of the position at time t, by summing independent steps, can be taken as 0 and $h^2 t/\delta$. In order that a limit will arise, take $\delta = h^2$, so that mean and variance are 0 and t. Let $S(t, h)$ denote the position of this random walk at time t.

As $h \downarrow 0$, so $\delta \downarrow 0$, the number of steps increases without bound, and the Central Limit Theorem implies that $S(t, h) \xrightarrow{D} N(0, t)$. This limiting process has other important properties: given $0 < u < v \le s < t$, consider what happens over the disjoint time intervals (u, v) and (s, t). For any given h, the random walks in these two intervals are independent, so the respective increments $S(v, h) - S(u, h)$ and $S(t, h) - S(s, h)$ are also independent. Moreover, if $t - s = v - u$, the statistical properties of these two increments are identical. We shall expect these properties to carry over to the limiting process, and to more than two disjoint time intervals.

Requiring a process to have certain properties is not enough to demonstrate that such a process can exist. That demonstration was accomplished by the mathematical prodigy Norbert Wiener (1894–1964), and the term "Brownian motion" is sometimes used interchangeably with "Wiener process". Grimmett and Stirzaker (1992) remark that Brownian motion is an observable physical phenomenon, and use the term in that sense only, reserving Wiener process for the mathematical model. We follow this lead.

Definition 8.27

A standard Wiener process $(W_t : t \ge 0)$ has the properties

(i) $W_0 = 0$ and $P(W_t$ is continuous for $t \ge 0) = 1$.

(ii) For any $u \ge 0$ and $t > 0$, the increment $W_{t+u} - W_u$ has the $N(0, t)$ distribution.

(iii) For all $n \ge 2$, and all choices of $0 \le t_1 \le t_2 \le \ldots \le t_{n+1} < \infty$, write $W^{(i)} = W_{t_{i+1}} - W_{t_i}$, for $1 \le i \le n$; then $W^{(1)}, \ldots, W^{(n)}$ are independent.

We reserve the notation W_t for a standard Wiener process henceforward.

Up to now, when we have met the Normal distribution, the context has been either one variable at a time, or a collection of independent variables. We will digress from Wiener processes to look at the properties of a collection of Normal variables that are not necessarily independent.

Definition 8.28

The *standard bivariate Normal* distribution has density

$$f(x,y) = \frac{1}{2\pi\sqrt{1-\rho^2}} \exp\left(-\frac{x^2 - 2\rho xy + y^2}{2(1-\rho^2)}\right)$$

for all real x and y.

Integrating out the other variable shows that when (X,Y) have this joint density, each of them is $N(0,1)$. Also

$$
\begin{aligned}
\mathrm{Cov}(X,Y) &= E(XY) = \int\int xy f(x,y)dydx \\
&= \frac{1}{2\pi\sqrt{1-\rho^2}} \int_{-\infty}^{\infty} x \left(\int_{-\infty}^{\infty} y \exp\left(-\frac{(x^2 - 2\rho xy + y^2)}{2(1-\rho^2)}\right) dy\right) dx.
\end{aligned}
$$

In the inner integral, put $y = \rho x + t\sqrt{1-\rho^2}$. This expression becomes

$$
\frac{1}{2\pi} \int_{-\infty}^{\infty} x \exp\left(-\frac{x^2}{2}\right) \left(\int_{-\infty}^{\infty} (\rho x + t\sqrt{1-\rho^2}) \exp\left(-\frac{t^2}{2}\right) dt\right) dx
$$

$$
= \frac{\rho}{2\pi} \int_{-\infty}^{\infty} x^2 \exp\left(-\frac{x^2}{2}\right) dx \int_{-\infty}^{\infty} \exp\left(-\frac{t^2}{2}\right) dt = \rho.
$$

Since each of X and Y has variance unity, the parameter ρ is the correlation between them. When $\rho = 0$, the density $f(x,y)$ factorises as the product $f_X(x).f_Y(y)$ showing that independence is the same as zero correlation for this distribution.

If X and Y have a standard bivariate Normal distribution, write $X_1 = \mu + \sigma X$, $Y_1 = \lambda + \tau Y$. Then X_1 is $N(\mu, \sigma^2)$, Y_1 is $N(\lambda, \tau^2)$, and $\mathrm{Cov}(X_1, Y_1) = \sigma\tau \mathrm{Cov}(X,Y)$, so that $\mathrm{Corr}(X_1, Y_1) = \rho$. Their joint density is

$$K \exp\left(\frac{-1}{2(1-\rho^2)}\left(\frac{(x-\mu)^2}{\sigma^2} - \frac{2\rho(x-\mu)(y-\lambda)}{\sigma\tau} + \frac{(y-\lambda)^2}{\tau^2}\right)\right), \quad (8.17)$$

where $K = 1/(2\pi\sigma\tau\sqrt{1-\rho^2})$. Let $\boldsymbol{x} = (x,y)^T$ and $\boldsymbol{\mu} = (\mu,\lambda)^T$ so that the expression in the exponent can be written as

$$-\frac{1}{2}(\boldsymbol{x} - \boldsymbol{\mu})^T V^{-1} (\boldsymbol{x} - \boldsymbol{\mu})$$

where $V = \begin{pmatrix} \sigma^2 & \rho\sigma\tau \\ \rho\sigma\tau & \tau^2 \end{pmatrix}$. The joint density is thus

$$\frac{1}{2\pi\sqrt{\det(V)}} \exp\left(-\frac{1}{2}(x-\mu)^T V^{-1}(x-\mu)\right).$$

Here $\det(\cdot)$ is the determinant. The diagonal entries of V give the variances of the components, the off-diagonal entries are their covariances, hence the terminology *variance–covariance matrix*. We use the more common name, *covariance matrix*.

This format indicates how to generalise to more than two components. Let V^{-1} be any invertible symmetric $n \times n$ matrix. A standard result from matrix algebra tells us that $V^{-1} = ADA^{-1}$, where $D = \text{Diag}(\lambda_1, \ldots, \lambda_n)$ is a diagonal matrix containing the eigenvalues of V^{-1}, and A is orthonormal. For notational simplicity, take $\mu = 0$, and write $x = (x_1, \ldots, x_n)^T$. Then

$$f(x) = K \exp\left(-\frac{1}{2}x^T V^{-1} x\right)$$

will be a density function provided that $K > 0$ and $\int f(x)dx = 1$. The latter condition can hold only if all the eigenvalues of V^{-1} are positive, otherwise the integral would be infinite when $K \neq 0$. Write $y = Ax$, so that the condition for a density function is

$$1 = K \int \exp\left(-\frac{1}{2}\sum \lambda_i y_i^2\right) dy = K \prod_{i=1}^{n} \sqrt{\frac{2\pi}{\lambda_i}},$$

hence $1/K^2 = (2\pi)^n / \det(D) = (2\pi)^n \det(V)$. Thus

$$f(x) = \frac{1}{\sqrt{(2\pi)^n \det(V)}} \exp\left(-\frac{1}{2}x^T V^{-1} x\right) \tag{8.18}$$

is a density function, so long as V^{-1} is positive-definite.

Definition 8.29

When $X = (X_1, \ldots, X_n)^T$ has the density function of Equation (8.18), it is said to have the *multivariate Normal distribution* $N(0, V)$. If μ is an n-vector, and $Y = \mu + X$, then Y has the distribution $N(\mu, V)$.

Theorem 8.30

If X is $N(0, V)$ and A is an $m \times n$ matrix of rank m, then $Y = AX$ is $N(0, AVA^T)$.

Proof

Suppose $m = n$. Then A is invertible, so for any $\mathcal{B}$,

$$P(\boldsymbol{Y} \in \mathcal{B}) = P(A\boldsymbol{X} \in \mathcal{B}) = P(\boldsymbol{X} \in A^{-1}\mathcal{B}) = \int_{A^{-1}\mathcal{B}} f(\boldsymbol{x})d\boldsymbol{x}.$$

In the integral, make the transform $\boldsymbol{y} = A\boldsymbol{x}$. We have

$$
\begin{aligned}
P(\boldsymbol{Y} \in \mathcal{B}) &= \int_{\mathcal{B}} f(A^{-1}\boldsymbol{y})(\det(A))^{-1}d\boldsymbol{y} \\
&= \frac{1}{\sqrt{(2\pi)^n \det(AVA^T)}} \int_{\mathcal{B}} \exp\left(-\frac{1}{2}\boldsymbol{y}^T(A^{-1})^T V^{-1} A^{-1} \boldsymbol{y}\right) d\boldsymbol{y} \\
&= \frac{1}{\sqrt{(2\pi)^n \det(AVA^T)}} \int_{\mathcal{B}} \exp\left(-\frac{1}{2}\boldsymbol{y}^T(AVA^T)^{-1}\boldsymbol{y}\right) d\boldsymbol{y}
\end{aligned}
$$

which establishes the result in this first case.

When $m < n$, the proof is more difficult, and is omitted. You can read it as Theorem 14.7 in Kingman and Taylor (1966). $\square$

Corollary 8.31

(i) X_i is $N(0, v_{ii})$; (ii) if $i \neq j$, the pair (X_i, X_j) have a bivariate Normal distribution with covariance matrix $\begin{pmatrix} v_{ii} & v_{ij} \\ v_{ji} & v_{jj} \end{pmatrix}$; (iii) if $\boldsymbol{c} = (c_1, \ldots, c_n)^T$ is a vector of constants, then $\boldsymbol{c}^T \boldsymbol{X}$ is $N(0, \boldsymbol{c}^T V \boldsymbol{c})$.

The statements in this corollary are special cases of the theorem, obtained when: (i) $A = \boldsymbol{e}_i^T = (0, \ldots, 1, \ldots, 0)$, the ith unit vector;
(ii) $A = \begin{pmatrix} \boldsymbol{e}_i^T \\ \boldsymbol{e}_j^T \end{pmatrix}$; (iii) $A = \boldsymbol{c}^T$.

This development brings out the central role of covariance for jointly Normal variables. From the definition of the Wiener process with $0 < u < v$, we have

$$
\begin{aligned}
\mathrm{Cov}(W_u, W_v) &= \mathrm{Cov}(W_u, W_u + (W_v - W_u)) \\
&= \mathrm{Cov}(W_u, W_u) + \mathrm{Cov}(W_u, W_v - W_u) = \mathrm{Var}(W_u) + 0 = u,
\end{aligned}
$$

i.e. for general s and t, $\mathrm{Cov}(X_s, X_t) = \min(s, t)$.

The scale change $X_t = \sigma W_t$ gives a process with all the properties of the standard Wiener process, except that the distribution is $N(0, \sigma^2 t)$. X_t is also referred to as a Wiener process, with σ^2 being its *diffusion coefficient*. In many applications, e.g. finance, there may be some expectation of a trend in the value of the asset being modelled, in which case $Y_t = \mu t + \sigma W_t$ has the $N(\mu t, \sigma^2 t)$ distribution, with μ being termed the *drift*. We shall look mainly at the standard

Wiener process, recognising that there is a simple transform to the cases when $\mu \neq 0$, or $\sigma^2 \neq 1$. The continuity of W_t shows that $P(W_t \geq x) = P(W_t > x)$ when $t > 0$, so we can afford to be a little careless in distinguishing strict and weak inequalities.

The reflection principle proved a useful tool with simple symmetric random walks, and there is a corresponding notion for the standard Wiener process. Given $t > 0$ and $x > 0$, define a new process $W_t^{(x)}$ as follows: if $W_s < x$ for $0 < s \leq t$, then $W_u^{(x)} = W_u$ on $0 \leq u \leq t$, while if $W_s = x$ for some s, $0 < s \leq t$, let $a = \inf\{s : W_s = x\}$ be the first time that W_s attains the value x. Write

$$W_u^{(x)} = \begin{cases} W_u & \text{for } 0 \leq u \leq a; \\ 2x - W_u & \text{for } a < u \leq t. \end{cases}$$

The process $(W_u^{(x)} : u \geq 0)$ is termed a Wiener process *reflected* in x.

The definition shows that if the value x is ever attained, then for $u \geq a$ we have $W_u^{(x)} = 2W_a - W_u = W_a - (W_u - W_a)$. The open intervals $(0, a)$ and (a, u) are disjoint, so the increment $W_u - W_a$ is independent of the process before time a; its distribution is symmetrical about zero, so $W_u^{(x)}$ has the same distribution as $W_a + (W_u - W_a) = W_u$ when $u \geq a$. Hence $W_u^{(x)}$ has the same statistical properties as W_u for all $u \geq 0$.

(Actually, we have glossed over a problem here. In the description of a Wiener process, independence of increments was assumed when the time points $\{u, v, s, t\}$ have been given as fixed, while the argument just deployed assumes we can also use this property when the non-overlapping intervals have been chosen by reference to the random time at which x is attained. The fact that this extension is permissible is the *strong Markov property* of the Wiener process, discussed in Section 13.4 of Grimmett and Stirzaker, 1992.)

Theorem 8.32

Let $M_t = \max\{W_s : 0 \leq s \leq t\}$. Then M_t has the same distribution as $|W_t|$.

Proof

Plainly $P(M_t \geq x) = P(M_t \geq x, W_t \geq x) + P(M_t \geq x, W_t < x)$. Since $W_t \geq x$ implies that $M_t \geq x$, the first term on the right is just $P(W_t \geq x)$. By the definition of a Wiener process reflected in x, the event $\{M_t \geq x, W_t < x\}$ is identical to $\{M_t \geq x, W_t^{(x)} > x\}$, so their probabilities are equal. But also $P(M_t \geq x, W_t^{(x)} > x) = P(W_t^{(x)} > x) = P(W_t > x)$; collecting these together, we have

$$P(M_t \geq x) = 2P(W_t \geq x).$$

Since W_t is $N(0, t)$, $P(|W_t| \geq x) = 2P(W_t \geq x)$, proving our claim. $\qquad\square$

This result is quite remarkable: it gives the distribution of the maximum value achieved by W_s over the period $0 \leq s \leq t$ in terms of the single quantity W_t. An easy integration shows that $E(M_t) = \sqrt{2t/\pi}$.

Corollary 8.33

Given $x > 0$, write $T_x = \inf\{t : W_t \geq x\}$ as the first time the Wiener process achieves level x. Then the density of T_x is

$$f(t) = \frac{x}{\sqrt{2\pi t^3}} \exp\left(-\frac{x^2}{2t}\right), \quad t > 0,$$

the so-called *inverse Gaussian distribution*.

Proof

$P(T_x \leq t) = P(M_t \geq x) = 2P(W_t \geq x)$, and W_t has the density $g(u) = \frac{1}{\sqrt{2\pi t}} \exp\left(\frac{-u^2}{2t}\right)$ of a $N(0, t)$ distribution. Thus $P(T_x \leq t) = 2\int_x^\infty g(u)du$; make the substitution $u = x\sqrt{t/y}$. Then

$$P(T_x \leq t) = 2\int_0^t g\left(x\sqrt{\frac{t}{y}}\right) \cdot \frac{x}{2}\sqrt{\frac{t}{y^3}}\,dy = \int_0^t f(y)dy,$$

after rearrangement, establishing our claim. $\qquad\square$

The theorem shows that $P(M_t \geq x) = 2P(W_t \geq x) = 2P(Z \geq x/\sqrt{t})$ where Z is $N(0, 1)$. Hence, whatever $x > 0$ is given, $P(M_t \geq x) \to 1$ as $t \to \infty$, i.e. the Wiener process is certain to reach any pre-assigned level sometime. On the other hand, the corollary shows that the mean time to reach x is $E(T_x) = \int_0^\infty \frac{x}{\sqrt{2\pi t}} \exp\left(-\frac{x^2}{2t}\right) dt$. Now when $t \geq x^2/2$, the term $\exp(-x^2/(2t))$ is at least e^{-1}, so $E(T_x) \geq e^{-1}\int_{x^2/2}^\infty \frac{x}{\sqrt{2\pi t}}dt$, which plainly diverges. The mean time to reach $x > 0$, no matter how small, is infinite.

The pair (W_t, M_t) are far from independent. We will find their joint density. When $m > w$ and $m > 0$, then

$$P(W_t \leq w) = P(W_t \leq w, M_t \leq m) + P(W_t \leq w, M_t > m).$$

By reflecting in the value of m, and recalling that we have taken $m > w$, we see that $P(W_t \leq w, M_t > m) = P(W_t^{(m)} \geq 2m - w)$, and this latter is the same as $P(W_t \geq 2m - w)$. Hence

$$P(W_t \leq w, M_t \leq m) = P(W_t \leq w) - P(W_t \geq 2m - w).$$

We know the density of W_t. Differentiation with respect to both m and w leads to the joint density

$$f(w, m) = \frac{2(2m - w)}{\sqrt{2\pi t^3}} \exp\left(-\frac{(2m - w)^2}{2t}\right) \qquad (8.19)$$

when $m > w$, $m > 0$ with, of course, $f(w, m) = 0$ if $m \leq 0$ or if $w > m$.

Example 8.34 (Hindsight)

Suppose John buys shares in Butterbean plc, intending to sell them at the highest price achievable before some time horizon T. By appropriate scale changes, take $T = 1$ and assume that $(W_t : t \geq 0)$ models the change in the share price after purchase at time zero.

Then M_1 is the maximum price that occurs while he holds the shares, and we know its distribution. We have calculated $E(M_t)$, so note that $E(M_1) = \sqrt{2/\pi} \approx 0.798$. But John has no way of knowing *at the time* that the maximum price over his time horizon has been reached: however well or badly his purchase is performing at the current time t, future changes have a mean value of zero. If he hangs on to the shares until $t = 1$, what can be said about his *regret*, the drop in value from the maximum?

Equation (8.19) gives the joint density of W_t and M_t. Write $Y_t = M_t - W_t$ so that Y_t is the regret at time t; we can deduce the joint density of W_t and Y_t, and then, by integrating out the variable w, we find a pleasing result: the density of Y_t is $\sqrt{2/(\pi t)} \exp(-y^2/(2t))$ on $y > 0$, i.e. Y_t has exactly the same density function as M_t!

Thus, at any time, the statistical properties of the maximum value achieved up to that time, and the drop from that maximum value, are identical. In particular, $E(Y_t) = \sqrt{2t/\pi}$, so $E(Y_1) = \sqrt{2/\pi} \approx 0.798$ is not only the expected value of his profit, had he had the foresight to sell at the peak price, it is also the unconditional mean value of his regret.

He may also be interested in the size of this regret for a given value of M_1. The definition of conditional density leads immediately to the expression

$$f(y|m) = f(y, m)/f_M(m) = \frac{m + y}{t} \exp\left(-\frac{y(2m + y)}{2t}\right), \quad y \geq 0.$$

for the density of $Y_t | M_t = m$. Thus

$$
\begin{aligned}
E(Y_t | M_t = m) &= \int_0^\infty y \cdot \frac{m + y}{t} \exp\left(\frac{-y(2m + y)}{2t}\right) dy \\
&= \left[-y \exp\left(\frac{-y(2m + y)}{2t}\right)\right]_0^\infty + \int_0^\infty \exp\left(\frac{-y(2m + y)}{2t}\right) dy \\
&= 0 + \int_0^\infty \exp\left(\frac{-y(2m + y)}{2t}\right) dy.
\end{aligned}
$$

This format shows clearly that $E(Y_t|M_t = m)$ is a decreasing function of m: the larger the maximum potential profit, the smaller the mean value of the regret. Simple manipulations on this last expression show that we can express $E(Y_t|M_t = m)$ as

$$\exp\left(\frac{m^2}{2}\right)\int_0^\infty \exp\left(\frac{-(m+y)^2}{2t}\right)\,dy = \exp\left(\frac{m^2}{2}\right)\sqrt{2\pi t}.P(Z \geq m\sqrt{t})$$

where Z is standard Normal. Some sample values in the case $t = 1$ may be interesting.

$m =$	0	0.5	1	1.5	2	3
$E(Y_1|M_1 = m) =$	1.253	0.876	0.656	0.516	0.421	0.372

As $E(W_1|M_1 = m) = m - E(Y_1|M_1 = m)$, there is a unique value m such that $E(W_1|M_1 = m) = 0$: that value turns out to be approximately 0.752. So if W_t ever reaches this level while he holds his shares, John can comfort himself with the thought that even if, in retrospect, this is the maximum value reached before $t = 1$, the final mean value will not be less than the purchase price.

Example 8.35 (Zeros of W_t)

The Wiener process has $W_0 = 0$, but what can be said about $P(u, v)$, the probability that W_t returns to its initial value over the time interval (u, v) when $0 \leq u < v$? Take $u > 0$, and condition on the value of W_u; write f_u as its density. Then

$$P(u, v) = \int_{-\infty}^\infty P(W_t \text{ has a zero in } (u, v)|W_u = x)f_u(x)du$$

$$= 2\int_{-\infty}^0 P(W_t \text{ has a zero in } (u, v)|W_u = x)f_u(x)du$$

by symmetry. Let $V_{t-u} = W_t - W_u$ for $t \geq u$, so that V_s is a Wiener process; given $W_u = x < 0$, the process W_t has a zero in $(u, v) \Leftrightarrow \max\{V_s : 0 \leq s \leq v - u\} \geq -x$. This maximum has the same distribution as $|V_{v-u}|$, hence $P(u, v)$ is

$$2\int_{-\infty}^0 \left(\int_{-x}^\infty \frac{2}{\sqrt{2\pi(v-u)}}\exp\left(\frac{-y^2}{2(v-u)}\right)dy\right)\frac{1}{\sqrt{2\pi u}}\exp\left(\frac{-x^2}{2u}\right)dx$$

$$= \frac{2}{\pi\sqrt{u(v-u)}}\int_{-\infty}^0\int_{-x}^\infty \exp\left(\frac{-y^2}{2(v-u)} - \frac{x^2}{2u}\right)dydx.$$

This double integral is over a wedge-shaped region of the second quadrant (draw it!), and we make the transformation from (x, y) to (r, θ) via

$$y = r\sqrt{2(v-u)}\sin(\theta), \quad x = r\sqrt{2u}\cos(\theta).$$

This gives

$$P(u,v) = \frac{2}{\pi} \int_0^\infty 2r \exp(-r^2)dr \int_{\pi/2}^{\pi/2+\alpha} d\theta$$

where $\tan(\alpha) = \sqrt{(v-u)/u}$, hence $\cos(\alpha) = \sqrt{u/v}$. We find

$$P(u,v) = \frac{2}{\pi} \arccos \sqrt{\frac{u}{v}} \qquad (8.20)$$

as the probability that W_t has a zero in the interval (u,v) where $0 < u < v$.

To extend this to the case $u = 0$, take $n \geq 2$ and $k \geq 1$, and let $A_{k,n}$ be the event that there is at least one zero in the interval $(1/n^{k+1}, 1/n)$. Equation (8.20) shows that $P(A_{k,n}) = \frac{2}{\pi} \arccos \sqrt{1/n^k}$. For fixed n, the events $(A_{k,n})$ form an increasing sequence, and their union A_n is the event that there is at least one zero in the open interval $(0, 1/n)$. Because $\arccos(1/n^k) \to \pi/2$ as $k \to \infty$, Theorem 1.5 shows that $P(A_n) = 1$. It is certain that the interval $(0, 1/n)$ contains a zero of W_t; and as this holds for all n, Equation (8.20) holds for $0 \leq u < v$. There is no "first" time the Wiener process returns to its start value.

A further appreciation of the erratic fluctuations of W_t (recalling that continuity is built in to its definition) arises if we attempt to differentiate it. The ratio $(W_{t+h} - W_t)/h$ has a $N(0, 1/h)$ distribution, so that, whenever $\epsilon > 0$ is given, $P(|W_{t+h} - W_t|/h > \epsilon) \to 1$ as $h \downarrow 0$. The ratio $(W_{t+h} - W_t)/h$ cannot possibly have a limit.

Example 8.36 (Brownian Bridge)

If we know the value of W_t for some $t > 0$, what can we say about the process W_s over $0 < s < t$? The unconditional joint distribution of (W_s, W_t) is bivariate Normal; in the expression (8.17) we have $\mu = \lambda = 0$, $\sigma^2 = s$, $\tau^2 = t$ and $\rho = \sqrt{s/t}$. Hence the density of W_s, conditional on $W_t = y$, is the ratio

$$\frac{1}{2\pi\sqrt{s(t-s)}} \exp\left(\frac{-t}{2(t-s)}\left(\frac{w^2}{s} - \frac{2wy}{t} + \frac{y^2}{t}\right)\right) \bigg/ \left(\frac{1}{\sqrt{2\pi t}} \exp\left(\frac{-y^2}{2t}\right)\right)$$

which simplifies to

$$\frac{1}{\sqrt{2\pi s(t-s)/t}} \exp\left(-\frac{(w - sy/t)^2}{2s(t-s)/t}\right), \quad -\infty < w < \infty.$$

This is the $N(sy/t, s(t-s)/t)$ distribution.

It will be no surprise that the distribution is Normal nor that, given $W_t = y$, the mean value of W_s is sy/t. Might we have guessed the expression $s(t-s)/t$

for the variance? Perhaps not that exact format, but it seems reasonable to expect the variance to be zero when $s = 0$ or $s = t$, maximal at $s = t/2$, and symmetrical about $s = t/2$.

Fixing $W_t = y$, and looking at W_s over $(0, t)$ is termed constructing a *Brownian bridge*; the other terminology used is *tied-down Wiener process*. One special case is when $y = 0$ and $t = 1$, where we note that $W_s | W_1 = 0$ is $N(0, s(1 - s))$ over $0 < s < 1$. Let $U_s = W_s - sW_1$ over this same interval. Plainly, U_s has a Normal distribution with mean zero, and $\text{Var}(U_s) = \text{Var}(W_s) - 2s\text{Cov}(W_s, W_1) + s^2\text{Var}(W_1) = s(1 - s)$, so U_s has the same distribution as $W_s | W_1 = 0$. It is an exercise to show that, if $0 < s < t < 1$, then $\text{Cov}(U_s, U_t) = s(1-t) = \text{Cov}(W_s | W_1 = 0, W_t | W_1 = 0)$. Now jointly distributed Normal variables are defined by their means and covariance matrices; the joint distribution of the collection $\{U_{s_i} : i = 1, \ldots, n\}$ is the same as that of the joint collection $\{W_{s_i} | W_1 = 0 : i = 1, \ldots, n\}$. Hence the statistical properties of $(U_s : 0 < s < 1)$ are the same as those of the tied-down Wiener process.

Example 8.37 (Drift and Diffusion)

Write $X_t = \mu t + \sigma W_t$, and suppose $\mu \neq 0$. Then the reflection principle does not apply to this process: we seek an alternative approach. One such is via the transformation $V_t \equiv \exp(-2\mu X_t / \sigma^2)$. The usefulness of this step stems from the following. Suppose $V_t = v$ corresponds to $X_t = x$, and take $s \geq 0$. Then

$$
\begin{aligned}
E(V_{t+s} | V_t = v) &= E(\exp\left(-\frac{2\mu X_{t+s}}{\sigma^2}\right) \Big| X_t = x) \\
&= E(\exp\left(-\frac{2\mu(X_t + (X_{t+s} - X_t))}{\sigma^2}\right) \Big| X_t = x) \\
&= \exp(-2\mu x/\sigma^2)E(\exp\left(-\frac{2\mu X_s}{\sigma^2}\right))
\end{aligned}
$$

because, conditional on $X_t = x$, the increment $X_{t+s} - X_t$ has the same distribution as X_s, and is independent of the value of x. Since X_s has the $N(\mu s, \sigma^2 s)$ distribution, its mgf $E(\exp(zX_s))$ is $\exp(z\mu s + z^2\sigma^2 s/2)$, and so this last expression is

$$
\exp(-2\mu x/\sigma^2)\exp(-2\mu^2 s/\sigma^2 + (-2\mu/\sigma^2)^2\sigma^2 s/2) = \exp(-2\mu x/\sigma^2) = v.
$$

Thus $E(V_{t+s} | V_t = v) = v$, i.e. $E(V_{t+s} | V_t) = V_t$. On average, V_u remains constant. As an application, suppose T is the first time this process reaches either the level a, where $a > 0$, or $-b$, where $b > 0$. Then (and this also needs an appeal to the strong Markov property), $E(V_T) = V_0 = 1$. If $p = P(X_t$ hits $+a$ before $-b)$, then V_T is either $\exp(-2a\mu/\sigma^2)$ or $\exp(2b\mu/\sigma^2)$, with respective probabilities p and $1 - p$. Hence

$$1 = p\exp(-2a\mu/\sigma^2) + (1 - p)\exp(2b\mu/\sigma^2),$$

from which we see that

$$p = \frac{\exp(2b\mu/\sigma^2) - 1}{\exp(2b\mu/\sigma^2) - \exp(-2a\mu/\sigma^2)}.$$

Consider the case when $\mu < 0$ so that, in the long run, X_t drifts to $-\infty$ with probability one. Let $b \to \infty$. Then p is interpreted as the probability the process ever reaches the value $+a$, and this limiting value of p is $\exp(2\mu a/\sigma^2)$. This is the same as saying that, for this process with negative drift, the probability its maximum value is at least $+a$ is $\exp(2\mu a/\sigma^2)$.

When $\mu > 0$, so that level $+a$ is certainly attained sometime, let T be the time required. To find the properties of T, let λ be a parameter, and define $U_t \equiv \exp(\lambda X_t - (\lambda\mu + \lambda^2\sigma^2/2)t)$. By a similar argument to that given for V_t above, we can show that, if $s \geq 0$, then

$$E(U_{t+s}|U_t = u) = u,$$

whatever the value of λ. Hence, since $U_0 = 1$,

$$
\begin{aligned}
1 &= E(U_T) = E(\exp(\lambda X_T - (\lambda\mu + \lambda^2\sigma^2/2)T)) \\
&= E(\exp(\lambda a - (\lambda\mu + \lambda^2\sigma^2/2)T)) = e^{\lambda a}E(e^{-\theta T})
\end{aligned}
$$

where $\theta = \lambda\mu + \lambda^2\sigma^2/2$. Thus $E(e^{-\theta T}) = e^{-\lambda a}$. The relationship between θ and λ can be written as a quadratic in λ, and the relevant solution is $\lambda = (\sqrt{\mu^2 + 2\theta\sigma^2} - \mu)/\sigma^2$. This gives

$$E(e^{-\theta T}) = \exp\left(-\frac{a}{\sigma^2}(\sqrt{\mu^2 + 2\theta\sigma^2} - \mu)\right)$$

as the Laplace transform of T. Tables of transforms give a neat result: the density of T is

$$\frac{a}{\sigma\sqrt{2\pi t^3}}\exp\left(-\frac{(a - \mu t)^2}{2\sigma^2 t}\right), \quad t > 0.$$

This is a variant on the standard Inverse Gaussian distribution, which we have already seen in this problem in the case $\mu = 0$.

EXERCISES

8.36. Show that both $(tW_{1/t} : t > 0)$ and $(W_{c^2t}/c : t > 0)$ are Wiener processes.

8.37. We have seen that $E(M_t) = \sqrt{2t/\pi}$. Now find $\text{Var}(M_t)$.

8.38. Write $Y_t = M_t - W_t$. Find the distribution of $M_t | Y_t = 0$, and hence show that $P(M_t > m | Y_t = 0) = \exp(-m^2/(2t))$ when $m > 0$.

8.39. Given $t > 0$, let A_t be the last time before t that $W_t = 0$, and let B_t be the first time after t that $W_t = 0$. Show that

$$P(A_t < x) = \frac{2}{\pi} \arcsin \sqrt{\frac{x}{t}}, \quad P(B_t < y) = \frac{2}{\pi} \arccos \sqrt{\frac{t}{y}}$$

when $0 \le x \le t$ and $y \ge t$. Find $P(A_t < x, B_t > y)$. Deduce the means of A_t, B_t.

8.40. Show that $P(M_t = W_t$ at least once in $u \le t \le v) = \frac{2}{\pi} \arccos \sqrt{\frac{u}{v}}$.

8.41. Consider Example 8.37, the Wiener process with drift and diffusion. Given a, μ and σ^2, what value of b makes it equally likely that X_t hits $+a$ or $-b$ first?

Find the distribution of $\min(X_t : t \ge 0)$ when $\mu = \sigma^2/2$.

8.42. Explain intuitively why it is not surprising that $E(Y_t | M_t = m)$ decreases with m, in Example 8.34.

9
Appendix: Common Distributions and Mathematical Facts

For handy reference, we collect together the definition and main properties of the probability distributions we have encountered, and some useful facts of mathematical life. The symbols μ and σ^2 denote the mean and variance.

9.1 Discrete Distributions

Here we take $0 < p < 1$, $q = 1 - p$, and $\lambda > 0$, and use the convention that $\binom{a}{b} = 0$ whenever $b < 0$ or $b > a$. Other restrictions will be specified. Where the pgf does not usefully simplify, it is omitted.

1. The Bernoulli distribution has $P(X = 1) = p$, $P(X = 0) = q$. It is $\text{Bin}(1, p)$, with $\mu = p$, $\sigma^2 = pq$ and pgf $q + pz$.

2. The Binomial distribution $\text{Bin}(n, p)$ has $P(X = k) = \binom{n}{k}p^k q^{n-k}$ for $0 \leq k \leq n$. $\mu = np$, $\sigma^2 = npq$ and the pgf is $(q + pz)^n$.

3. The Geometric distribution $G_0(p)$ has $P(X = k) = q^k p$ for $k = 0, 1, \ldots$. $\mu = q/p$, $\sigma^2 = q/p^2$ and the pgf is $p/(1 - qz)$, valid for $|z| < 1/q$.

4. The Geometric distribution $G_1(p)$ has $P(X = k) = q^{k-1}p$ for $k = 1, 2, \ldots$. $\mu = 1/p$, $\sigma^2 = q/p^2$ and the pgf is $pz/(1 - qz)$, valid for $|z| < 1/q$.

5. The Hypergeometric distribution with parameters M, N and n has $P(X =$

$k) = \binom{M}{k}\binom{N}{n-k}/\binom{M+N}{n}$. $\mu = Mn/(M+N)$, $\sigma^2 = MNn(M+N-n)/((M+N)^2(M+N-1))$.

6. The Negative Binomial distribution $\mathrm{NB}_0(r,p)$ has $P(X = k) = \binom{k+r-1}{r-1}p^r q^k$ for $k \geq 0$. Here $\mu = rq/p$, $\sigma^2 = rq/p^2$ and the pgf is $p^r/(1 - qz)^r$ when $|z| < 1/q$.

7. For $r \geq 1$, the Negative Binomial distribution $\mathrm{NB}_r(r,p)$ has $P(X = k) = \binom{k-1}{r-1}p^r q^{k-r}$ for $k \geq r$. Here $\mu = r/p$, $\sigma^2 = rq/p^2$, and the pgf is $p^r z^r/(1 - qz)^r$ when $|z| < 1/q$.

8. The Poisson distribution $\mathrm{Poiss}(\lambda)$ has $P(X = k) = \exp(-\lambda)\lambda^k/k!$. $\mu = \lambda$, $\sigma^2 = \lambda$, and the pgf is $\exp(\lambda(z - 1))$.

9. The (discrete) Uniform distribution $U(m,n)$ has $P(X = k) = 1/(n-m+1)$ for $m \leq k \leq n$. $\mu = (m + n)/2$, $\sigma^2 = ((n - m + 1)^2 - 1)/12$, and the pgf is $z^m(1 - z^{n-m})/(1 - z)$ when $z \neq 1$.

9.2 Continuous Distributions

The density is defined on the whole real line unless otherwise stated. Again, the mgf is omitted when its form is not useful, as well as when it does not exist.

1. The Beta distribution $B(\alpha, \beta)$ with $\alpha > 0$ and $\beta > 0$ has density $Kx^{\alpha-1}(1-x)^{\beta-1}$ for $0 < x < 1$, where $K = \Gamma(\alpha + \beta)/(\Gamma(\alpha)\Gamma(\beta))$. $\mu = \alpha/(\alpha + \beta)$, $\sigma^2 = \alpha\beta/((\alpha + \beta)^2(\alpha + \beta + 1))$.

2. The Cauchy distribution has density $a/(\pi(a^2 + (x - \theta)^2))$. Its mean does not exist.

3. The Chi-square distribution on n degrees of freedom, χ_n^2, is the Gamma $\Gamma(n/2, 1/2)$ distribution (see below). $\mu = n$, $\sigma^2 = 2n$, and the mgf is $(1 - 2t)^{-n/2}$ when $|t| < 1/2$.

4. The Exponential distribution $E(\lambda)$ has density $\lambda \exp(-\lambda x)$ on $x > 0$. $\mu = 1/\lambda$, $\sigma^2 = 1/\lambda^2$, and the mgf is $\lambda/(\lambda - t)$ for $t < \lambda$.

5. The Gamma distribution $\Gamma(\alpha, \lambda)$ with $\alpha > 0$ and $\lambda > 0$ has density $Kx^{\alpha-1}\exp(-\lambda x)$ on $x > 0$, where $K = \lambda^\alpha/\Gamma(\alpha)$. $\mu = \alpha/\lambda$, $\sigma^2 = \alpha/\lambda^2$, and the mgf is $\lambda^\alpha/(\lambda - t)^\alpha$ for $|t| < \lambda$.

6. The Inverse Gaussian distribution has density $\frac{a}{\sqrt{2\pi x^3}} \exp\left(\frac{-a^2}{2x}\right)$ on $x > 0$. Its mean is infinite.

7. The Laplace (or double exponential) distribution has density $K \exp(-K|x - \theta|)/2$. $\mu = \theta$, $\sigma^2 = 1/K^2$.

8. The Normal distribution $N(\mu, \sigma^2)$ has density $\frac{1}{\sigma\sqrt{2\pi}} \exp\left(-\frac{(x-\mu)^2}{2\sigma^2}\right)$. The parameters are the mean and variance, the mgf is $\exp(\mu t + \sigma^2 t^2/2)$.

9. The Pareto distribution $\text{Par}(\beta, c)$ has density $c\beta/(1 + \beta x)^{c+1}$ on $x > 0$. $\mu = 1/(\beta(c-1))$ if $c > 1$, $\sigma^2 = 1/(\beta^2(c-1)^2(c-2))$ if $c > 2$.

10. The (continuous) Uniform distribution $U(a, b)$ has density $1/(b - a)$ on $a < x < b$. $\mu = (a+b)/2$, $\sigma^2 = (b-a)^2/12$, the mgf is $\frac{e^{tb} - e^{ta}}{t(b-a)}$ if $t \neq 0$.

11. The Weibull distribution has density $\alpha\rho(\rho x)^{\alpha-1} \exp(-(\rho x)^\alpha)$ on $x > 0$. $\mu = \Gamma(1 + 1/\alpha)/\rho$, $\sigma^2 = \Gamma(1 + 2/\alpha)/\rho^2 - \mu^2$.

9.3 Miscellaneous Mathematical Facts

The title speaks for itself. Any student who has hard-wired the items in this list alongside their multiplication tables will save *hours* of time, over one for whom this is uncertain territory.

A. Finite Sums

(i) $1 + 2 + 3 + \cdots + n = n(n+1)/2$.

(ii) $1^2 + 2^2 + 3^2 + \cdots + n^2 = n(n+1)(2n+1)/6$.

(iii) $1 + x + x^2 + \cdots + x^n = (1 - x^{n+1})/(1 - x)$ if $x \neq 1$.

(iv) $1 + 2x + 3x^2 + \cdots + nx^{n-1} = (1 - (n+1)x^n + nx^{n+1})/(1 - x)^2$ if $x \neq 1$.

(v) $(a + b)^n = \sum_{r=0}^{n} \binom{n}{r} a^{n-r} b^r = a^n + na^{n-1}b + \binom{n}{2}a^{n-2}b^2 + \cdots + b^n$.

B. Series

(i) $1 + x + x^2 + x^3 + \cdots = (1 - x)^{-1}$ if $|x| < 1$.

(ii) $1 + 2x + 3x^2 + 4x^3 + \cdots = (1 - x)^{-2}$ if $|x| < 1$.

(iii) $x - \frac{x^2}{2} + \frac{x^3}{3} - \frac{x^4}{4} + \cdots = \ln(1 + x)$ if $-1 < x \leq 1$.

(iv) $-x - \frac{x^2}{2} - \frac{x^3}{3} - \frac{x^4}{4} - \cdots = \ln(1 - x)$ if $-1 \leq x < 1$.

(v) $1 + x + \frac{x^2}{2!} + \frac{x^3}{3!} + \cdots = e^x = \exp(x)$ for all x.

(vi) $(1 - x)^{-\alpha} = 1 + \alpha x + \frac{\alpha(\alpha+1)x^2}{2!} + \frac{\alpha(\alpha+1)(\alpha+2)x^3}{3!} + \cdots$ if $|x| < 1$.

(vii) $(1 - 2x)^{-\beta/2} = 1 + \beta x + \frac{\beta(\beta+2)x^2}{2!} + \frac{\beta(\beta+2)(\beta+4)x^3}{3!} + \cdots$ if $|x| < 1/2$.

(viii) $1 + \frac{1}{2^2} + \frac{1}{3^2} + \frac{1}{4^2} + \cdots = \frac{\pi^2}{6} \approx 1.645$.

C. Integration

(i) $\int_0^\infty x^n e^{-x} dx = n!$ for $n = 0, 1, 2, \ldots$.

(ii) $\int_0^\infty x^n e^{-\alpha x} dx = n!/\alpha^{n+1}$ for $n = 0, 1, 2, \ldots$.

(iii) $\int_0^\infty x^{\alpha-1} e^{-x} dx = \Gamma(\alpha)$ for $\alpha > 0$.

(iv) $\int_{-\infty}^\infty \exp(-x^2) dx = \sqrt{\pi}$.

D. Approximations and Limits

We write $a(n) \sim b(n)$ to mean that $a(n)/b(n) \to 1$ as $n \to \infty$.

(i) Stirling's formula is $n! \sim \sqrt{2\pi} e^{-n} n^{n+1/2}$. Among its consequences are $\binom{2n}{n} \frac{1}{2^n} \sim \frac{1}{\sqrt{\pi n}}$, and $\binom{2n}{n} p^n q^n \sim \frac{(4pq)^n}{\sqrt{\pi n}}$. Stirling's formula is excellent, even for low values of n.

(ii) The harmonic series $1 + \frac{1}{2} + \frac{1}{3} + \cdots + \frac{1}{n} \sim \ln(n)$.

Even better, $1 + \frac{1}{2} + \frac{1}{3} + \cdots + \frac{1}{n} - \ln(n) \to \gamma = 0.5772\ldots$ as $n \to \infty$.

(iii) $\left(1 + \frac{x}{n}\right)^n \to \exp(x)$ as $n \to \infty$. Extending this, if $\alpha(n) \to \alpha$ and $\beta(n) \to \beta$, then $\left(1 + \frac{\alpha(n)}{n}\right)^{n\beta(n)} \to \exp(\alpha\beta)$.

(iv) The statement $f(t) = o(g(t))$ means that the function $f(t)$ is of a smaller order of magnitude than the function $g(t)$, as $t \to$ some limit (often zero), in the sense that the ratio $f(t)/g(t) \to 0$.

(v) Similarly, we write $f(t) = O(g(t))$ to mean that $f(t)$ is "of the order of" $g(t)$, in the sense that the ratio $|f(t)/g(t)|$ is bounded above as $t \to$ some limit. (If $f(t) = o(g(t))$, then $f(t) = O(g(t))$, but not necessarily conversely.)

E. Notation

(i) $(n)_k = n(n-1)\cdots(n-k+1)$.
(ii) iidrv means independent, identically distributed random variables.
(iii) CLT means Central Limit Theorem.
(iv) WLLN (and SLLN) are the Weak (and Strong) Laws of Large Numbers.

Bibliography

Anděl, J (2001) Mathematics of Chance. Wiley.

Box, GEP and Muller, ME (1958) A note on the generation of random normal deviates. Annals of Mathematical Statistics 29, 610–11.

Breiman, L (1961) Optimal gambling systems for favorable games. Proceedings of the Fourth Berkeley Symposium on Mathematical Statistics and Probability, Vol 1, 65–78.

Burke, PJ (1956) The output of a queuoing system Operations Research 4, 699–704.

Capiński, M and Kopp, PE (1999) Measure, Integral and Probability. Springer, SUMS.

Chung, KL (1960) Markov Chains with Stationary Transition Probabilities. Springer. (Second edition 1967.)

Daley, DJ and Kendall, DG (1965) Stochastic rumours. Journal of the Institute of Mathematics and its Applications 1, 42–55.

Goldie, CM (1977) Lanchester square-law battles: transient and terminal distributions. Journal of Applied Probability 14, 604–10.

Grimmett, GR and Stirzaker, DR (1992) Probability and Random Processes. Second edition. Oxford University Press. (Third edition 2001.)

Grinstead, CM and Snell, JL (1997) Introduction to Probability. McGraw-Hill.

Harmer, GP and Abbott, D (1999) Parrondo's paradox. Statistical Science 14, 206–13.

Harper, LH (1967) Stirling behavior is asymptotically normal. Annals of Mathematical Statistics 38, 410–14.

Kelly, JL Jr (1956) A new interpretation of information rate. Bell Systems Technical Journal 35, 917–26.

Kendall, DG (1951) Some problems in the theory of queues. Journal of the Royal Statistical Society, Series B, 13, 151–85.

Kingman JFC and Taylor, SJ (1966) Introduction to Measure and Probability. Cambridge University Press.

Lindley, DV (1952) The theory of queues with a single server. Proceedings of the Cambridge Philosophical Society 48, 277–89.

Maehara, H and Ueda, S (2000) On the waiting time in a janken game. Journal of Applied Probability 37, 601–5.

Norris, JR (1997) Markov Chains. Cambridge University Press.

Ross, S (1998) A First Course in Probability, Fifth edition. Prentice-Hall.

Simpson, EH (1951) The interpretation of interaction in contingency tables. Journal of the Royal Statistical Society, Series B, 13 238–41.

von Bortkewitsch (1898) Das Gesetz der Kleinen Zahlen. Teubner. (For a more accessible version of the data, see page 46 of "Biostatistics: concepts and applications for biologists", Brian Williams (1993). Chapman and Hall.)

Solutions

Some of these solutions are skeletal, and the notation is occasionally taken for granted. But enough detail is (believed to be) included for a full solution to be concocted.

Chapter 1

1.1. $P(A) = P(A \cap B) + P(A \cap B^c)$ and $P(B) = P(A \cap B) + P(A^c \cap B)$. But $P(\text{Exactly one}) = P((A \cap B^c) \cup (A^c \cap B)) = P(A \cap B^c) + P(A^c \cap B)$ (disjoint). Hence $P(\text{Exactly one}) = P(A) + P(B) - 2P(A \cap B)$.

1.2. $P(A \cap B) = P(A) + P(B) - P(A \cup B) = 0.6 + 0.7 - 0.8 = 0.5$. City have a 50% chance of being champions.

1.3. It holds for $n = 2$. Assuming it holds for N, then

$$P\left(\bigcup_{n=1}^{N+1} A_n\right) = P\left(\bigcup_{n=1}^{N} A_n \cup A_{N+1}\right) \leq P\left(\bigcup_{n=1}^{N} A_n\right) + P(A_{N+1})$$

using the known result for the union of two events. Now use the inductive hypothesis for N, to complete the induction step. For the other result, apply Boole's inequality to $\{A_i^c\}$, and note that $1 - P(\cap A_i) = P((\cap A_i)^c) = P(\cup A_i^c)$.

1.4. $P(\cap_{n=1}^{\infty} B_n) = 1 - P((\cap_{n=1}^{\infty} B_n)^c) = 1 - P(\cup_{n=1}^{\infty} B_n^c) = 1 - \lim P(B_n^c)$ using Theorem 1.5. The last term is $1 - \lim(1 - P(B_n)) = \lim P(B_n)$.

1.5. Write $A_n = B_n^c$, and use a similar argument to the last exercise.

1.6. The case $n = 2$, and all values of k, is just the last part of Theorem 1.2. Suppose inductively we have proved the case of N and all values of k. Then

$$P\left(\bigcup_{n=1}^{N+1} A_n\right) = P\left(\bigcup_{n=1}^{N} A_n\right) + P(A_{N+1}) - P\left(\bigcup_{n=1}^{N} (A_n \cap A_{N+1})\right).$$

Fix the value of k; apply the inductive assumption for the union of N terms to the two expressions on the right side, and watch the result emerge.

1.7. Ω has eight outcomes such as HHH, HTT etc., each with probability $1/8$. Two of these – HHH, TTT – have all results the same, so the chance is $1/4$. The nub of the flaw is the imprecision of the phrase "the third coin"; not until we have seen all the outcomes do we know which is this third coin.

1.8. The chance they are all different is $(100)_n/100^n$. When $n = 16$, the value is 0.2816, so the chance of at least one duplicate is 0.7184. Peter is strongly favoured to win; when $n = 12$, the chances are very nearly equal.

1.9. We have $-\ln(1-x) = x + x^2/2 + x^3/3 + \cdots$. Thus the lower bound is clear; and if $n \geq 3$ and $x < 1/2$, then $x^n/n < x^2/2^{n-1}$, which gives the upper bound. Write $(n)_K/n^K$ as the product of terms $(1 - i/n)$, take logs and sum for the bounds on $-\ln(p_K)$. Finally, when K is near $\sqrt{-2n\ln(p)}$, so $K(K-1)/(2n) \approx -\ln(p)$, and differs from $-\ln(p_K)$ by terms of order $1/\sqrt{n}$.

1.10. $P(r \text{ correct }) = \binom{8}{r}\binom{4}{6-r}/\binom{12}{6}$. So $P(4) = 420/924$, $P(5) = 224/924$ and $P(6) = 28/924$, and $P(4+) = 672/924 = 8/11$.

1.11. (a) There are $4\binom{13}{5} = 5148$ flushes, of which 40 are Straight Flushes.
(b) Plainly $13 \times 48 = 624$ hands have Four of a Kind.
(c) $\binom{13}{2} = 78$ ways to select the ranks of the pairs, 6 ways each to choose two cards from any rank. 44 possibilities for the last card, so $78 \times 6 \times 6 \times 44 = 123\,552$ hands with Two Pairs.

1.12. (a) There are $\binom{49}{6}$ ways to select six numbers, $\binom{49-r}{6}$ to select them, avoiding the first r. The probability none of the first r are among the winning numbers is the ratio of these, which reduces to $x(r)$.
(b) Whatever the value of i, $P(A_i) = x(1)^n$; similarly, if $i < j$, $P(A_i \cap A_j) = x(2)^n$. Similar expressions hold for three and more numbers; as there are $\binom{49}{k}$ ways to select k numbers, the value of S_k in Corollary 1.3 is $\binom{49}{k}x(k)^n$. This corollary is precisely what we seek here – the chance that at least one number does not appear within n draws. When $n = 50$, $S_1 = 0.0714$, $S_2 = 0.00216$, $S_3 = 0.000037$ so Bonferroni's inequalities show the chance is 0.0693.

1.13. Ω is the unit disc, $P(A) = (\text{Area of } A)/\pi$. Area within 0.5 of centre is $\pi/4$, so $p = 1/4$. The chord with $y = 1/\sqrt{2}$ has angle $\pi/2$ at the centre, so the total area of the segment when $y > 1/\sqrt{2}$ is $\pi/4 - 1/2$. Hence the probability is $1/4 - 1/(2\pi) \approx 0.0908$. The region $|x - y| < 1$, $|x + y| < 1$ is a square of side $\sqrt{2}$, so the probability is $1 - 2/\pi \approx 0.3634$.

1.14. The sources of help are x, $1 - x$ and $|x - y|$ away. Draw the diagram in the unit square to see that, given $0 \leq t \leq 1/2$, the area of the region when at least one of these occurs is $1 - (1 - 2t)^2 = 4t(1 - t)$.

1.15. Our model is that the centre $C = (x, y)$ of the coin has a uniform distribution over the square in which it falls. To win a prize, C must be no closer than $d/2$ to any side of the square, i.e. both $d/2 < x < s - d/2$ and $d/2 < y < s - d/2$. The chance of a prize is $(s - d)^2/s^2$.

Chapter 2

2.1. (a) $P(B, \text{then } R) = (8/14) \times (6/13) = 24/91$.
(b) The answer is the same - think about it.
(c) $P(\text{Same}) = P(BBB) + P(RRR) = (8 \times 7 \times 6 + 6 \times 5 \times 4)/(14 \times 13 \times 12) = 19/91$.

2.2. Expand the right side using the definition of $P(E|F)$.
Let $A = \{1\}$ and $B = \{1, 2, 3\}$. Then $P(A|B) = 1/3$, also $P(A^c|B^c) = 1$, so (a) is false. Also $P(A|B^c) = 0$, so (b) is false. But (c) is true, by Theorem 2.2.

2.3. Let A = See Red, B = Other side is Red. Plainly $P(A) = 1/2$. So $P(B|A) = P(A \cap B)/P(A) = (1/3)/(1/2) = 2/3$.

The argument given fails as seeing Red does more than just eliminate the BB card – it makes the RR card more likely than the RB card.

2.4. Write $P(A) = P(A|C)P(C) + P(A|C^c)P(C^c)$, similarly $P(B)$, and rearrange $P(A) - P(B)$.

2.5. Best tactics are to put one Black ball in one urn, all other balls in the other urn. Then $P(\text{Black}) = 0.5(1 + (b-1)/(w+b-1)) = x$, say. To prove this is best, take $k \le (w+b)/2$ and put k balls (r Black) in one urn, the rest in the other. Then

$$P(\text{Black}) = \frac{1}{2}\left(\frac{r}{k} + \frac{b-r}{w+b-k}\right) = \frac{kb + r(w+b-2k)}{2k(w+b-k)}.$$

Given k, this is largest when r is as large as possible. Plainly $r \le k$ and $r \le b$; taking $r = b$ (if possible) makes $P(\text{Black})$ no more than 0.5, so not as much as x. With $r = k$, the expression is $(2b + w - 2k)/(2b + 2w - 2k)$, which is maximised when k is minimised – i.e. $k = 1$, as claimed.

2.6. If she sticks, $P(\text{Win}) = 1/3$ (guessing). Suppose she picks A, and swaps. If A were the winning box (probability 1/3), she loses. Otherwise (probability 2/3), the host must open the only other losing box, so swapping inevitably gets the winning box. Swapping doubles her winning chances from 1/3 to 2/3. (N.B. The precise wording of this question is important!)

2.7. $P(\text{Score}) = (2/3) \times (3/4) + (1/3) \times (1/4) = 7/12$. Hence $P(\text{Good}|\text{Score}) = P(\text{Score}|\text{Good}) \times P(\text{Good})/P(\text{Score}) = (2/3) \times (3/4)/(7/12) = 6/7$.

2.8. $P(\text{Formal}) = (1/2) \times (2/3) + (1/3) \times (1/2) + (1/6) \times (1/3) = 5/9$. Then $P(\text{North.}|\text{Formal}) = P(\text{North.} \cap \text{Formal})/P(\text{Formal}) = (1/18)/(5/9) = 1/10$.

2.9. $P(R) = n/1000$, so $P(S) = 0.95(n/1000) + 0.05(1000 - n)/1000 = (50 + 0.9n)/1000$. Thus $P(R|S) = (0.95n/1000)/((50 + 0.9n)/1000) = 19n/(1000 + 18n)$.

2.10. $P(\text{All } H) = 1/n + ((n-1)/n) \times 2^{-k} = ((n-1)2^{-k} + 1)/n$. Hence $P(\text{Double H}|\text{All } H) = P(\text{Double H and All } H)/P(\text{All } H) = 2^k/(2^k + n - 1)$.

2.11. Let A = Mother a carrier, B = Information on Penny, and C = Karen's first child has the haemophilia gene. Then

$$P(C) = P(C|A)P(A) + P(C|A^c)P(A^c) = (1/4)P(A) + 0 = P(A)/4.$$

Hence we evaluate $P(A|B)/4$.

(a) $P(B) = P(B|A)P(A) + P(B|A^c)P(A^c) = 0.5(P(B|A) + P(B|A^c))$. Here $P(B|A) = (1/2)(1/2) + (1/2).1 = 3/4$ and $P(B|A^c) = 1$, so $P(B) = 7/8$. Hence $P(A|B) = P(B|A)P(A)/P(B) = (3/4)(1/2)/(7/8) = 3/7$. So $P(C) = 3/28$.

(b) Now $P(B|A) = (1/2)(1/4) + (1/2).1 = 5/8$, again $P(B|A^c) = 1$, so $P(B) = 13/16$. Thus $P(A|B) = (5/8)(1/2)/(13/16) = 5/13$, and $P(C) = 5/52$.

(c) But now $P(A|B) = 1$, so $P(C) = 1/4$.

2.12. Write A = It was £10 and G = Gina says it was £10. Then $P(G) = P(G|A)P(A) + P(G|A^c)P(A^c) = 0.95 \times 0.4 + 0.05 \times 0.6 = 0.41$, and Bayes gives $0.95 \times 0.4/P(G) = 38/41$. Let D =Derek says it was £20; then $P(D) = 0.10 \times (38/41) + 0.80 \times (3/41) = 6.2/41$, and Bayes gives $P(A) = 0.10 \times (38/41)/(6.2/41) = 19/31$. The chance it was £10 is 19/31.

2.13. If $P(A) = 0$, then $P(A \cap B) = 0$ since $A \cap B \subset A$, so $P(A \cap B) = P(A)P(B)$. If $P(A) = 1$, then $P(A^c) = 0$, so $P(A \cap B) = P(A \cap B) + P(A^c \cap B) = P(B) = P(B)P(A)$.

2.14. Always $P(A \cup B) = P(A) + P(B) - P(A \cap B)$. Mutually exclusive means that $P(A \cup B) = P(A) + P(B)$, hence $P(A \cap B) = 0$. For independence, $P(A \cap B) = P(A)P(B)$, so either $P(A) = 0$ or $P(B) = 0$, i.e. one of them is trivial.

2.15. Suppose $P(A) = a/n$, $P(B) = b/n$ and $P(A \cap B) = c/n$ with a, b and c integers, with $0 < a, b < n$. For independence, $c/n = (a/n) \times (b/n)$, i.e. $ab = cn$. Hence c cannot be zero, but since n is prime, the product ab has no factor n, so independence is impossible.

2.16. Let $n = 6k + r$ with $0 \le r \le 5$. Then $P(A \cap B) = k/n$. When $r = 0$, $P(A) = 1/2$, $P(B) = 1/3$ and $P(A \cap B) = 1/6$, so A and B are independent. When $r = 2$, $P(A) = 1/2$ and $P(B) = 2k/n$, which makes them independent again. But for $r = 1, 3, 4, 5$, they are not independent.

2.17. (a) $P(\text{No Six}) = (5/6)^4 = 625/1296 < 1/2$. Hence $P(\text{Six}) > 1/2$.
(b) $P(\text{No Double Six}) = (35/36)^{24} \approx 0.5086 > 1/2$, so $P(\text{Double Six}) < 1/2$.

2.18. Left diagram is equivalent to two parallel components with failure chances 0.2 and $1 - 0.95^2 = 0.0975$. Hence $P(\text{Left fails}) = 0.2 \times 0.0975 = 0.0195$. Also $P(\text{Right fails}) = 0.3^3 = 0.027$. So $P(\text{System Fails}) = 1 - (1 - 0.0975) \times (1 - 0.027) = 0.0459735$. Let $A = $ System does not fail, and $B = $ Neither * component fails. Given B, left system must work, so $P(A|B) = P(\text{Right works}) = 0.973$. Thus $P(B|A) = P(A|B)P(B)/P(A) = 0.973 \times 0.95^2/0.9540265 = 0.9223$ or so.

2.19. $P(\text{Five}) = (1/6)^4 = 1/1296$; $P(\text{Four}) = 5 \times (1/6)^3 \times (5/6) = 25/1296$.
$P(\text{Full House}) = 10 \times (1/6)^2 \times (5/6) \times (1/6) = 50/1296$.
$P(\text{Threes}) = 10 \times (1/6)^2 \times (5/6) \times (4/6) = 200/1296$.
$P(\text{Two Pairs}) = 15 \times (1/6) \times (5/6) \times (1/6) \times (4/6) = 300/1296$.
$P(\text{One Pair}) = 10 \times (1/6) \times (5/6) \times (4/6) \times (3/6) = 600/1296$.
$P(\text{No Pair}) = (5/6) \times (4/6) \times (3/6) \times (2/6) = 120/1296$. (Check sum is $1296/1296$.) For best tactics, leave Fives or a Full House alone. Chance of improving Fours is plainly $1/6$. From Threes, $P(\text{Fives}) = 1/36$, $P(\text{Fours}) = 10/36$, $P(\text{Full House}) = 5/36$. From Two Pairs, $P(\text{Full House}) = 1/3$. From One Pair, $P(\text{Fives}) = 1/216$, $P(\text{Fours}) = 15/216$, $P(\text{Full House}) = (5/216 + 15/216) = 20/216$, $P(\text{Threes}) = 60/216$, and $P(\text{Two Pairs}) = 60/216$. From No Pair, chances of improvement are the unconditional chances of the six better hands.

Chapter 3

3.1. $p_k = \binom{M}{k}\binom{N}{n-k}/\binom{M+N}{n} = \frac{(M)_k(N)_{n-k}n!}{k!(n-k)!(M+N)_n}$. Rewrite as the product

$$p_k = \binom{n}{k} \frac{M}{M+N} \cdots \frac{M-k+1}{M+N-k+1} \cdot \frac{N}{M+N-k} \cdots \frac{N-n+k+1}{M+N-n+1},$$

which plainly converges to the binomial probability (recall that n, k are fixed). Not a surprise, as when M, N are enormous, it makes little difference whether or not the sampled balls are replaced.

3.2. The number who vote to convict has a $\text{Bin}(12, 0.9)$ distribution. The probability of conviction is $P(10) + P(11) + P(12)$. $P(10) = 66(0.9)^{10}(0.1)^2 = 0.2301$, $P(11) = 12(0.9)^{11}(0.1) = 0.3766$ and $P(12) = (0.9)^{12} = 0.2824$, making a total of 0.8891.

3.3. To win, the server has to win the last point; use the binomial for the previous points. $P(4 - 0) = p^4$; $P(4 - 1) = 4p^4q$; $P(4 - 2) = 10p^4q^2$; $P(3 - 3) = 20p^3q^3$. To win from Deuce, win the next two points (p^2), or win one, lose one ($2pq$) to return to Deuce, and win subsequently. Hence $D = p^2 + 2pqD$, and $D = p^2/(1 - 2pq)$. Adding up over the disjoint ways of winning, chance of winning is $p^4(1 + 4q + 10q^2) + 20p^3q^3.p^2/(1 - 2pq)$ which simplifies to the given expression.

3.4. $p_{k+1}/p_k = \frac{e^{-\lambda}\lambda^{k+1}}{(k+1)!} \cdot \frac{k!}{e^{-\lambda}\lambda^k} = \frac{\lambda}{k+1}$, so the (p_k) increase so long as $k+1 < \lambda$ and then decrease to zero. There is a unique maximum if λ is NOT an integer.

3.5. (a) The correct key is equally likely to be in K places, hence $U(1,K)$.
(b) $P(\text{Success}) = 1/K$ independently each try, so $G_1(1/K)$.

3.6. Write $B = \{n : p_1(n) > p_2(n)\}$, $C = \{n : p_1(n) < p_2(n)\}$. Then $|P_1(A) - P_2(A)| = |x - y|$, where $x = P_1(A \cap B) - P_2(A \cap B)$ and $y = P_2(A \cap C) - P_1(A \cap C)$ are both positive. So $|P_1(A) - P_2(A)|$ is maximised by taking $A = B$ or $A = C$, and $d(P_1, P_2) = P_1(B) - P_2(B) = P_2(C) - P_1(C) = \sum_n |p_1(n) - p_2(n)|/2$. For the Bin$(3, 1/3)$, the respective chances are $8/27$, $12/27$, $6/27$ and $1/27$; for the Poisson, the first few are 0.3679, 0.3679, 0.1839, 0.0613, with $P(X \geq 4) = 0.01899$. Use the sum formula to see that the distance is 0.1148.

3.7. The pgf of a $G_1(p)$ is $\sum_{k=1}^{\infty} pq^{k-1}z^k = pz \sum_{k=1}^{\infty} (qz)^{k-1} = pz/(1-qz)$ if $|z| < 1/q$.

The pgf of a $U(1, K)$ is $\sum_{j=1}^{K} z^j/K = \frac{z}{K}.(1 + z + \cdots + z^{K-1})$.

The pgf of the NB$_0$ is $\sum_{k=0}^{\infty} \binom{k+r-1}{r-1}p^r q^k z^k = (p/(1-qz))^r$ if $|z| < 1/q$.

3.8. Write $g(z) = c\ln(1-qz)$. For $g(1) = 1$, we need $c = 1/\ln(1-q)$. The coefficient of z^k in the expansion of $g(z)$ is $-cq^k/k$, hence $p_k = -q^k/(k\ln(p))$ when $k \geq 1$.

3.9. Let $Y = $ residual time with $P(Y > y) = \exp(-\lambda y)$, and let $s > 0$. Then $P(Y > T + s|Y > T) = P(Y > T + s)/P(Y > T) = \exp(-\lambda(T+s))/\exp(-\lambda T) = \exp(-\lambda s)$ as required.

3.10 (a) $k = 1/2$, Prob. $= 3/4$ and $F(x) = x^2/4$ for $0 \leq x \leq 2$.
(b) $k = 1$, Prob. $= \int_1^{\pi/2} \sin(x)dx = \cos(1) \approx 0.5403$, and $F(x) = 1 - \cos(x)$ for $0 \leq x \leq \pi/2$.
(c) Since $\cos(0) > 0$ and $\cos(2) < 0$, no k allows this to be a density function.
(d) $k = 1$, Prob. $= 0$, $F(x) = (1-x^2)/2$ on $-1 \leq x \leq 0$, $F(x) = (1+x^2)/2$ on $0 \leq x \leq 1$.

3.11. Take $c = 12$ for $f(x) \geq 0$ and $\int_0^1 f(x)dx = 1$. Then $F(x) = 6x^2 - 8x^3 + 3x^4$ over $0 \leq x \leq 1$, last part is $F(0.5) - F(0.2) \approx 0.5067$.

3.12. Mr C's position is Bin$(10, p) + 1$. $P(\text{Win}) = p^{10}$, and $P(\text{Place}) = p^{10} + 10p^9 q + 45p^8 q^2$. All have the same winning chance when $p^{10} = 1/11$, leading to $p \approx 0.7868$. Then $P(\text{Place})$ adds up to 0.6377. The winning chances are all equal, but the distribution of Mr.C's position differs from that of the other ten horses; no real surprise the answer is not 3/11.

3.13. For $0 \leq x \leq 1$, $f(x) \geq 0$ and $F(x) = -\int_0^x \ln(u)du = x - \ln(x)$, so $F(1) = 1$. $P(A) = F(3/4) - F(1/4) \approx 0.3692$, $P(A \cap B) = F(1/2) - F(1/4) = 0.25$ and $P(B) = F(1/2) \approx 0.8466$. Hence $P(A|B) = 0.25/0.8466 \approx 0.2953 < P(A)$.

3.14. When $f(x)$ is Gamma, its derivative is $Ke^{-\lambda x}x^{\alpha-2}(\alpha - 1 - \lambda x)$. If $\alpha \leq 1$, this is always negative, f is decreasing, and has its maximum at $x = 0$. (If $\alpha < 1$, f asymptotes to $+\infty$ as $x \to 0$.) If $\alpha > 1$, the maximum is when $x = (\alpha - 1)/\lambda$. When $f(x)$ is Beta, differentiate. For $\alpha < 1$, then $f \to \infty$ as $x \to 0$, and when $\beta < 1$, the same occurs as $x \to 1$. If $\alpha = \beta = 1$, f is constant over $(0, 1)$. If $\alpha = 1$ and $\beta > 1$, sole maximum at $x = 0$; if $\alpha > 1$ and $\beta = 1$, sole maximum at $x = 1$. When both $\alpha > 1$ and $\beta > 1$, sole maximum is at $(\alpha - 1)/(\alpha + \beta - 2)$.

Chapter 4

4.1. The pgf of $G_0(p)$ is $g(z) = p/(1 - qz)$. So $g'(z) = pq(1 - qz)^{-2}$, $g''(z) = 2pq^2(1 - qz)^{-3}$. Hence $\mu = g'(1) = q/p$, $\sigma^2 = 2q^2/p^2 + q/p - q^2/p^2 = q/p^2$. $G_1(p)$ has pgf $h(z) = pz/(1 - qz)$. Thus $h'(z) = p/(1 - qz) + pqz(1 - qz)^{-2}$, $h''(z) = 2pq(1 - qz)^{-2} + 2pq^2z(1 - qz)^{-3}$ and $\mu = 1/p$, $\sigma^2 = q/p^2$ (as with $G_0(p)$, of course). The $NB_r(r, p)$ has pgf $k(z) = p^r z^r (1 - qz)^{-r} = (h(z))^r$. Hence $k'(z) = r(h(z))^{r-1}h'(z)$ and $k''(z) = r(r-1)(h(z))^{r-2}(h'(z))^2 + r(h(z))^{r-1}h''(z)$. Here $\mu = rh'(1) = r/p$ and $\sigma^2 = rq/p^2$ after simplification.

4.2. $E(X) = \sum_{r=1}^{K} r/K = K(K + 1)/(2K) = (K + 1)/2$ (or use symmetry).

$E(X^2) = \sum_{r=1}^{K} r^2/K = K(K + 1)(2K + 1)/(6K)$ and $\sigma^2 = E(X^2) - (E(X))^2$ reduces to $(K^2 - 1)/12$. (The sum formulae are in Appendix 9.3A.)

4.3. $E(X) = \sum_{k=0}^{n} k\binom{M}{k}\binom{N}{n-k}/\binom{M+N}{n} = M \sum_{k=1}^{n} \binom{M-1}{k-1}\binom{N}{n-k}/\binom{M+N}{n}$.

Use $\binom{M+N}{n} = \frac{M+N}{n}\binom{M+N-1}{n-1}$ to obtain $E(X) = \frac{Mn}{M+N} \sum_{k=1}^{n} \binom{M-1}{k-1}\binom{N}{n-k}/\binom{M+N-1}{n-1}$ and note that the sum is unity, as the summands are the entries of the hypergeometric with parameters $M - 1, N, n - 1$. For the variance, evaluate $E(X(X - 1))$ using a similar trick. $E(X(X - 1)) = \sum_{k=0}^{n} k(k - 1)\binom{M}{k}\binom{N}{n-k}/\binom{M+N}{n}$. For $k \geq 2$ the numerator in the sum is $M(M - 1)\binom{M-2}{k-2}\binom{N}{n-k}$, and write the denominator as $\frac{(M+N)(M+N-1)}{n(n-1)}\binom{M+N-2}{n-2}$ to find $E(X(X - 1)) = \frac{M(M-1)n(n-1)}{(M+N)(M+N-1)}$. Thus $\sigma^2 = E(X(X - 1)) + E(X) - (E(X))^2$ which reduces to the given expression.
(a) Plainly $(Mn)/(M + N) \to np$, and the variance is $n.\frac{M}{M+N}.\frac{N}{M+N}.\frac{M+N-n}{M+N-1} \to n.p.(1 - p).1$ as needed.
(b) Here $M = 6$, $N = 43$ and $n = 6$ so $\mu = 36/49$ and $\sigma^2 = 5547/9604 \approx 0.5776$.

4.4. $E(X) = \sum_{k=0}^{\infty} kP(X = k) = \sum_{k=1}^{\infty} P(X = k) \sum_{n=1}^{k} 1 = \sum_{n=1}^{\infty} \sum_{k=n}^{\infty} P(X = k) = \sum_{n=1}^{\infty} P(X \geq n)$. Interchange of order of the summations is justified as all terms are non-negative.

4.5. $E(\frac{1}{1+X}) = \sum_{k=0}^{\infty} \frac{1}{1+k} \frac{e^{-\lambda}\lambda^k}{k!} = \frac{e^{-\lambda}}{\lambda} \sum_{k=0}^{\infty} \frac{\lambda^{k+1}}{(k+1)!}$. The sum is $e^\lambda - 1$, so the answer is $(1 - e^{-\lambda})/\lambda$. For the binomial, the value of $E(1/(1 + X))$ is

$$\sum_{k=0}^{n} \frac{1}{k+1}\binom{n}{k}p^k q^{n-k} = \frac{1}{p(n+1)} \sum_{k=0}^{n} \binom{n+1}{k+1}p^{k+1}q^{n-k}.$$

The sum term is $(p + q)^{n+1} - q^{n+1}$, so $E(1/(1 + X)) = (1 - q^{n+1})/(p(n + 1))$.

4.6. Take X as $Bin(100, 0.6)$, so the profit is $£P$ where $P = 40X + 5(100 - X) - 1500 = 35X - 1000$. Since $E(X) = 60$ and $Var(X) = 24$, the mean profit (as in the Trixie example) is $£1100$ and the standard deviation is $£35\sqrt{24} = £171.46$.

4.7. Take W as $G_1(0.2)$ (waiting time for the first Success), X as $NB_4(4, 0.2)$ (time for the fourth Success) and Y as $Bin(20, 0.2)$. Then $E(W) = 5$, $Var(W) = 20$, $E(X) = 20$, $Var(X) = 80$ (using Exercise 4.1), $E(Y) = 4$, $Var(Y) = 3.2$. Hence the standard deviations are, in order, $4.472, 8.944$ and 1.789.

4.8. In each tube, $X=$Number of bacteria is Poiss(20θ), so $P(X = 0) = \exp(-20\theta)$. As 70 out of 100 tubes are clear, the estimate of $P(X = 0)$ is $70/100$, so solve $\exp(-20\theta) = 0.7$ to estimate θ as $-\ln(0.7)/20 \approx 0.0178$.

4.9. Let $f(x) = \sum\limits_{n=0}^{\infty} \frac{B_n x^n}{n!}$, so that $f'(x) = \sum\limits_{n=0}^{\infty} \frac{B_{n+1} x^n}{n!} = \sum\limits_{n=0}^{\infty} \frac{x^n}{n!} \sum\limits_{k=0}^{n} \binom{n}{k} B_k$. Changing

the order of summation, $f'(x) = \sum\limits_{k=0}^{\infty} \frac{B_k x^k}{k!} \sum\limits_{n=k}^{\infty} \frac{x^{n-k}}{(n-k)!} = f(x)e^x$. Thus $\int \frac{f'(x)}{f(x)} dx =$

$\int e^x dx$, i.e. $\ln(f(x)) = e^x + c$. Now $f(0) = 1$, so $c = -1$, hence the result.

4.10. The algorithm returns the value n if, and only if, $q_{n-1} < u \leq q_n$. As the length of that interval is $q_n - q_{n-1} = p_n$, so $P(X = n) = p_n$.

4.11. $E(X^k) = \int\limits_{0}^{1} x^k dx = 1/(k+1)$. Hence $E(X) = 1/2$, $E(X^2) = 1/3$, so $\mathrm{Var}(X) = 1/12$. Plainly $a \leq Y \leq b$, so let $a \leq y \leq b$; then $P(Y \leq y) = P(a + (b-a)X \leq y) = P(X \leq (y-a)/(b-a)) = (y-a)/(b-a)$, so Y has density $1/(b-a)$ over (a,b). Thus Y is $U(a,b)$, $E(Y) = a + (b-a)/2 = (a+b)/2$ and $\mathrm{Var}(Y) = (b-a)^2/12$.

4.12. $E(X^k) = \int\limits_{0}^{\infty} x^k \lambda e^{-\lambda x} dx = (\int\limits_{0}^{\infty} y^k e^{-y} dy)/\lambda^k = k!/\lambda^k$. Hence $E(X) = 1/\lambda$, $E(X^2) = 2/\lambda^2$, so $\mathrm{Var}(X) = 1/\lambda^2$, and the standard deviation is $1/\lambda$.

4.13. We require $1 = \int f(x)dx = \int\limits_{1}^{\infty}(c/x^3)dx = [-c/(2x^2)]_1^{\infty} = c/2$, so $c = 2$ and

then plainly $f(x) \geq 0$ also. Then $E(X) = \int\limits_{1}^{\infty}(2x/x^3)dx = [-2/x]_1^{\infty} = 2$, but

$E(X^2) = \int\limits_{1}^{\infty}(2x^2/x^3)dx = [2\ln(x)]_1^{\infty}$ which diverges. Hence $\mathrm{Var}(X) = \infty$ (there

is no ambiguity in the integral, as $X \geq 1$).

4.14. The distribution function of X is $F(x) = 1 - \exp(-\lambda x)$ on $x \geq 0$. For $y \geq 0$, $P(Y \leq y) = P(aX \leq y) = F(y/a)$, so Y has density $f(y/a)/a = \theta \exp(-\theta y)$ on $y > 0$, where $\theta = \lambda/a$. For $w \geq 0$, $P(W \leq w) = P(X^2 \leq w) = P(X \leq \sqrt{w}) = 1 - \exp(-\lambda\sqrt{w})$, so W has density $\frac{\lambda}{2\sqrt{w}} \cdot \exp(-\lambda\sqrt{w})$ on $w > 0$. For $v \geq 0$, $P(V \leq v) = P(|X - 1| \leq v) = P(1 - v \leq X \leq 1 + v) = F(1+v) - F(1-v)$. For $0 \leq v \leq 1$, this is $\exp(-\lambda(1-v)) - \exp(-\lambda(1+v))$, so V has density $\lambda\exp(-\lambda)[\exp(-\lambda v) + \exp(\lambda v)]$, but for $v > 1$ it is just $F(1+v)$, so the density here is $\lambda\exp(-\lambda(1+v))$.

4.15. $E(X^n) = \int\limits_{-\infty}^{\infty}(x^n/2)\exp(-|x|)dx$. Split the range into $(-\infty, 0)$ and $(0, \infty)$. When

n is odd, these are finite and cancel, when n is even they double up. Thus $E(X) =$

0, $\mathrm{Var}(X) = E(X^2) = \int\limits_{0}^{\infty} x^2 \exp(-x)dx = 2$. For $y \geq 0$, $P(|X| \leq y) = P(-y \leq$

$X \leq y) = 2P(0 \leq X \leq y) = \int\limits_{0}^{y} \exp(-x)dx$, so $|X|$ is $E(1)$.

4.16. (This function arose in Exercise 3.10(a).) Since $F(x) = x^2/4$ for $0 \leq x \leq 2$, we solve $u = x^2/4$ to find $x = 2\sqrt{u}$ for the simulations, where u is from $U(0,1)$. $P(2X \leq y) = F(y/2) = y^2/16$ on $0 \leq y \leq 4$, so the density is $y/8$ on that interval. $P(\ln(X) \leq y) = F(e^y) = e^{2y}/4$ on $-\infty < y \leq \ln(2)$, so the density is $e^{2y}/2$ on that interval. $P(1/X \leq y) = P(X \geq 1/y) = 1 - F(1/y)$, making the density $f(1/y)/y^2$, i.e. $1/(2y^3)$ on $y > 1/2$. Plainly $0 \leq (X-1)^2 \leq 1$, so let $0 \leq y \leq 1$. Then $P((X-1)^2 \leq y) = F(1+\sqrt{y}) - F(1-\sqrt{y}) = \sqrt{y}$, so the density is $1/(2\sqrt{y})$ on $0 < y < 1$.

4.17. $P(Y \leq y) = P(\tan(X) \leq y) = P(X \leq \arctan(y)) = (\arctan(y) + \pi/2)/\pi$.

Differentiate, the given density emerges. Then $E(Y) = \int\limits_{-\infty}^{\infty} \frac{y\,dy}{\pi(1+y^2)}$ is not defined,

since $\int\limits_{-\infty}^{\infty} \frac{|y|\,dy}{\pi(1+y^2)} = \infty$. So the mean (and hence the variance) does not exist. If U is $U(0,1)$, then $\pi U - \pi/2$ is $U(-\pi/2, \pi/2)$, so $\tan(\pi U - \pi/2)$ will simulate.

4.18. $E(X) = \int_0^\infty x f(x)\,dx = \int_0^\infty f(x)\left(\int_0^x 1\,dy\right)dx = \int_0^\infty 1.\left(\int_y^\infty f(x)\,dx\right)dy$ which is equal to $\int_0^\infty P(X > y)\,dy$.

4.19. $E(X^k) = \int\limits_0^\infty x^k.Kx^{\alpha-1}e^{-\lambda x}\,dx$, where $K = \lambda^\alpha/\Gamma(\alpha)$. But $\int\limits_0^\infty x^{k+\alpha-1}e^{-\lambda x}\,dx = \Gamma(\alpha+k)/\lambda^{\alpha+k}$, so $E(X^k) = \Gamma(\alpha+k)/(\lambda^k\Gamma(\alpha)) = (\alpha+k-1)_k/\lambda^k$ by properties of the Gamma function. Hence $E(X) = \alpha/\lambda$, $E(X^2) = \alpha(\alpha+1)/\lambda^2$ so $\mathrm{Var}(X) = \alpha/\lambda^2$. For the Beta, $E(X^k) = \int\limits_0^1 x^k.Kx^{\alpha-1}(1-x)^{\beta-1}\,dx$, where $K = \Gamma(\alpha+\beta)/(\Gamma(\alpha)\Gamma(\beta))$. But $\int\limits_0^1 x^{k+\alpha-1}(1-x)^{\beta-1}\,dx = \Gamma(\alpha+k)\Gamma(\beta)/\Gamma(\alpha+\beta+k)$, so $E(X^k) = \frac{\Gamma(\alpha+\beta).\Gamma(\alpha+k)}{\Gamma(\alpha)\Gamma(\alpha+\beta+k)} = \frac{(\alpha+k-1)_k}{(\alpha+\beta+k-1)_k}$. Hence $E(X) = \alpha/(\alpha+\beta)$, $E(X^2) = \alpha(\alpha+1)/((\alpha+\beta)(\alpha+\beta+1))$, so $\mathrm{Var}(X) = \alpha\beta/((\alpha+\beta)^2(\alpha+\beta+1))$.

4.20. $0 \le E((X+\theta Y)^2) = E(X^2) + 2\theta E(XY) + \theta^2 E(Y^2)$. This quadratic in θ cannot have distinct real roots, or it would be negative for some θ, so ("$b^2 \le 4ac$") $(E(XY))^2 \le E(X^2)E(Y^2)$. Thus $(\mathrm{Cov}(X,Y))^2 \le \mathrm{Var}(X).\mathrm{Var}(Y)$, so $-1 \le \mathrm{Corr}(X,Y) \le 1$.

4.21. $f_X(x) = \int\limits_{-\infty}^{\infty} f(x,y)\,dy = \int\limits_x^\infty 2\exp(-x-y)\,dy = 2\exp(-2x)$ on $x > 0$. $f_Y(y) = \int\limits_{-\infty}^{\infty} f(x,y)\,dx = \int\limits_0^y 2\exp(-x-y)\,dx = 2e^{-y}(1-e^{-y})$ on $y > 0$. $f(x|Y=4) = f(x,4)/f_Y(4) = 2e^{-x-4}/(2e^{-4}(1-e^{-4})) = e^{-x}/(1-e^{-4})$ on $0 < x < 4$. $f(y|X=4) = f(4,y)/f_X(4) = 2e^{-4-y}/(2e^{-8}) = e^{4-y}$ on $y > 4$. X and Y are not independent. Firstly, $Y \ge X$, more formally $f(x,y)$ is not the product of the marginal densities. When $U = X+Y$ and $V = X/Y$, the worked example in the text shows $J = u/(1+v)^2$, so $g(u,v) = 2e^{-u}u/(1+v)^2$ on $0 < u < \infty$, $0 < v < 1$. Integrating, U has density ue^{-u} on $u > 0$, V has density $2/(1+v)^2$ on $0 < v < 1$, so $g(u,v) = g_U(u)g_V(v)$ and U, V are independent.

4.22. Write $f(x,y) = P(X=x, Y=y)$, $f_X(x) = P(X=x)$, $f_Y(y) = P(Y=y)$. Suppose $f(x,y) = f_X(x)f_Y(y)$ for all x, y. Then $F(x,y) = \sum f(x',y') = \sum f_X(x')f_Y(y') = \sum f_X(x')\sum f_Y(y') = F_X(x)F_Y(y)$, the sums being over all $x' \le x$ and $y' \le y$. Conversely, suppose $F(x,y) = F_X(x)F_Y(y)$ for all x, y. Take $x' < x$, $y' < y$ and note $P(x' < X \le x, y' < Y \le y) = F(x,y) - F(x',y) - F(x,y') + F(x',y') = F_X(x)F_Y(y)-$ etc, reducing to $\sum f_X(x_i)\sum f_Y(y_j)$, the sums being over $x' < x_i \le x$ and $y' < y_j \le y$. If there is a maximal $x' < x$ with $f_X(x') > 0$, the x-sum is $f_X(x)$. Otherwise there are $(x_m) \uparrow x$ with $f_X(x_m) > 0$; take $x' = x_m$, let $(x_m) \uparrow x$ to get the same result. Similarly for the y-sum.

4.23. Use symmetry, or: $E(XY) = \int\limits_0^1 y\,dy \int\limits_{y-1}^{1-y} x\,dx = \int\limits_0^1 0\,dy = 0$. The whole region has unit area, so, for $-1 < x < 0$, $P(X \le x) = (x+1)^2/2$, and the density is $1+x$; for $0 < x < 1$, $P(X \le x) = 1 - (1-x)^2/2$ with density $1-x$. Thus $E(X) = \int\limits_{-1}^0 x(1+x)\,dx + \int\limits_0^1 x(1-x)\,dx = 0$, so $E(XY)$ and $E(X)E(Y)$ are equal, both being zero. X, Y are not independent, as $Y - 1 \le X \le 1 - Y$.

4.24. $P(W > w) = P(X > w, Y > w) = (1 - w)^2$ by independence, hence W has density $2(1 - w)$ over $(0, 1)$. $P(V \leq v) = P(X \leq v, Y \leq v) = v^2$ by independence, so V has density $2v$ over $(0, 1)$. They cannot be independent, as $W \leq V$. And since $VW = XY$ (yes?), so $E(VW) = E(XY) = E(X)E(Y) = 1/4$.

4.25. μ has density $1/80$ on $10 < \mu < 90$. Also $f(x|\mu) = 1/20$ on $\mu - 10 < x < \mu + 10$. Hence $f(x, \mu) = 1/1600$ on $10 < \mu < 90$, $\mu - 10 < x < \mu + 10$. We seek $f_X(x) = \int\limits_{-\infty}^{\infty} f(x, \mu)d\mu$. For $0 < x < 20$, $f_X(x) = \int\limits_{10}^{10+x} d\mu/1600 = x/1600$; for $20 < x < 80$, then $f_X(x) = \int\limits_{x-10}^{x+10} d\mu/1600 = 1/80$; for $80 < x < 100$, $f_X(x) = \int\limits_{x-10}^{90} d\mu/1600 = (100 - x)/1600$. And $f(\mu|x) = f(x, \mu)/f_X(x)$. So for $0 < x < 20$, $f(\mu|x) = 1/x$ on $10 < \mu < 10 + x$. For $20 < x < 80$, $f(\mu|x) = 1/20$ on $x - 10 < \mu < x + 10$. For $80 < x < 100$, $f(\mu|x) = 1/(100 - x)$ on $x - 10 < \mu < 90$.

4.26. Sum along rows and columns. The marginal mass functions are

Goals	0	1	2	3	≥ 4
X	0.261	0.322	0.237	0.094	0.086
Y	0.339	0.356	0.210	0.079	0.016

Given $Y = 1$, rescale the second row of the given table by the factor 0.356 so that the conditional X-scores have probabilities $0.185, 0.371, 0.222, 0.140$ and 0.081. Since this set differs from the marginal distribution of X, X and Y are not independent. Calculate $E(X) = 1.422$, $E(Y) = 1.077$ from the table above. From the original table, $E(XY) = \sum xyP(X = x, Y = y) = 1.564$, so $Cov(X, Y) = 1.564 - 1.422 \times 1.077 = 0.033$. Also $E(X^2) = 3.492$ so $Var(X) = 1.999$; $E(Y^2) = 2.163$ so $Var(Y) = 1.003$, making $Corr(X, Y) = 0.033/\sqrt{1.999 \times 1.003} = 0.023$.

4.27. $P(W > w) = P(X_1 > w, \ldots, X_n > w) = P(X_1 > w) \cdots P(X_n > w)$ by independence, so that $P(W > w) = \exp(-\lambda_1 w) \cdots \exp(-\lambda_n w) = \exp(-w \sum \lambda_i)$. Hence W is $E(\lambda_1 + \cdots + \lambda_n)$.

4.28. $P(T \leq t|N = n) = P(X_1 + \cdots + X_n \leq t)$ where $\{X_i\}$ are independent, each $E(\lambda)$, so conditionally T is $\Gamma(n, \lambda)$ (a worked example), with density $f_n(t) = \lambda^n t^{n-1}/n!$. Hence $P(T \leq t) = \sum\limits_{n=1}^{\infty} pq^{n-1} \int\limits_0^t f_n(u)du$; differentiate to get the density of T as

$$\sum\limits_{n=1}^{\infty} pq^{n-1} f_n(t) = \frac{pe^{-\lambda t}}{qt} \sum\limits_{n=1}^{\infty} \frac{(qt\lambda)^n}{n!} = pe^{-\lambda t}(e^{qt\lambda} - 1)/(qt).$$

4.29. We use $h(t) = \int\limits_{-\infty}^{\infty} f(x)g(t - x)dx$, with $f(x) = 1$ over $(0, 1)$, $g(u) = u$ when $0 < u < 1$ and $g(u) = 2 - u$ when $1 < u < 2$. For $0 < t < 1$ then $h(t) = \int\limits_0^t 1.(t - x)dx = t^2/2$. For $1 < t < 2$ then $h(t) = \int\limits_0^{t-1} 1.(2 - t + x)dx + \int\limits_{t-1}^1 1.(t - x)dx = 3t - t^2 - 3/2$.

For $2 < t < 3$ then $h(t) = \int\limits_{t-2}^1 1.(2 - t + x)dx = (3 - t)^2/2$.

4.30. Take X in the first half, Y in the second so that X is $U(0, 1/2)$ and Y is $U(1/2, 1)$, independently. Hence $f(x, y) = 4$ on $0 < x < 1/2$, $1/2 < y < 1$; using the usual diagram, the probability a triangle can be formed is $1/2$.

4.31. $E(X|p) = np$. When p is $U(0, 1)$, $E(p) = 1/2$, so $E(X) = E(E(X|p)) = n/2$. When p is Beta, $E(p) = \alpha/(\alpha + \beta)$ (Exercise 4.19) so $E(X) = n\alpha/(\alpha + \beta)$. Always $E(X^2|p) = Var(X|p) + (E(X|p))^2 = np(1 - p) + n^2p^2 = np + n(n - 1)p^2$. When p is $U(0, 1)$, $E(p^2) = 1/3$ so $E(X^2) = n/2 + n(n - 1)/3$, hence

$\text{Var}(X) = n/6 + n^2/12$. When p is Beta, $E(p^2) = \alpha(\alpha+1)/((\alpha+\beta)(\alpha+\beta+1))$, so $\text{Var}(X) = n\alpha\beta(\alpha+\beta+n)/((\alpha+\beta)^2(\alpha+\beta+1))$.

4.32. $E(W_n^2) = E(E(W_n^2|Y_n)) = E(E(X_n^2+2X_nU_n+U_n^2|Y_n))$. Split this up as $E(X_n^2)+2E(X_n)E(E(U_n|Y_n)) + E(E(U_n^2|Y_n))$. Now $E(U_n^r|Y_n = k) = E(W_k^r)$, hence, if $p_k = P(Y_n = k)$, then $\sigma_n^2 = \text{Var}(W_n) = E(X_n^2) + 2E(X_n)\sum p_k\mu_k + \sum p_k(\sigma_k^2 + \mu_k^2) - (E(X_n))^2 - 2E(X_n)\sum p_k\mu_k - (\sum p_k\mu_k)^2$. Thus $\sigma_n^2 = \text{Var}(X_n) + \sum p_k\sigma_k^2 + \sum p_k\mu_k^2 - (\sum p_k\mu_k)^2$, and we know the values of μ_k, p_k and $\text{Var}(X_n)$.

4.33. The variance of X, conditional on $Y = y$ should be the (conditional) expectation of the square of the difference between $X|Y = y$ and $E(X|Y = y)$, as the expression states. We also have $\text{Var}(X|Y) = E(X^2|Y) - (E(X|Y))^2$, hence $E(\text{Var}(X|Y)) = E(X^2) - E((E(X|Y))^2)$. Now $\text{Var}(E(X|Y)) = E((E(X|Y))^2) - (E(E(X|Y)))^2 = E((E(X|Y))^2) - E(X^2)$. Eliminate $E((E(X|Y))^2)$ to obtain the result stated.

Chapter 5

5.1. $x = P(X = k|X + Y = n) = P(X = k \cap X + Y = n)/P(X + Y = n) = P(X = k \cap Y = n - k)/P(X + Y = n)$. Now $X + Y$ is $\text{Poiss}(\lambda + \mu)$, X and Y are independent, so

$$x = \frac{e^{-\lambda}\lambda^k}{k!} \cdot \frac{e^{-\mu}\mu^{n-k}}{(n-k)!} / \frac{e^{-(\lambda+\mu)}(\lambda+\mu)^n}{n!} = \binom{n}{k}\left(\frac{\lambda}{\lambda+\mu}\right)^k\left(\frac{\mu}{\lambda+\mu}\right)^{n-k}$$

as claimed.

5.2. An application of the previous exercise with $\lambda = 8$, $\mu = 4$. Conditional on the total being 16, the number that are sports-related is $\text{Bin}(16, 2/3)$. Hence the mean number is $32/3$, and if the total is 4, sports-related is $\text{Bin}(4, 2/3)$ so $P(\text{All sports}) = (2/3)^4 = 16/81$.

5.3. $x = P(X = k|X + Y = r) = P(X = k \cap X + Y = r)/P(X + Y = r) = P(X = k \cap Y = r - k)/P(X + Y = r)$. Now $X + Y$ is $\text{Bin}(m + n, p)$ (easy), X and Y are independent, so x is the ratio of $\binom{m}{k}p^kq^{m-k}\binom{n}{r-k}p^{r-k}q^{n-r+k}$ and $\binom{m+n}{r}p^rq^{m-r}$. The terms in p and q cancel, leaving $\binom{m}{k}\binom{n}{r-k}/\binom{m+n}{r}$, a hypergeometric distribution. The absence of p is because of the conditioning on the total: once we know that $X + Y = r$, what p was is irrelevant.

5.4.

$$\text{Var}\left(\sum c_i X_i\right) = E\left(\left(\sum c_i X_i\right)^2\right) - \left(E\sum c_i X_i\right)^2$$

$$= E\left(\sum c_i^2 X_i^2 + 2\sum_{i<j} c_i c_j X_i X_j\right) - \left(\sum c_i\mu_i\right)^2$$

$$= \sum c_i^2(E(X_i^2) - \mu_i^2) + 2\sum_{i<j} c_i c_j(E(X_i X_j) - \mu_i\mu_j)$$

$$= \sum c_i^2\text{Var}(X_i) + 2\sum_{i<j} c_i c_j\text{Cov}(X_i, X_j)$$

5.5. Let X be the number of purchases to obtain at least one copy of each. We use $E(X) = \sum_{n\geq 1} P(X \geq n)$ from Exercise 4.4. Let $I(i, n)$ be the indicator variable taking the value 1 if toy i is NOT found in the first $n - 1$ purchases, and zero otherwise. Define the event $A(i, n) = \{I(i, n) = 1\}$. Plainly $P(A(i, n)) = (1 - p_i)^{n-1}$, $P(A(i, n) \cap A(j, n)) = (1 - p_i - p_j)^{n-1}$ when $i \neq j$, and similarly for more than two events. Key step: $X \geq n \Leftrightarrow \bigcup_i A(i, n)$ occurs. Hence

$$P(X \geq n) = S_1(n) - S_2(n) + S_3(n) - \cdots + (-1)^{N-1} S_N(n)$$

using Corollary 1.3, where $S_j(n)$ is the sum, over all sets of j toys, of the probabilities that none of those j toys are found in the first $n - 1$ purchases. A typical term is $(1 - p_1 - p_2 - \cdots - p_j)^{n-1}$, so

$$E(X) = \sum_{n=1}^{\infty} \sum_{r=1}^{N} (-1)^{r-1} S_r(n) = \sum_{r=1}^{N} (-1)^{r-1} \sum_{n=1}^{\infty} S_r(n).$$

But $\sum_{n=1}^{\infty} (1 - x)^{n-1} = (1 - (1 - x))^{-1} = x^{-1}$ here, so $E(X) = \sum_{r=1}^{N} (-1)^{r-1} U_r$ where the U_r are as defined.

5.6. Let I_k be 1 or 0 according as the ith pair match or not. Then $S = \sum_{k=1}^{n} I_k$ is the total number of matches. Plainly $P(I_k = 1) = 1/n$, and since $P(I_j = 1|I_k = 1) = 1/(n - 1)$ if $j \neq k$, then $P(I_j = 1 \cap I_k = 1) = 1/(n(n - 1))$ when $j \neq k$. Hence $E(S) = n.(1/n) = 1$, and $\text{Var}(S) = n\text{Var}(I_1) + n(n - 1)\text{Cov}(I_1, I_2)$ (symmetry), i.e.

$$\text{Var}(S) = n\frac{1}{n}\left(1 - \frac{1}{n}\right) + n(n - 1)\left(\frac{1}{n(n - 1)} - \frac{1}{n^2}\right) = 1.$$

Both the mean and the variance of the number of matches are unity, when $n \geq 2$.

5.7. Put $\alpha = P(I = 1)$, $\beta = P(J = 1)$. The given condition is equivalent to $P(I = 1, J = 1) = \alpha\beta$. Thus $\alpha = P(I = 1) = P(I = 1, J = 1) + P(I = 1, J = 0) = \alpha\beta + P(I = 1, J = 0)$, hence $P(I = 1, J = 0) = \alpha(1 - \beta) = P(I = 1)P(J = 0)$. The other two conditions for independence follow in the same way.

5.8. Let I_k be 1 or 0 according as the kth married couple are adjacent so that the required total is $S = I_1 + \cdots + I_n$. Wherever wife k sits, there are two choices for husband k to be next to her, so $P(I_k = 1) = 2/(2n - 1)$. Thus $E(S) = 2n/(2n - 1)$. Take $j \neq k$. To find $P(I_j = 1|I_k = 1)$, we must distinguish the cases where wife j sits adjacent to one of pair k, or does not; the chances of these are $2/(2n - 2)$ and $(2n - 4)/(2n - 2)$ respectively. If she does, the value is $1/(2n - 3)$, otherwise it is $2/(2n - 3)$, so overall

$$P(I_j = 1|I_k = 1) = \frac{2}{2n - 2}\cdot\frac{1}{2n - 3} + \frac{2n - 4}{2n - 2}\cdot\frac{2}{2n - 3} = \frac{1}{n - 1}.$$

Hence $P(I_j = 1 \cap I_k = 1) = 2/((n - 1)(2n - 1))$ if $j \neq k$. Thus

$$
\begin{aligned}
\text{Var}(S) &= n\text{Var}(I_1) + n(n - 1)\text{Cov}(I_1, I_2) \\
&= n.\frac{2}{2n - 1}\cdot\frac{2n - 3}{2n - 1} + n(n - 1).\left(\frac{2}{(n - 1)(2n - 1)} - \frac{4}{(2n - 1)^2}\right)
\end{aligned}
$$

which simplifies to $4n(n - 1)/(2n - 1)^2$.

With the same notation for the second problem, $P(I_k = 1) = 2/n$ so $E(S) = 2$. Again, splitting the cases according as wife j is or is not adjacent to husband k,

$$P(I_j = 1|I_k = 1) = \frac{1}{n - 1}\cdot\frac{1}{n - 1} + \frac{n - 2}{n - 1}\cdot\frac{2}{n - 1} = \frac{2n - 3}{(n - 1)^2},$$

hence $P(I_j = 1 \cap I_k = 1) = (4n - 6)/(n(n-1)^2)$. This time $\text{Var}(S) = n.\frac{2}{n}(1 - \frac{2}{n}) + n(n-1)(\frac{4n-6}{n(n-1)^2} - \frac{4}{n^2}) = (2n-4)/(n-1)$.

5.9. Imagine the noodle chopped finely up, without position disturbance, into N pieces each of length y/N. "Smoothness" means that the pieces are effectively straight, each short enough to cross at most one line. Let $I_k = 1$ when the kth piece crosses a line, so that $\sum I_k$ is the total number of crossings. We know $P(I_k = 1) = 2y/(N\pi)$ from the needle problem, hence $E(S) = N.(2y/(N\pi)) = 2y/\pi$.

5.10. Take $P(X = t) = P(X = -t) = 1/2$.

5.11. $M(t) = E(e^{tX}) = \int_0^1 e^{tx}dx = (e^t - 1)/t$ when $t \neq 0$. Now $Y = a + (b-a)X$ is $U(a,b)$, its mgf is $e^{at}M((b-a)t) = (e^{bt} - e^{at})/(t(b-a))$ when $t \neq 0$.

5.12. We will prove by induction that the density of $X_1 + \cdots + X_n$ is $f_n(t) = n/(\pi(n^2 + t^2))$. The assertion is then immediate. The result holds for $f_1(t)$, by definition of Cauchy. Suppose it holds up to n: then

$$f_{n+1}(t) = \int f_n(x)f_1(t-x)dx = \frac{n}{\pi^2}\int_{-\infty}^{\infty} \frac{dx}{(n^2 + x^2)(1 + (t-x)^2)}.$$

The partial fractions expansion of the integrand is

$$\frac{1}{(t^2 + (n-1)^2)(t^2 + (n+1)^2)} \left(\frac{2tx + 1 + t^2 - n^2}{n^2 + x^2} + \frac{t^2 - 1 + n^2 - 2t(x-t)}{1 + (x-t)^2}\right).$$

Now

$$\int_{-\infty}^{\infty} \left(\frac{2tx}{n^2 + x^2} - \frac{2t(x-t)}{1 + (x-t)^2}\right)dx = \left[t\ln\frac{n^2 + x^2}{1 + (x-t)^2}\right]_{-\infty}^{\infty} = 0.$$

Also

$$\int_{-\infty}^{\infty} \frac{1 + t^2 - n^2}{n^2 + x^2}dx = \frac{1 + t^2 - n^2}{n}\left[\arctan(\frac{x}{n})\right]_{-\infty}^{\infty} = \frac{(1 + t^2 - n^2)\pi}{n}$$

and

$$\int_{-\infty}^{\infty} \frac{t^2 - 1 + n^2}{1 + (x-t)^2}dx = (t^2 - 1 + x^2)[\arctan(x-t)]_{-\infty}^{\infty} = (t^2 - 1 + n^2)\pi$$

so their sum reduces to $\frac{\pi}{n}((n^2 - 1)(n-1) + (n+1)t^2)$. Hence $f_{n+1}(t)$ is

$$\frac{n}{\pi^2}\cdot\frac{1}{(t^2 + (n-1)^2)(t^2 + (n+1)^2)}\cdot\frac{\pi}{n}.(n+1)(t^2 + (n-1)^2) = \frac{n+1}{t^2 + (n+1)^2}$$

as claimed.

5.13. Write down the formal expression for $E(X)$ as an integral, and make the substitution $y = (x - \mu)/\sigma$. This quickly gives $E(X) = \mu$. Use $\text{Var}(X) = E((X-\mu)^2)$, and the same substitution.

5.14. (P1) Let X be $N(\mu, \sigma^2)$ with density $f(x) = \frac{1}{\sigma\sqrt{2\pi}} \exp(-\frac{(x-\mu)^2}{2\sigma^2})$ and distribution function $F(x)$. Let $W = aX + b$. Then if $a > 0$, $P(W \le w) = P(aX + b \le w) = P(X \le (w - b)/a) = F((w - b)/a)$. Hence W has density $f((w - b)/a)/a$ which reduces to the density of a $N(a\mu + b, a^2\sigma^2)$ variable. A similar argument holds when $a < 0$.

For (P2), we first show that if U, V are independent each $N(0, 1)$, then their sum is $N(0, 2)$. The result then follows, using (P1) and $X = \mu + \sigma U$, $Y = \lambda + \tau V$. The density of $U + V$ is

$$\frac{1}{2\pi} \int_{-\infty}^{\infty} \exp\left(-\frac{x^2}{2}\right) \exp\left(-\frac{(t-x)^2}{2}\right) dx = \frac{1}{2\pi} \int_{-\infty}^{\infty} \exp\left(-\frac{t^2}{4} - \left(x - \frac{t}{2}\right)^2\right) dx.$$

Now put $x - t/2 = y$; the density of $U + V$ is seen to be proportional to $\exp(-t^2/4)$, which is enough to show that $U + V$ is $N(0, 2)$.

5.15. The mgf is $\exp(t^2/2)$, which we expand as a power series. Hence all odd moments are zero, and the term in t^{2m} is $t^{2m}/(2^m.m!)$; the $(2m)$th moment is then $(2m)!/(2^m.m!)$.

5.16. From the previous exercise, the fourth moment of a $N(0, 1)$ variable is 3, which is hence the kurtosis of a Standard Normal. But $Z = (X - \mu)/\sigma$ simply standardizes X, so the kurtosis is always 3.

5.17. (Solution by Tom Liggett, Chance magazine, 2000.) Let X and Y be iidrv defined over $(0, 1)$. Then $-1 \le X - Y \le 1$, so $(X - Y)^4 \le (X - Y)^2$ with equality only when $(X - Y)^2$ is 0 or 1. Take expectations as $E((X - \mu) - (Y - \mu))^4 \le E((X - \mu) - (Y - \mu))^2$, which reduces to $2m_4 + 6\sigma^4 \le 2\sigma^2$. Equality is when $P(X - Y = 1 \text{ or } 0) = 1$, i.e. the common distribution is degenerate or Bernoulli.

5.18. Write $X_n = Z_1^2 + \cdots + Z_n^2$, the $\{Z_i\}$ being independent $N(0, 1)$ variables. Thus $E(X_n) = nE(Z^2) = n$, and $\text{Var}(X_n) = n\text{Var}(Z^2) = 2n$, since $\text{Var}(Z^2) = E(Z^4) - (E(Z^2))^2 = 3 - 1 = 2$. Note that, using $\chi_n^2 \equiv \Gamma(n/2, 1/2)$,

$$\int_0^{\infty} \exp(-x/2).x^{n/2-1}dx = \Gamma(n/2)/(1/2)^{n/2} = 1/K,$$

say. The mgf of X_n is

$$M(t) = K \int_0^{\infty} \exp(tx).\exp(-x/2).x^{n/2-1}dx = K \int_0^{\infty} \exp(-x(1/2 - t)).x^{n/2-1}dx$$

which is $(1/2)^{n/2}/\Gamma(n/2).\Gamma(n/2)/(1/2 - t)^{n/2} = (1 - 2t)^{-n/2}$. Thus $Y_n = (X_n - n)/\sqrt{2n}$ has mgf $H(t) = \exp(-nt/\sqrt{2n})M(t/\sqrt{2n})$, hence

$$-\ln(H(t)) = t\sqrt{\frac{n}{2}} - \ln\left(1 - \frac{2t}{\sqrt{2n}}\right)^{-n/2} = t\sqrt{\frac{n}{2}} + \frac{n}{2}\ln\left(1 - t\sqrt{\frac{2}{n}}\right).$$

Use $-\ln(1 - x) = x + x^2/2 + \cdots$ when x is small, so that

$$-\ln(H(t)) = t\sqrt{\frac{n}{2}} - \frac{n}{2}\left(t\sqrt{\frac{2}{n}} + \frac{t^2}{2}.\frac{2}{n} + O\left(\frac{1}{n^{3/2}}\right)\right) = -\frac{t^2}{2} + O\left(\frac{1}{\sqrt{n}}\right),$$

and $H(t) \to \exp(t^2/2)$ as claimed. (For the O-notation, see Appendix 9.3D.)

5.19. Assume the times (X_i) in minutes are iid with mean unity and variance $(1/3)^2$, and let $S = X_1 + \cdots + X_{110}$. Then the CLT justifies taking S as approximately $N(110, 110/9)$. Thus

$$P(S \geq 120) = P((S - 110)/\sqrt{110/9} \geq (120 - 110)/\sqrt{110/9}) \approx P(Z \geq 2.86),$$

where Z is $N(0, 1)$. The probability is about 0.002.

5.20. Use the CLT. Each throw has mean 3.5 and variance 35/6, so the total T over 1200 throws is well approximated by $N(4200, 7000)$.

$$P(T \geq t) = P((T - 4200)/\sqrt{7000} \geq (t - 4200)/\sqrt{7000}) \approx P(Z \geq z)$$

where $z = (t - 4200)/\sqrt{7000}$. (a) $t = 4000 \Rightarrow z = -2.39 \Rightarrow p \approx 0.92$; (b) $t = 4250 \Rightarrow z = 0.60 \Rightarrow p \approx 0.27$; (c) $t = 4400 \Rightarrow z = 2.39 \Rightarrow p \approx 0.08$.

5.21. (a) $P(|X - 60| \geq 10) \leq \sigma^2/10^2 = 1/4$, so the chance of a score between 50 and 70 is at least 0.75.

(b) $\bar{X}$ will have mean 60 and variance $25/n$; so $P(|\bar{X} - 60| \geq 5) \leq (\sigma^2/n)/25 = 1/n$. To have this at most 0.10, take $n \geq 10$.

(c) If $\bar{X}$ is $N(60, 25/n)$, then $P(|\bar{X} - 60| \geq 5) = P(|Z| \geq \sqrt{n})$, with $Z = (\bar{X} - 60)\sqrt{n}/5 \approx N(0, 1)$. For this to be 0.10, we take $\sqrt{n} \approx 1.65$, and so $n \approx 2.72$. Hence take $n = 3$ (a little low for the CLT!)

5.22. If X is Poiss(n), then $P(X \leq n) = \exp(-n) \sum_{k=0}^{n} n^k/k!$. For large n, X is approximately $N(n, n)$, so $P(X \leq n) = P((X - n)/\sqrt{n} \leq 0) \to 1/2$ as $n \to \infty$. And $P(X \leq n + \sqrt{n}) = P((X - n)/\sqrt{n}) \leq 1) \to P(Z \leq 1) \approx 0.84$.

5.23. The errors are $X_1, \ldots, X_{12}$, with sum $S = X_1 + \cdots + X_{12}$. Since $\mu_i = 0$ and $\sigma_i^2 = 1/12$, so $E(S) = 0$ and $\text{Var}(S) = 1$. Hence $P(|S| \geq 1) = 2P(S \geq 1)$ (symmetry) $\approx 2P(Z \geq 1) \approx 0.32$.

5.24. In Example 5.7, take $N = 49$. Then $E(S_N) = 49 \sum_{n=1}^{49} 1/n = 219.48$, and $\text{Var}(S_N) = 49^2 \sum_{n=1}^{49} 1/n^2 - E(S_N) \approx 61^2$, so the mean time is about 220 draws with a s.d. of about 60. The actual result – 262 draws – is about 2/3 of a s.d. away from the mean, not at all surprising.

5.25. The inverse transform is

$$x = \frac{u}{\sqrt{u^2 + v^2}} \exp\left(-\frac{u^2 + v^2}{4}\right), \quad y = \frac{v}{\sqrt{u^2 + v^2}} \exp\left(-\frac{u^2 + v^2}{4}\right).$$

Let E denote the term $\exp(-(u^2 + v^2)/4)$. Then the respective values of $\frac{\partial x}{\partial u}$ and $\frac{\partial x}{\partial v}$ are

$$\frac{E}{\sqrt{u^2 + v^2}}\left(1 - \frac{u^2}{u^2 + v^2} - \frac{u^2}{2}\right) = \frac{2v^2 - u^2 v^2 - u^4}{2(u^2 + v^2)^{3/2}}.E,$$

$$\frac{E}{\sqrt{u^2 + v^2}}\left(-\frac{uv}{u^2 + v^2} - \frac{uv}{2}\right) = -\frac{uv(u^2 + v^2 + 2)}{2(u^2 + v^2)^{3/2}}.E,$$

and the partial derivatives of y can be written down by symmetry. The Jacobian is the absolute value of

$$\frac{E^2}{4(u^2 + v^2)^3}((2v^2 - u^2 v^2 - u^4)(2u^2 - v^2 u^2 - v^4) - u^2 v^2(u^2 + v^2 + 2)^2)$$

which reduces to $\frac{1}{2}\exp(-\frac{u^2+v^2}{2})$. The density of (X,Y) is $f(x,y) = 1/\pi$ on $x^2+y^2 \le 1$, so $g(u,v) = \frac{1}{\pi}\cdot\frac{1}{2}\cdot\exp(-\frac{u^2+v^2}{2})$ on $-\infty < u, v < \infty$, which factorises as the product of $N(0,1)$ densities for U and V, so U and V are independent, each Standard Normal.

Given Y which is $U(0,1)$, then $X = 2Y - 1$ is $U(-1,1)$, so first generate pairs (X_1, X_2) that are independent $U(-1,1)$. Reject the pair unless $W = X_1^2 + X_2^2 \le 1$, if not rejected compute $T = \sqrt{-2\ln(W)/W}$, and return $X_1.T$, $X_2.T$ as two independent $N(0,1)$ values.

5.26. (a) $g_i(z) = (1 + z^i)/2$ hence $\mathrm{Var}(X_i) = i^2/4$ and $\sigma_n^2 = n(n+1)(2n+1)/24$, so $R(n)/\sigma_n = O(1/\sqrt{n}) \to 0$. Hence $\sum X_i$ is approximately $N(n(n+1)/4, \sigma_n^2)$.

(b) $g_i(z) = (1 + z + \cdots + z^i)/(i+1)$, so $\mathrm{Var}(X_i) = ((i+1)^2 - 1)/12$, hence $\sigma_n^2 = n(n+1)(2n+7)/72$. Again $R(n)/\sigma_n \to 0$, and $\sum X_i$ is approximately $N(n(n+1)/4, \sigma_n^2)$.

5.27. $p_n = P(T_3 = n) = P(U_{n-1} = 2)/n$, and U_{n-1} has pgf $z(z+1)\cdots(z+n-2)/(n-1)!$. Hence read off the coefficient of z^2. Calculate $p_3 = 1/6$, $p_4 = 1/8$, $p_5 = 11/120$, $p_6 = 5/72$ and $p_7 = 137/2520$ to get the median.

5.28. We know that $P(T_{r+1} = k | T_r = m) = m/(k(k-1))$ for $k \ge m+1$, hence if $n \ge m+1$ then $P(T_{r+2} > n | T_r = m) = \sum_k P(T_{r+2} > n \cap T_{r+1} = k | T_r = m) = \sum_{k>m+1} P(T_{r+2} > n | T_{r+1} = k, T_r = m)\frac{m}{k(k-1)} = \sum_{k>m+1} P(T_{r+2} > n | T_{r+1} = k)\frac{m}{k(k-1)} = \sum_{k=m+1}^{n-1}\frac{k}{n}\frac{m}{k(k-1)} + \sum_{k=n}^{\infty}\frac{m}{k(k-1)} = \frac{m}{n}\sum_{k=m+1}^{n-1}\frac{1}{k-1} + \frac{m}{n-1} = \frac{m}{n}\sum_{k=m}^{n-2}\frac{1}{k} + \frac{m}{n-1}$.

5.29. Write $m/n = x$, so that $P(T_{r+2} > n | T_r = m) \approx x\ln(1/x) + x = x(1 - \ln(x))$. The solution of $x(1 - \ln(x)) = 1/2$ is $x = 0.1867$, and $n = m/x \approx 5.36m$.

5.30. Plainly $E(W_n) = 2E(U_n) = 2\sum_{i=1}^n 1/i$. Also $\mathrm{Var}(W_n) = \mathrm{Var}(U_n) + \mathrm{Var}(V_n) + 2\mathrm{Cov}(U_n, V_n)$. Now $\mathrm{Cov}(U_n, V_n) = \sum_{i,j}\mathrm{Cov}(I_i, J_j) = \sum_{i=1}^n \mathrm{Cov}(I_i, J_i)$ by independence if $i \ne j$. This sum is $\sum_{i=2}^n(-1/i^2)$, hence $\mathrm{Var}(W_n) = 2\mathrm{Var}(U_n) - 2\sum_{i=2}^n(1/i^2) = \sum_{i=1}^n((2/i) - (4/i^2)) + 2$.

Chapter 6

6.1. For $0 < a < 1$, $P(|X| > a) = 1 - a$, for $a \ge 1$ the probability is zero. Also $E(X) = 0$, $\mathrm{Var}(X) = 1/3$ and $E(|X|) = 1/2$. Formally Markov gives $P(|X| > a) \le 1/(2a)$, which is useless when $a \le 1/2$, and Chebychev gives $P(|X| > a) \le 1/(3a^2)$ which is no help when $a \le 1/\sqrt{3}$. Here Markov is better than Chebychev over $(1/2, 2/3)$, Chebychev is better over $(2/3, 1)$.

6.2. $P(X > 30) = \sum_{n=31}^{\infty} \exp(-25)25^n/n!$, nearly 14%. The CLT takes X as approximately $N(25, 25)$, so $P(X > 30) = P((X - 25)/5 \ge (30.5 - 25)/5) \approx P(Z \ge 1.1) \approx 0.135$, splendid. Markov gives $P(X > 30) \le 25/30 = 5/6$, very poor.

6.3. (i) Stick to first choice, $P(\text{Injury-free}) = \sum_i \alpha_i p_i^n = A_1$ say.
(ii) New choice daily, $P(\text{Injury-free}) = (\sum_i \alpha_i p_i)^n = A_2$ say.
Let X take the value p_i with probability α_i, so that $A_1 = E(X^n)$ and $A_2 = (E(X))^n$. The corollary to Jensen's inequality says that $A_1 \ge A_2$, whatever the values of α, p, i.e. choice (i) is always less hazardous. Numerical values are: $n = 10$, 0.9887 v 0.9852; $n = 100$, 0.9552 v 0.8610; $n = 1000$, 0.8596 v 0.2240.

6.4. (a) Take $X = \sqrt{V}$ and $Y = 1\sqrt{V}$, so that $XY = 1$. Cauchy–Schwarz says $1 \le E(V)E(1/V)$, i.e. $E(1/V) \ge 1/E(V)$.

(b) $E(1/V) = E(1/(\mu - (\mu - V))) = \frac{1}{\mu}E((1 - \frac{\mu-V}{\mu})^{-1})$. Since the distribution is symmetrical about μ, V cannot exceed 2μ, and we expand $(1-x)^{-1}$ in the usual fashion to get $E(1/V) = E(\frac{1}{\mu}(1 + \frac{\mu-V}{\mu} + \frac{(\mu-V)^2}{\mu^2} + \cdots))$. But the odd central moments are all zero, the even ones are non-negative, so $E(1/V) \geq \frac{1}{\mu}(1 + \frac{\sigma^2}{\mu^2})$ as claimed.

6.5. Let $£X_i$ be the price on day i, so that $\bar{X} = (X_1 + \cdots + X_{250})/250$ is the mean price over the year. Under daily purchase she gets $M = \sum(1000/X_i)$ shares, under bulk purchase she gets $N = 250\,000/\bar{X}$. Let Y be a random variable taking the value X_i with probability $1/250$ for $1 \leq i \leq 250$. Then $M = 250\,000E(1/Y)$, and $N = 250\,000/E(Y)$, so we know that $M \geq N$. *Daily purchase buys more shares for the same outlay.*
Note: we expect $\bar{X}$ to be close to the overall mean price μ, and part (b) of the previous exercise indicates roughly the expected excess of M over N.

6.6. The flaw is that the step $E(X/(X+Y)) = E(X)/E(X+Y)$ has no justification (we have just seen that $E(1/V) \neq 1/E(V)$ in general). A valid argument is $1 = E((X+Y)/(X+Y)) = E(X/(X+Y)) + E(Y/(X+Y))$ using $E(V+W) = E(V) + E(W)$. Symmetry shows the last sum is $2E(X/(X+Y))$.

6.7. For $0 \leq y \leq 1$, $P(Y_n \leq y) = P(X_n \leq ny) = [ny]/n$, where $[x]$ is the largest integer that does not exceed x. Thus $ny - 1 \leq [ny] \leq ny$, so $y - 1/n \leq [ny]/n \leq y$, and $[ny]/n \to y$, i.e. $Y_n \xrightarrow{D} U(0,1)$.

6.8. For $y > 0$, $P(Y_n > y) = P(\theta_n X_n > y) = P(X_n > y/\theta_n)$. When r is an integer, $P(X_n > r) = (1-p_n)^r$, so we can write

$$(1-p_n)^{y/\theta_n} \geq P(Y_n > y) \geq (1-p_n)^{y/\theta_n + 1}.$$

But $(1-p_n)^{y/\theta_n} \to \exp(-cy)$ as $n \to \infty$, since $p_n/\theta_n \to c$ and $p_n \to 0$, hence $Y_n \xrightarrow{D} E(c)$.

6.9. We claim that each Y_i has the ten-point discrete uniform distribution $U(0,9)$. For, $Y_i = j \Leftrightarrow x + j/10^i \leq X < x + (j+1)/10^i$ for some $x = n/10^{i-1}$, where n is an integer in the range $0 \leq n < 10^{i-1}$. There are 10^{i-1} such integers, giving 10^{i-1} intervals each of length $1/10^i$, hence the result. Moreover, given the values of $Y_{k(1)}, \ldots, Y_{k(m)}$, the distribution of any other Y_j is plainly the same, so the $\{Y_i\}$ are iidrv. Given j, $0 \leq j \leq 9$, let $I_k = 1$ if $Y_k = j$ and let $I_k = 0$ otherwise. Then the $\{I_k\}$ are iid, $P(I_k = 1) = 1/10$, and the SLLN shows that $(I_1 + \cdots + I_n)/n \xrightarrow{a.s.} 1/10$, which means that $P(\text{Digit } j \text{ has frequency } 1/10) = 1$. Thus $P(\text{All digits have frequency } 1/10) = 1$, so $P(X \text{ is normal}) = 1$.

6.10. $E(|X|^r) = E(|X - X_n + X_n|^r) \leq C_r(E(|X - X_n|^r) + E(|X_n|^r)) < \infty$ since $X_n \xrightarrow{r} X$. Minkowski says $(E(|X+Y|^r))^{1/r} \leq (E(|X|^r))^{1/r} + (E(|Y|^r))^{1/r}$. Put $Y = X_n - X$; then $(E(|X_n|^r))^{1/r} - (E(|X|^r))^{1/r} \leq (E(|X_n - X|^r))^{1/r} \to 0$. Interchange X, X_n in this; the result follows.

6.11. The answers are that if $a_n \to a$, and $X_n \to X$ in some mode, then $a_n X_n \to aX$ in that same mode, be it a.s., rth mean, P or D. The proofs follow in that order.
(a) Clearly $a_n X_n(\omega) \to aX(\omega)$ by the properties of convergence of real numbers whenever $X_n(\omega) \to X(\omega)$. So when $P(\omega : X_n(\omega) \to X(\omega)) = 1$, so $P(\omega : a_n X_n(\omega) \to aX(\omega)) = 1$, proving almost sure convergence.
(b) Write $(a_n X_n - aX) = (a_n X_n - a_n X) + (a_n X - aX)$ to obtain

$$E(|a_n X_n - aX|^r) \leq C_r(E(|a_n X_n - a_n X|^r) + E(|a_n X - aX|^r))$$

as in Theorem 6.5. But $E(|a_n|^r|X_n - X|^r) \to 0$ if $X_n \xrightarrow{r} X$, and $|a_n - a|^r E(|X|^r) \to 0$ since $a_n \to a$. Hence we have convergence in rth mean.

(c) Always $|a_n X_n - aX| \le |a_n X_n - a_n X| + |a_n X - aX|$. Also $P(|a_n X_n - a_n X| > \epsilon/2) = P(|a_n||X_n - X| > \epsilon/2)$ can be made as small as you like by choice of n, if we have convergence in probability. We ensure that $P(|a_n - a||X| > \epsilon/2)$ is small by first choosing K so that $P(|X| > K)$ is small, then choosing n large enough to ensure that $|a_n - a| < \epsilon/(2K)$. Since $P(|a_n X_n - aX| > \epsilon) \le P(|a_n X_n - a_n X| > \epsilon/2) + P(|a_n X - aX| > \epsilon/2)$, the result follows.

(d) The result is clear when $a = 0$, so suppose $a > 0$, and hence (for n large enough) $a_n > 0$ also. (The case $a < 0$ is similar.) Now

$$|P(a_n X_n \le t) - P(aX \le t)| \le |F_n(t/a_n) - F_n(t/a)| + |F_n(t/a) - F(t/a)|.$$

If F is continuous at t/a, the second term on the right is small, from convergence in distribution; we work on the first term. Take $u < t < v$, and see that

$$\begin{aligned}(a_n X_n \le t) &= (a_n X_n \le t \cap aX_n \le u) \cup (a_n X_n \le t \cap aX_n > u)\\ &\subset (aX_n \le u) \cup (u/a < X_n \le t/a_n),\end{aligned}$$

i.e. $F_n(t/a_n) \le F_n(u/a) + P(u/a < X_n \le t/a_n)$. In a similar fashion,

$$\begin{aligned}(aX_n \le v) &= (aX_n \le v \cap a_n X_n \le t) \cup (aX_n \le v \cap a_n X_n > t)\\ &\subset (a_n X_n \le t) \cup (t/a_n < X_n \le v/a),\end{aligned}$$

i.e. $F_n(v/a) \le F_n(t/a_n) + P(t/a_n < X_n \le v/a)$. Combining these,

$$F_n(v/a) - P_n(t/a_n, v/a) \le F_n(t/a_n) \le F_n(u/a) + P_n(u/a, t/a_n).$$

This holds for all $u < t < v$, and $a_n \to a$, so $F_n(t/a_n)$ is arbitrarily close to $F_n(t/a)$, proving convergence in distribution.

6.12. Suppose $P(X = \mu) = 1$. Then $P(|X_n - \mu| > \epsilon) = P(X_n \le \mu - \epsilon) + P(X_n > \mu + \epsilon)$ which does not exceed $F_n(\mu - \epsilon) + (1 - F_n(\mu + \epsilon))$. But $F_n(\mu - \epsilon) \to 0 \to F(\mu - \epsilon)$ and $F_n(\mu + \epsilon) \to 1 = F(\mu + \epsilon)$, so $P(|X_n - \mu| > \epsilon) \to 0$.

6.13. Fix n. By Markov's inequality, $P(S_n > t) \le n\mu/t \to 0$ as $t \to \infty$. But $S_n > t \Leftrightarrow N(t) < n$, so $N(t) \stackrel{a.s.}{\to} \infty$ as $t \to \infty$. Now $S_{N(t)} \le t < S_{N(t)+1}$, so $\frac{S_{N(t)}}{t} \le 1 < \frac{S_{N(t)+1}}{t} = \frac{S_{N(t)}}{t} + \frac{X_{N(t)+1}}{t}$. Plainly $X_{N(t)+1}/t \stackrel{a.s.}{\to} 0$, so $S_{N(t)}/t \stackrel{a.s.}{\to} 1$. But $\frac{N(t)}{t} = \frac{N(t)}{S_{N(t)}} \cdot \frac{S_{N(t)}}{t}$. The SLLN shows that $S_{N(t)}/N(t) \stackrel{a.s.}{\to} \mu$ since $N(t) \to \infty$, so (using arguments similar to Exercise 6.11) the result follows.

6.14. In the notation of Example 6.21,

$$X_{k+1} = (1 - x)X_k(1 + \theta) + \begin{cases} xX_k(1 + u) & \text{with probability } p\\ xX_k(1 - d) & \text{with probability } q,\end{cases}$$

and $X_n = X_0((1 + \theta)(1 - x) + (1 + u)x)^Y((1 + \theta)(1 - x) + (1 - d)x)^{n-Y}$, so that

$$\frac{1}{n}\ln\left(\frac{X_n}{X_0}\right) \stackrel{a.s.}{\to} p\ln(1 + \theta + (u - \theta)x) + q\ln(1 + \theta - (d + \theta)x).$$

This is maximised at some x with $0 < x < 1$ provided there is such an x that solves

$$\frac{p(u - \theta)}{1 + \theta + (u - \theta)x} = \frac{q(d + \theta)}{1 + \theta - (d + \theta)x}.$$

However, if, when $x = 1$, the left side exceeds the right side, $x = 1$ maximises the growth rate, and all funds should go into the risky investment. The condition for this reduces to $\theta < (pu - ud - dq)/(1 + uq - dp)$. And if, when $x = 0$, the right side exceeds the left, then $x = 0$ maximises and all funds should be in the safe investment: the condition here reduces to $\theta > pu - qd$, i.e. the safe investment gives a higher mean return than the risky one. Otherwise, the optimal action is to put a proportion $(1 + \theta)(pu - qd - \theta)/((d + \theta)(u - \theta))$ in the risky investment.

Chapter 7

7.1. Let $p_n = P(\text{Become extinct in generation } n)$. The event "Extinct by generation n" is the union of the disjoint events "Extinct by generation $n - 1$" and "Extinct in generation n". Hence $x_n = x_{n-1} + p_n$.

7.2. The pgf of X_1 is the given pgf, that of X_2 is $0.4 + 0.2(0.4 + 0.2z + 0.4z^2) + 0.4(0.4 + 0.2z + 0.4z^2)^2 = 0.544 + 0.104z + 0.224z^2 + 0.064z^3 + 0.064z^4$. Hence $x_1 = 0.4$, $x_2 = 0.544$; so $p_2 = 0.144$.

7.3. Solve $z = g(z)$, i.e. $2z^3 - 5z + 3 = 0$. This cubic factorises as $(z - 1)(2z^2 + 2z - 3)$, so the roots other than unity are $(-1 \pm \sqrt{7})/2 = 0.8229$ or -1.8229. The smallest positive root is $x = 0.8229$, the chance of extinction; with probability 0.1771, the process explodes. $\mu = 1.1$, $\sigma^2 = 1.09$ so $E(X_{100}) = 1.1^{100} \approx 13\ 781$ and $\text{Var}(X_{100}) \approx 43378^2$. $E(X_{100}|X_{100} \neq 0) \approx E(X_{100}/(1 - x)) \approx 77\ 800$.

7.4. The sum of the coefficients is always unity, but we need $0.4 \leq \lambda \leq 1.2$ for a pgf. Since $g'(1) = 1.2 - \lambda + 3(\lambda - 0.4) = 2\lambda$, it certainly dies out $\Leftrightarrow 0.4 \leq \lambda \leq 0.5$.

7.5. We use $h_{n+1}(z) = E(E(z^{S_{n+1}}|X_1))$. Conditional on $X_1 = k$, $S_{n+1} = 1 + \sum_{r=1}^{k} Y_r$, where $Y_r = Y_{r0} + Y_{r1} + \cdots + Y_{rn}$ is the total number in the first n generations of that branching process consisting of the descendants of the rth member of the first generation. These Y_r are iidrv, each with pgf $h_n(z)$, so $E(z^{S_{n+1}}|X_1 = k) = z(h_n(z))^k$, and $h_{n+1}(z) = \sum p_k z(h_n(z))^k = zg(h_n(z))$, taking $h_0(z) = z$. When $g(z) = q + pz^2$ and $h_n \to h$, then $h = z(q + ph^2)$, with formal solutions $h(z) = (1 \pm \sqrt{1 - 4pqz^2})/(2pz)$. Only the minus sign can lead to a pgf, and then $h(1) = (1 - \sqrt{1 - 4pq})/(2p)$. Now $\sqrt{1 - 4pq} = |1 - 2p|$, so if $p \leq 1/2$, then $h(1) = 1$ and we have a genuine pgf. Here, the process dies out a.s., and $S = \lim S_n$ is an honest random variable. When $p > 1/2$, there is positive probability the process does not die out, so then $S = \infty$. Formally, $h(1) = q/p = P(\text{Process dies out}) < 1$, and $h(z)$ is that part of the pgf of S corresponding to S when the process dies out. The residual probability $1 - q/p$ is the chance of explosion.

7.6. We use induction. The formula is easily seen to hold when $n = 1$, and suppose it holds for n. Then $g_{n+1}(z) = g(g_n(z)) = p/(1 - qg_n(z))$. Hence

$$g_{n+1}(z) = \frac{p(q^{n+1} - p^{n+1} - qz(q^n - p^n))}{q^{n+1} - p^{n+1} - qz(q^n - p^n) - qp(q^n - p^n - qz(q^{n-1} - p^{n-1}))}$$

which simplifies to the desired expression. Since $P(X_n = 0) = g_n(0)$, we have $P(Y_n = k) = P(X_n = k)/(1 - P(X_n = 0))$ when $k \geq 1$, so Y_n has pgf $(g_n(z) - g_n(0))/(1 - g_n(0))$. When $p = q = 1/2$, $g(z) = (2 - z)^{-1}$; we prove $g_n(z) = (n - (n-1)z)/(n + 1 - nz)$ in a similar fashion. Thus $x_n = g_n(0) = n/(n+1)$, so $p_n = 1/(n(n+1))$ in the notation of Exercise 7.1. The pgf of Y_n is $z/(n+1-nz)$, so let $u = n/(n+1)$. Then $P(Y_n > N) = u^N$. Hence $P(Y_n/n > x) = P(Y_n > nx) \sim (1 - \frac{1}{n+1})^{nx} \to \exp(-x)$, as required.

7.7. Use the expression for $\text{Var}(X_n)$ in Theorem 7.4 to get the limit. Its corollary shows that $\text{Var}(Y_n) = \sigma^2(\mu^n - 1)/(\mu^{n+1}(\mu - 1)) \to \sigma^2/(\mu^2 - \mu)$. Let $W_n \equiv$

$Y_n|X_n \neq 0$, so that $E(W_n^r) = E(X_n^r/\mu^{rn})/P(X_n \neq 0)$. Thus $E(W_n) \to 1/(1-x)$, and $E(W_n^2) = (\mathrm{Var}(X_n) + (E(X_n))^2)/(\mu^{2n}(1-x_n))$, so $\mathrm{Var}(W_n) \to (\sigma^2(1-x) - \mu x(\mu-1))/((\mu^2-\mu)(1-x)^2)$.

7.8. Write $X_{n+1} = W_{n+1} + Y_{n+1}$ where W_{n+1} is the sum of X_n variables, each with pgf $g(z)$ and the pgf of Y_{n+1} is (independently) $h(z)$. So $k_{n+1}(z) = k_n(g(z))h(z)$. Find k_1 and k_2 to obtain $k_3(z) = g(g(g(z)))h(g(g(z)))h(g(z))h(z)$.

7.9. Use $x = P(\text{Ruin}) = (\theta^c - \theta^a)/(\theta^c - 1)$ where $\theta = q/p$. Las Vegas has $\theta = 20/18$; (a) $x = 1$; (b) here $a = 1, c = 4$ so $x = 0.7880$; (c) $x = 1 - P(2 \text{ Wins}) = 1 - (18/38)^2 = 0.7756$. UK has $\theta = 75/73$; (a) $x = 0.9997$; (b) $x = 0.7600$; (c) $x = 1 - (73/148)^2 = 0.7567$.

7.10. Condition on the first step, as in the Gambler's Ruin argument. Mean profit to unit stake is $5x - (1-x) = 6x - 1$, which favours us when $x > 1/6$. By the branching process analogy, $p_a = 1$ in cases (a) and (b); in case (c), we seek the smallest positive root of $z = 0.8 + 0.2z^6$, i.e. $x = 0.9265$. Then $p_a = x^a$.

7.11. $S_n = \sum X_i$ with $E(X_i) = p - q$ and $\mathrm{Var}(X_i) = 4pq$, so $E(S_n) = n(p-q)$ and $\mathrm{Var}(S_n) = 4npq$. If $p = 0.6$ and $n = 100$, mean is 20, variance is 96, so $P(S_n > 10) \approx P((S_n - 20)/\sqrt{96} \geq (10.5 - 20)/\sqrt{96}) \approx P(Z \geq -0.97) \approx 0.83$.

7.12. $\alpha(2k, 2n)/\alpha(2k+2, 2n) = u_{2k}u_{2n-2k}/(u_{2k+2}u_{2n-2k-2}) = (k+1)(2n - 2k - 1)/((2k+1)(n-k))$ which exceeds unity iff $2k < n - 1$. Thus the sequence decreases from $k = 0$ to the middle value then, by symmetry, increases. For $0 < x < 1$, $F_n(x) = P(T_n/(2n) \leq x) = \sum_{k \leq nx} \alpha(2k, 2n - 2k)$. Stirling's formula shows that $u_{2r} = \binom{2r}{r}/2^{2r} \sim 1/\sqrt{\pi r}$, and so

$$F_n(x) \sim \sum_{k \leq nx} \frac{1}{\sqrt{\pi k}} \cdot \frac{1}{\sqrt{\pi(n-k)}} \sim \frac{1}{\pi} \int_0^{nx} \frac{du}{\sqrt{u(n-u)}}.$$

Substitute $x = n\sin^2(\theta)$ to find $F_n(x) \sim \frac{2}{\pi} \arcsin \sqrt{x}$.

7.13. (a) Since $\binom{30}{15} = 155\,117\,520$, we require $\binom{30}{16+a}$ to be at most 5% of this. Trial and error shows that $a = 5$ is not enough, but the chance when $a = 6$ is 0.9623. (b) Here $\binom{30}{10} = 30\,045\,015$, and we want $\binom{30}{11+a}$ to be at most 5% of this. If $a = 12$, this is not enough, but when $a = 13$ the chance is 0.9802.

7.14. The entries in P^2 are $(5/12, 7/12; 7/18, 11/18)$ and the $p_{ii}^{(3)}$ are $29/72$ and $65/108$. The vector π is $(2/5, 3/5)$. For the other matrix, the entries are $(2/3, 1/6, 1/6; 0, 1/2, 1/2; 4/9, 5/18, 5/18)$, the diagonal entries in P^3 are $2/9$, $1/6$ and $17/54$, and $\pi = (2/5, 3/10, 3/10)$.

7.15. Many possible solutions, e.g. make all (p_{ij}) zero except as follows: $p_{12} = p_{23} = p_{24} = p_{31} = p_{41} = 1/2$ will make $\{1, 2, 3, 4\}$ transient with period 3; $p_{56} = p_{65} = 1$ makes $\{5, 6\}$ recurrent, period 2; $p_{78} = p_{79} = p_{87} = p_{89} = p_{97} = p_{98} = 1/2$ makes $\{7, 8, 9\}$ aperiodic recurrent. $\{10, 11, 12\}$ can be the last class, aperiodic transient if $p_{10,10} = p_{10,11} = p_{11,12} = p_{12,10} = 1/3$. Then $p_{10,1} = 1/3$ makes the first transient class accessible from the second; make the other row sums unity, using positive entries in column 7 only.

7.16. With states $\{0, 1, 2\}$, the row-by-row entries are $(q, p, 0; 0, q, p; 1, 0, 0)$. Plainly irreducible, aperiodic, with stationary vector $(1/(2+p))(1, 1, p)$ so he misses a proportion $p/(2+p)$ matches. Work out P^3; the entries $p_{i2}^{(3)}$ are $2p^2q$, pq^2 and p^2, which give the chances he is suspended for the Brazil match, given his current state.

7.17. If it is either transient or null, $p_{ij}^{(n)} \to 0$ for all i, j, which contradicts $\sum_j p_{ij}^{(n)} = 1$ for a *finite* sum.

7.18. $q_{ij} = P(X_n = j, X_{n+1} = i)/P(X_{n+1} = i) = P(X_{n+1} = i|X_n = j)P(X_n = j)/\pi_i = p_{ji}\pi_j/\pi_i$. Then $\sum_j q_{ij} = \sum_j p_{ji}\pi_j/\pi_i = 1$ since $\pi_i = \sum_j \pi_j p_{ji}$. Plainly every $q_{ij} \geq 0$, so Q is a transition matrix. Also $(\pi Q)_j = \sum_i \pi_i q_{ij} = \sum_i \pi_j p_{ji} = \pi_j$ since $\sum_i p_{ji} = 1$.

7.19. From any state i we can reach 0 via $i-1, i-2$, etc., and from 0 we can go straight to any state j, hence irreducible. Aperiodic since $p_{00} > 0$. $\pi P = \pi$ reads: $\pi_0 = \pi_0 a_0 + (1-\theta)\pi_1$; $\pi_n = \pi_0 a_n + \pi_n \theta + (1-\theta)\pi_{n+1}$ for $n \geq 1$. Hence $\pi_n(1-\theta) = \pi_0(1 - a_0 - \cdots - a_{n-1}) = \pi_0 \sum_{k \geq n} a_k$ for $n \geq 1$. We want $(1-\theta) = \pi_0(1 - \theta + \sum_{n=1}^{\infty} \sum_{k \geq n} a_k) = \pi_0(1 - \theta + \mu)$. So there is a stationary probability vector, and the chain is positive recurrent, iff μ is finite. Foster's criterion looks at $Py = y$, i.e. $\theta y_1 = y_1$ and $(1-\theta)y_{i-1} + \theta y_i = y_i$ when $i \geq 2$, so $y_1 = 0$ and $y_i = y_{i-1}$ for $i \geq 2$, hence $y_i \equiv 0$ and the chain is never transient. Thus it is null recurrent when $\mu = \infty$.

The other chain is plainly irreducible and aperiodic, and we look at the system: $\pi_0 = \pi_0 a_0 + (1-\theta)\sum_{i \geq 1} \pi_i$; $\pi_n = \pi_0 a_n + \theta \pi_n$ for $n \geq 1$. Here $\pi_n = a_n \pi_0/(1-\theta)$ for $n \geq 1$, so we need $1 - \theta = \pi_0(1 - \theta + \sum_{n \geq 1} a_n) = \pi_0(2 - a_0 - \theta)$ which can always be satisfied. This chain is always positive recurrent.

Intuitively, μ is the mean distance "jumped" when we leave zero; in the first chain, we return to zero a step at a time, in the second we jump straight back to zero, so how far we went is irrelevant.

7.20. The chain is plainly irreducible, and now aperiodic since $r = p_{00} > 0$. Look at $Qy = y$, Q and $y = (\ldots, y_{-2}, y_{-1}, y_1, y_2, \ldots)$ as in Foster's criteria. Let $x = q/(1-r)$; for positive subscripts, the system is $(1-x)y_2 = y_1$, and $(1-x)y_{i+1} = y_i - x y_{i-1}$ when $i \geq 2$. Write $Y(\theta) = \sum_{i \geq 1} y_i \theta^i$, and obtain $Y = y_1 \theta(1-\theta)^{-1}(1 - x\theta/(1-x))^{-1}$. Similarly, let $Z(\theta) = \sum_{i \geq 1} y_{-i}\theta^i$, and find $Z = y_{-1}\theta(1-\theta)^{-1}(1 - \theta(1-x)/x)^{-1}$. If $x = 1/2$, both of these contain $(1-\theta)^{-2}$, so (y_n) is unbounded, hence the chain is not transient. But if $x \neq 1/2$, choose one of $\{y_1, y_{-1}\}$ to be zero, the other non-zero, and get a bounded non-zero solution – hence transient. Take $\mathbf{u}$ as the vector all of whose entries are unity. Then $\mathbf{u}P = \mathbf{u}$, and $\sum_i u_i = \infty$ so the chain is null recurrent when not transient, i.e. when $p = q > 0$.

7.21. Let $\pi = (1/N, 1/N, \ldots, 1/N)$. Plainly $\pi P = \pi$ so it is positive recurrent. The previous exercise gives examples for transient and null chains.

7.22. The $\{p_{ij}\}$ are zero except for $p_{00} = 1$ and $p_{i,i-1} = 33/37$, $p_{i,i+8} = 4/37$ for $i \geq 1$. Then $\{0\}$ is a closed aperiodic recurrent class, the rest form a transient class of period 9. As the mean winnings are $-1 \times 33/37 + 8 \times 4/37 = -1/37$, negative, the state inevitably hits zero eventually. Meantime, you lose one unit most of the time, with occasional increases of eight units.

7.23. Let $a_i = \sum_j a_{ij} > 0$ be the number of doors out of (and into) room i, $a = \sum_i a_i$ and $\pi_i = a_i/a$. Write $p_{ij} = a_{ij}/a_i$. Independence of steps means this is a Markov chain, plainly irreducible. Now $(\pi P)_j = \sum_i \pi_i p_{ij} = \sum_i \frac{a_i}{a} \cdot \frac{a_{ij}}{a_i} = (\sum_i a_{ij})/a = (\sum_i a_{ji})/a$ (two-way doors), i.e. $(\pi P)_j = a_j/a = \pi_j$. In the long run, the distribution converges to π, showing the chance of being in room i is proportional to its number of doors. $P(\text{Same room}) = \sum_i \pi_i^2 = (\sum_i a_i^2)/a^2$.

7.24. Plainly $p_{i,i-1} = i/N$ and $p_{i,i+1} = 1 - i/N$ for $1 \leq i \leq N-1$, and $p_{01} = 1 = p_{N,N-1}$. This is a finite irreducible chain of period 2 (hence it is positive recurrent), and induction verifies the unique solution of $\pi P = \pi$ is $\pi_k = \binom{N}{k}/2^N$.

7.25. We have $p_{ii}^{(2n)} = \binom{2n}{n}p^n q^n$ which, using Stirling's formula, is asymptotic to $(4pq)^n/\sqrt{n\pi}$. When $p \neq 1/2$, $|4pq| < 1$, so the sum is convergent and the chain is transient. When $p = 1/2$, $4pq = 1$, so the sum diverges, so the chain is recurrent. But since $p_{ii}^{(2n)} \to 0$, it is null recurrent.

In two dimensions, $p_{ii}^{(2n)} = \frac{1}{4^{2n}}\sum_{r=0}^{n} \frac{(2n)!}{r!r!(n-r)!(n-r)!}$ which can be written as

$\binom{2n}{n}\sum_{r=0}^{n}\binom{n}{r}\binom{n}{n-r}/4^{2n}$. The sum is $\binom{2n}{n}$, so the whole expression is $\left(\binom{2n}{n}/2^{2n}\right)^2$. Stirling shows this is asymptotic to $1/(n\pi)$, which means that $p_{ii}^{(2n)} \to 0$, while the sum diverges – null recurrence.

Chapter 8

8.1. $p_{ij}(t + h) - p_{ij}(t) = (p_{ii}(h) - 1)p_{ij}(t) + \sum_{k \neq i} p_{ik}(h)p_{kj}(t)$. The first term is negative and does not exceed $1 - p_{ii}(h)$ in modulus. The second term is positive and at most $\sum_{k \neq i} p_{ik}(h) = 1 - p_{ii}(h)$. Hence the result.

8.2. $p_{ii}(t) \geq (p_{ii}(t/n))^n$, so $\ln(p_{ii}(t)) \geq n\ln(p_{ii}(t/n))$. Write $x_n = 1 - p_{ii}(t/n)$, and recall that for small x, $\ln(1 - x) \geq -x - x^2$. Thus $\ln(p_{ii}(t))/n \geq \ln(1 - x_n) \geq -x_n - x_n^2$. Since $q_i = \lim_{n \to \infty} (nx_n/t)$, given any $\epsilon > 0$, take n large enough to ensure $nx_n/t \leq q_i + \epsilon$. Thus $\ln(p_{ii}(t))/n \geq -t(q_i + \epsilon)/n - t^2(q_i + \epsilon)^2/n^2$, i.e.

$$\ln(p_{ii}(t)) \geq -t(q_i + \epsilon) - t^2(q_i + \epsilon)^2/n.$$

Let $n \to \infty$ to eliminate the last term, what is left holds for all $\epsilon > 0$, so $\ln(p_{ii}(t)) \geq -q_i t$. The rest is immediate.

8.3. Let $x = p_{11}(t)$ and $y = p_{13}(t)$, and use the forward equations for x and y to find $x' = -x + y$, $y' = 1 - x - 2y$. Differentiate the first again, eliminate y', then eliminate y to obtain $x'' + 3x' + 3x = 1$. This solves by standard methods to give the stated expression, using $p_{11}(0) = 1$ and $p'_{11}(0) = -1$. Obtain $p_{13}(t) = y = x' + x$, then $p_{12}(t) = 1 - x - y$. The other terms of $P(t)$ follow from symmetry.

8.4. Using the same method as the previous exercise, the second order ODE for x is

$$x'' + (\alpha + \beta + \gamma)x' + (\beta\gamma + \gamma\alpha + \alpha\beta)x = \beta\gamma.$$

The solution will contain trigonometric terms if the roots of the auxiliary equation have imaginary parts – which is the condition stated. The last part comes from the specific solution $x = \beta\gamma/(\beta\gamma + \gamma\alpha + \alpha\beta)$ of the non-homogeneous equation – the general solution of the homogeneous equation has a term $\exp(-Kt)$.

8.5. If the rates at which the Home team scores Goals and Behinds are λ_G and λ_B, and for the Away team we have μ_G and μ_B, then the non-zero off-diagonal entries are $q_{i,i-6} = \mu_G$, $q_{i,i-1} = \mu_B$, $q_{i,i+1} = \lambda_B$ and $q_{i,i+6} = \lambda_G$. The diagonal entries then make the row sums zero, the state space is all integers, positive and non-positive.

8.6. One approach is to decide whether the inter-event intervals are iid Exponential variables. This is clearly false in (a), (b), (c) or (e), true for (d) and (f).

8.7. The forward equations are $x'_n = -\beta n x_n + \beta(n-1)x_{n-1}$, so the PDE becomes $\frac{\partial G}{\partial t} = \beta z(z - 1)\frac{\partial G}{\partial z}$. For the given expression, $G = \sum e^{-\beta t}(1 - e^{-\beta t})^{n-1}z^n = ze^{-\beta t}/(1 - z + ze^{-\beta t})$, which satisfies the PDE and initial conditions (check!). Differentiate the PDE twice with respect to z, put $z = 1$, and write $m(t) = E(X_t)$, $u(t) = E(X_t(X_t - 1))$. We find $m'(t) = \beta m$, hence $m(t) = \exp(\beta t)$, and $u' = 2\beta m + 2\beta u$. Using an integrating factor we find $u(t) = 2\exp(\beta t)(\exp(\beta t) - 1)$, leading to $\text{Var}(X_t) = \exp(\beta t)(\exp(\beta t) - 1)$. To find t_n, differentiate $x_n(t) = e^{-\beta t}(1 - e^{-\beta t})^{n-1}$; maximum at $t = \ln(n)/\beta$.

8.8. The method of characteristics looks at $\frac{dt}{1} = -\frac{dz}{(\lambda z - \mu)(z - 1)} = \frac{dG}{0}$. The first pair solve as $(\lambda - \mu)t = \ln(\frac{\lambda z - \mu}{z - 1}) + \text{const}$, hence $G = f((\lambda - \mu)t + \ln\frac{z-1}{\lambda z - \mu})$. The initial condition gives $z = f(\ln\frac{z-1}{\lambda z - \mu})$, hence $f(u) = \frac{1 - \mu e^u}{1 - \lambda e^u}$. The displayed result follows. Note that $G(t, 0) = (\mu(1 - E))/(\lambda - \mu E)$, where $E = \exp(-t(\lambda - \mu))$. Thus the pgf of Y_t is $(G(t,z) - G(t,0))/(1 - G(t,0))$, which simplifies to $[(\lambda - \mu)zE]/[\lambda - \mu E - \lambda z(1 - E)]$. Differentiate with respect to z, put $z = 1$ and find the mean size is $(\lambda - \mu E)/[(\lambda - \mu)E]$.

8.9. Write down the forward equations ($n = 0$ is a special case), multiply by z^n and sum. The Lagrange equation is

$$\frac{\partial G}{\partial t} = \alpha(z-1)G + \beta z(z-1)\frac{\partial G}{\partial z}.$$

Solve to obtain const$= \beta t + \ln((1-z)/z)$ and const$= \alpha \ln z + \beta \ln G$. Then $G(0,z) = 1$ leads to $G = \exp(-\alpha t)(1 - z(1 - e^{-\beta t}))^{-\alpha/\beta}$.

8.10. The forward equations for the immigration–death process lead to the given PDE. The method of characteristics gives $\frac{dt}{1} = \frac{dz}{\mu(z-1)} = \frac{dG}{\alpha(z-1)G}$; taking the first two and the last two, we find $\mu t = \ln(z-1)$+const and $\rho z = \ln(G)$+const, where $\rho = \alpha/\mu$. Thus $\ln(G) - \rho z = f(\mu t - \ln(z-1))$, and the initial conditi on $G(0,z) = 1$ shows $-\rho z = f(-\ln(z-1))$, hence $f(u) = -\rho(1 + e^{-u})$. This leads to $\ln(G) = \rho(z-1)(1 - \exp(-\mu t))$, so G is the claimed Poisson pgf.

8.11. Given $T = t$, the mean size of the V-population is $\exp(\phi t)$. Hence, using conditional expectations, the overall mean size is $\int_0^\infty \theta \exp(-\theta t).\exp(\phi t)dt$, which is infinite if $\phi \geq \theta$, and $\theta/(\theta - \phi)$ if $\theta > \phi$. Complete sentence to "also 2". And when $\theta = \phi$, the mean size of either population at the time of the first birth in the other is infinite!

8.12. The respective probabilities of $0, 1, \ldots, 4$ infections are $(144, 36, 28, 53, 459)/720$.

8.13. Keeping as close a parallel with epidemics as we can, let $X(t)$, $Y(t)$ and $Z(t) = N - X(t) - Y(t)$ be the numbers of Ignorants, Spreaders and Stiflers. Given $(X(t), Y(t)) = (x, y)$, the non-zero transition rates out of that state are to $(x - 1, y + 1)$ at rate βxy, to $(x, y - 2)$ at rate $\beta y(y-1)/2$ and to $(x, y-1)$ at rate $\beta y(N-x-y)$. Here β is related to the rate of mixing, and the likelihood a spreader begins to tell the rumour – there is no threshold. For an exact analysis via random walks, the transition rates are plain. Taking the deterministic equations as with epidemics, $\frac{dx}{dt} = -\beta xy$ and $\frac{dy}{dt} = \beta(xy - y(y-1) - y(N-x-y))$, giving $\frac{dy}{dx} = (N - 2x - 1)/x$. Taking $y = 1$ when $x = N - 1$, integrate to obtain $y = (N-1)\ln(x/(N-1)) - 2x + 2N - 1$. The epidemic is over when $y = 0$; take N large, and write $u = x/(N-1)$. This leads to $2u = \ln(u) + 2$, approximately, i.e. $u = 0.2032$. The rumour is expected to spread to about 80% of the population.

8.14. The forward equations are $p_0'(t) = -\lambda p_0(t) + \mu p_1(t)$, and, for $n \geq 1$, $p_n'(t) = \lambda p_{n-1}(t) - (\lambda + \mu)p_n(t) + \mu p_{n+1}(t)$. Multiply by z^n, sum, simplify.

8.15. Use Equation (8.10) with $\lambda_n = \lambda$ and $\mu_n = \mu$ to see that $\pi_n = \rho^n \pi_0$. Then $F(w) = P(W \leq w) = \sum_n P(W \leq w \cap Q = n)$. When $Q = n$, W is the sum of n independent variables, each $E(\mu)$, so is $\Gamma(n, \mu)$. Thus $F(w) = \pi_0 + \sum_{n \geq 1} \pi_n \int_0^w e^{-\mu t}\frac{\mu^n t^{n-1}}{(n-1)!}dt = \pi_0[1 + \lambda \int_0^w e^{(\lambda - \mu)t}dt]$ which simplifies as required. Busy and Idle periods alternate, their mean lengths are $1/(\mu - \lambda)$ and $1/\lambda$, so in a long time T the mean number of Busy periods is $N = T/(1/(\mu - \lambda) + 1/\lambda)$. The mean number of arrivals in T is λT, all get served in the long run, so the mean number served per Busy period is $\lambda T/N$ which simplifies to $1/(1 - \rho)$.

8.16. For $M/M/2$, we find $\pi_0 = (2-\rho)/(2+\rho)$, and the pgf reduces to $\pi_0(2+\rho z)/(2-\rho z)$. The mean is then $4\rho/(4-\rho^2)$, the variance is $4\rho(4+\rho^2)/(4-\rho^2)^2$. For the $M/M/1$ bank, the chance of immediate service is $1 - \rho$; for $M/M/2$, recalling the arrival rate is 2λ, $\pi_0 = (1-\rho)/(1+\rho)$ and $\pi_1 = 2\rho\pi_0$, so the chance of immediate service is $(1 - \rho)(1 + 2\rho)/(1 + \rho)$ which is plainly higher. The mean times in the systems reduce to $1/(\mu(1 - \rho))$ and $1/(\mu(1 - \rho^2))$. Again, the latter is shorter.

8.17. When $\mu = 2$, then $\rho = 2$ and $\pi_0 = 3/23$, $\pi_1 = \pi_2 = 6/23$, $\pi_3 = 4/23$, so (a)$= 19/23$, (b)$= 2$ and (c)$= 25/46$. When $\mu = 4/3$, then $\rho = 3$, $\pi_0 = 2/53$, $\pi_1 = 6/53$, $\pi_2 = \pi_3 = 9/53$, and the answers are $26/53$, 1 and $60/53$.

8.18. Per hour, queue is $M(180)/M(30)/4$. Here $k\mu = 120 < 180 = \lambda$, so the queue would grow without bound. With the given limit, it is $M/M/4/10$, so Equation (8.10) implies $\pi_n = \lambda^n \pi_0/(n!\mu^n) = 6^n\pi_0/n!$ when $n \le 4$, and $\pi_n = \pi_4(3/2)^{n-4}$ for $4 \le n \le 10$. This yields $\pi_0 = 1/1798.28$, hence $\pi_1, \ldots, \pi_{10}$. $P(\text{Forecourt full}) = \pi_{10} = P(\text{Arrival drives on})$, so mean number lost per hour is $180\pi_{10} \approx 61.6$. (a) leads to $M/M/4/12$, from which we find $\pi_{12} = 0.3371$, and (b) is $M/M/5/10$, now $\pi_n = 6^n\pi_0/n!$ for $n \le 5$, and $\pi_n = \pi_5(6/5)^{n-5}$ for $5 \le n \le 10$. Then $\pi_{10} = 0.2126$, so the respective forecourt losses are about 60.7 and 38.3. An extra pump is much better than extending the forecourt.

8.19. Per minute, $\lambda = 3/10$, $\mu = 1/3$, so $\rho = 0.9$. The respective values of $\text{Var}(S)$ are 0, 9 and 3, so the mean queue lengths are 4.95, 9 and 6.3.

8.20. Differentiate to obtain $M''_B(0) = M''_S(0)(1 + \lambda E(B))^2 + E(S)\lambda M''_B(0)$. This gives $M''_B(0) = E(B^2)$, use the known expression for $E(B)$ to find $\text{Var}(B) = (\text{Var}(S) + \lambda(E(S))^3)/(1 - \lambda E(S))^3$.

8.21. The arrival rate is state-dependent, being $\lambda(N - n)$ when n machines are in the "queue". With Exponential service time, we have a birth–death process with states $0 \le n \le N$, $\lambda_n = \lambda(N - n)$ and $\mu_n = \mu$ for $n \ge 1$. Equation (8.10) shows $\pi_n = \pi_0 N!\lambda^n/(\mu^n(N - n)!)$.

8.22. With no restrictions, $M/M/K$ is stable provided that $\lambda < \mu K$, and then Equation (8.11) is used to find π_0, and then all π_n. Calls arriving when the state is at least K are lost, so the proportion of lost calls is $\sum_{r \ge K} \pi_r = \pi_0\rho^K K\mu/(K!(K\mu - \lambda))$.

8.23. For $M/E_r/c$ with $c \ge 2$, suppose $Y_t = c$. An arriving customer C could find all servers occupied (when each customer has one stage left), or a free server (if some customer has at least two stages left). So we do not know whether C could go straight to service. Modify Y_t to be the vector $(w, a_1, \ldots, a_c)$, where $w \ge 0$ is the number of waiting customers, and server i has a_i stages of service left.

8.24. Suppose the first K in the queue give up the option of using the slower server. The next customer has a mean total time of $(K + 2)/\mu_1$ with the faster server, or $1/\mu_2$ with the slower. She should wait if, and only if, $K < \mu_1/\mu_2 - 2$.

8.25. $K > 4$, otherwise the queue is unstable. When $K = 4$, compute $\pi_0 = 2/53$, and then $P(\text{At most 3 others waiting}) = \pi_0 + \cdots + \pi_7 \approx 41.6/53$, not good enough. When $K = 5$, then $\pi_0 = 16/343$, and even $P(\text{At most 1 waiting}) = 313.84/343 > 0.9$. So $K = 5$ suffices; then the output is Markov, so the cashier queue is $M/M/n$ with the same arrival rate. We need $n \ge 3$ for stability; and when $n = 3$, $\pi_0 = 1/9$, $\pi_0 + \cdots + \pi_3 = 19/27$ which meets the criterion.

8.26. As a birth and death process, Equation (8.11) shows that $\pi_n = \lambda^n\pi_0/(n!\mu^n)$, hence the distribution is $\text{Poiss}(\lambda/\mu)$.

8.27. (a) $F(t) = t$ and $\mathcal{F}(t) = 1 - t$ so $\phi(t) = 1/(1 - t)$, all on $0 < t < 1$.
(b) For $t \ge 1$, $F(t) = 1 - 1/t^2$, $\mathcal{F}(t) = 1/t^2$, and $\phi(t) = 2/t$. For $E(\lambda)$, we know $f_k(t)$; use Theorem 8.18 to find $H(t) = \lambda t$ and $h(t) = \lambda$.

8.28. Let the dead processes be iidrv $\{Y_i\}$; assuming the emissions form a Poisson process, times after dead periods until recording are iidrv, $U_i \sim E(\lambda)$. Then $X_i \equiv Y_i + U_i$ is the interval between recordings.

8.29. W_k is $N(k\mu, k\sigma^2)$, so $P(X_t \ge k) = P(W_k \le t) = P((W_k - k\mu)/(\sigma\sqrt{k}) \le (t - k\mu)/(\sigma\sqrt{k})) = P(Z \le (t - k\mu)/(\sigma\sqrt{k}))$, Z being $N(0, 1)$.

8.30. $P(T > s + t | T > s) = P(T > s + t \cap T > s)/P(T > s) = \mathcal{F}(t + s)/\mathcal{F}(s)$. So if $\mathcal{F}(s)\mathcal{F}(t) > \mathcal{F}(s + t)$, then $P(T > s + t | T > s) < P(T > t)$, i.e. if the lightbulb has survived for time s, it is less likely to survive a further time t than is a new lightbulb. So $\mathcal{F}(s)\mathcal{F}(t) > \mathcal{F}(s + t)$ could be NBU. Similarly for NWU.

8.31. $\mathcal{F}_z(t) \equiv P(T_z > t) = P(T > t + z | T > z) = (1 + \beta z)^c / (1 + \beta(t + z))^c$, and $E(T_z) = \int_0^\infty \mathcal{F}_z(t) dt$. In the integral, put $t = (1 + \beta z)u$; it transforms to $(1 + \beta z) \int_0^\infty 1/(1 + \beta u)^c du = (1 + \beta z) E(T)$.

8.32. (a) Let Y_i be the cost of replacement i, so that $P(Y_i = c_2) = 1 - F(\xi)$. Then, since f is the density of failure time, $E(Y_i) = c_2(1 - F(\xi)) + c_1 \int_0^\xi f(u) du = c_1 F(\xi) + c_2(1 - F(\xi)) = C$. By the SLLN, $K_n/n \to E(Y_i)$ almost surely.
(b) We have $K_{X(t)}/t = (K_{X(t)}/X(t)).(X(t)/t)$. As $X(t) \to \infty$, the first term converges a.s. to C, and the second to $1/\mu$, where μ is the mean lifetime. The lifetime is ξ with probability $\mathcal{F}(\xi)$, otherwise has density f, so $\mu = \xi \mathcal{F}(\xi) + \int_0^\xi t f(t) dt$ which simplifies to $\int_0^\xi \mathcal{F}(t) dt$. Hence the result. "The mean cost of replacement, per unit time, is this last expression."

8.33. Let Y_i be the cost of replacements in the interval $i\tau < t \le (i+1)\tau$; then its mean is $c_1 H(\tau) + c_3$, since $H(\tau)$ is the mean number of unscheduled replacements. The result is immediate.

8.34. $F(z) = (8z + 7z^2 + z^3)/16$, so $U(z) = 16/((1 - z)(4 + z)^2)$. Expanding, $U(z) = \sum_0^\infty z^n \sum_0^\infty (k+1)(-1)^k (z/4)^k$. Thus $u_n = 1 - 2/4 + 3/4^2 - 4/4^3 + \cdots + (-1)^n(n+1)/4^n \to (1 + 1/4)^{-2} = 16/25$. $E(T) = 1/2 + 7/8 + 3/16 = 25/16$.

8.35. A Geometric distribution arises when we carry out Bernoulli trials with Success probability p, independent on each trial. So in a Renewal Process, Geometric lifetime is equivalent to "New is exactly as good as Used", and we can ignore how old the current component is.

8.36. The joint distributions of any finite collection are clearly Gaussian in each case. We check covariances: if $s < t$, then $\text{Cov}(sW_{1/s}, tW_{1/t}) = st\text{Cov}(W_{1/s}, W_{1/t}) = st.(1/t) = s = \min(s,t)$ as required.
Also $\text{Cov}(W_{c^2t}/c, W_{c^2s}/c) = (1/c^2)\min(c^2t, c^2s) = \min(t,s)$ as required.

8.37. $E(M_t^2) = E(|W_t|^2) = t$, so $\text{Var}(M_t) = t(1 - 2/\pi)$.

8.38. Equation (8.19) gives the joint density of W and M; rewrite it as the joint density of M_t and $Y_t = M_t - W_t$. Y_t has the same density as M_t, so $P(M_t > m | Y_t = 0) = \int_m^\infty \sqrt{2/(\pi t^3)} u \exp(-u^2/(2t)) du / \sqrt{2/(\pi t)}$ which simplifies to the given formula.

8.39. $P(A_t < x) = P(\text{No zeros in } (x, t)) = 1 - (2/\pi) \arccos \sqrt{x/t} = (2/\pi) \arcsin \sqrt{x/t}$. $P(B_t < y) = P(\text{At least one zero in } (t, y)) = (2/\pi) \arccos \sqrt{t/y}$. $P(A_t < x, B_t > y) = P(\text{No zeros in } (x, y)) = 1 - (2/\pi) \arccos \sqrt{x/y}$. By differentiation, A_t has density $1/(\pi\sqrt{x(t - x)})$ over $0 < x < t$, its mean is $t/2$. Also B_t has density $\frac{1}{\pi y}\sqrt{\frac{t}{y-t}}$ on $y > t$, its mean is infinite.

8.40. The event occurs iff, given $Y_u = y$, the maximum gain over the interval (u, v) is at least y. Hence $P = \int_0^\infty \sqrt{\frac{2}{\pi u}} \exp(-y^2/(2u)) \int_y^\infty \sqrt{\frac{2}{\pi(v-u)}} \exp(-m^2/(2(v - u))) dm dy$. Treat this in the same way as the similar integral in Example 8.35; the result drops out.

8.41. Put $p = 1/2$ and $2\mu/\sigma^2 = \lambda$, so that $1/2 = (\exp(b\lambda) - 1)/(\exp(b\lambda) - \exp(-a\lambda))$, leading to $\exp(b\lambda) = 2 - \exp(-a\lambda)$. Thus $b = \ln(2 - \exp(-a\lambda))/\lambda$. When $\lambda = 1$, $\mu > 0$, the process drifts to $+\infty$, and $V \equiv \min(X_t : t \ge 0)$ is defined. Let $a \to \infty$ in the expression for p; note that $1 - p$ is then $P(\text{Ever hit } -b)$. But $p = 1 - \exp(-b)$, so $P(V \le -b) = \exp(-b)$, so V is distributed as $-E(1)$.

8.42. Intuitively, the larger the value of m, the later the maximum is reached, so the less time it has to decrease later.

Index